69

Fortschritte der Chemie organischer Naturstoffe

Progress in the Chemistry of Organic Natural Products

Founded by
L. Zechmeister

Edited by
W. Herz, G. W. Kirby,
R. E. Moore, W. Steglich,
and Ch. Tamm

Authors:
D. Deepak, J. F. Grove, E. Haslam,
A. Khare, N. K. Khare,
S. Srivastava

Springer-Verlag Wien GmbH

List of Contributors

DEEPAK, Dr. D., Department of Chemistry, Lucknow University, Lucknow 226 007, India.

GROVE, Dr. J. F., 3 Homestead Court, Welwyn Garden City, Herts AL7 4LY, United Kingdom.

HASLAM, Prof. E., Department of Chemistry, The University of Sheffield, Sheffield S3 7HF, United Kingdom.

KHARE, Prof. Dr. A., Department of Chemistry, Lucknow University, Lucknow 226 007, India.

KHARE, Dr. N. K., Department of Chemistry, Lucknow University, Lucknow 226 007, India.

SRIVASTAVA, Dr. S., Department of Chemistry, Lucknow University, Lucknow 226 007, India.

Non-Macrocyclic Trichothecenes, Part 2 (*1*)*

J. F. GROVE
3 Homestead Court, Welwyn Garden City, Herts AL7 4LY,
United Kingdom

Contents

* Reviewing the literature published between January 1987 and December 1995.

1. Introduction and Nomenclature

A total of 182 trichothecenes, based on the trichothecane skeleton (**1**), have now been isolated from natural sources. They are made up of 113 non-macrocyclic and 69 macrocyclic compounds (*2*). Thirty-four more naturally-occurring trichothecenes have therefore been described since 1986 (*1*) and of these 30 are non-macrocyclic compounds.†

In some papers, *e.g.* (*3*), stereo-diagrams of type (**2A**), initially used in 1982 for the unsubstituted nucleus, have been used to represent ring A-substituted 12,13-epoxytrichothec-9-enes, drawn conventionally as (**2**). These "inverted ring A" diagrams can be misleading since, for example, 8β-substituents are, perforce, denoted by broken lines which are normally used to indicate α-substituents in the conventional structure (**2**).

Apotrichothecane (**3**), by definition (*4*), has *cis* fusion of rings A/B with the 11-H α-oriented. A number of naturally occurring apotrichothecene relatives have been described (see Sect. 5.2) with rings A/B *trans* fused and the 11-H β-oriented (**4**). These relatives are 11-epiapotrichothec-9-enes.

Some confusion has been caused by authors (*5, 6*) using the name trichothecene for its 12,13-epoxide (**2**; R=H). The last named authors (*6, 7*) have also used the trivial name trichoene for the skeleton (**5**) (biosynthetic numbering, no position 1) when 12,13-epoxytrichoene, based on skeleton (**6**), would have been more appropriate. In this review trichoene is used for skeleton (**6**), whence 9,12-trichodiene (**7**), the biosynthetic precursor of the trichothecenes, is correctly derived.

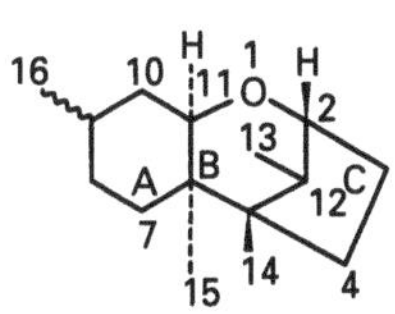

(**1**)

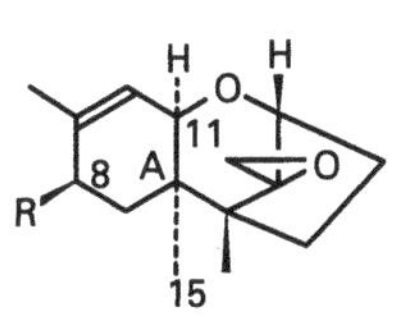

(**2**)

≡

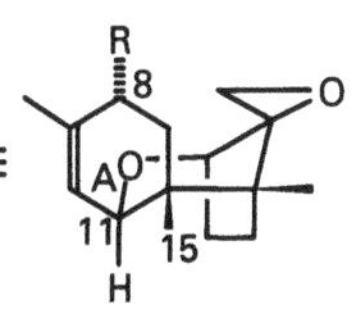

(**2A**)

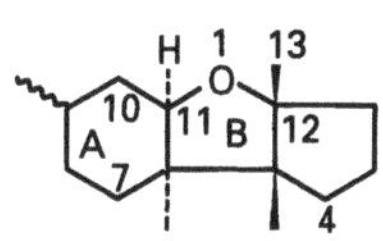

(**3**)

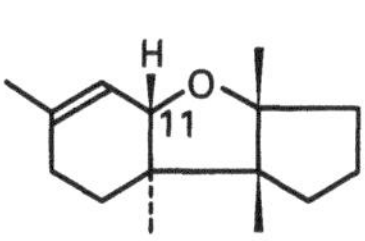

(**4**)

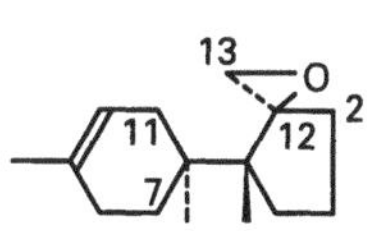

(**5**)

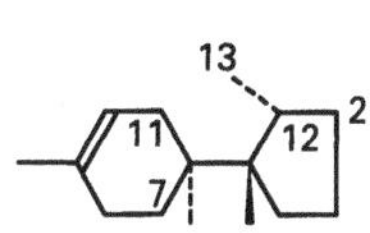

(**6**)

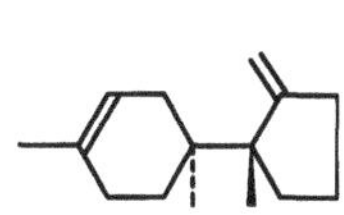

(**7**)

2. Naturally Occurring Compounds

Twenty eight simple 12,13-epoxytrichothecenes, isolated from natural sources since the publication (*1*) of Part 1, are listed in Tables 1 and 2.

---

† Hereafter, in this review, called "simple trichothecenes".

Table 1. *Naturally Occurring Simple 12,13-Epoxytrichothec-9-enes That Have Been Described Since 1986*

Substituents in (**8**)								
R^1	R^2	R^3	R^4	R^5	Trivial name	Formula	Organism[a, b]	Refs.
H	H	H	H	OH	Isotrichodermol	$C_{15}H_{22}O_3$	*F. crookwellense, F. graminearum*	*(9, 10)*
OH	H	H	H	OH	8-Hydroxyisotrichodermol	$C_{15}H_{22}O_4$	*F. crookwellense*	*(9)*
H	OH	H	H	OH	7-Hydroxyisotrichodermol	$C_{15}H_{22}O_4$	*F. crookwellense*	*(9)*
OH	OH	OH	H	OH		$C_{15}H_{22}O_6$	*F. graminearum*	*(11)*
OH	OH	H	H	OAc	7,8-Dihydroxyisotrichodermin	$C_{17}H_{24}O_6$	*F. crookwellense*	*(9)*
H	H	OH	OH	OAc	3-Acetoxyscirpene-4,15-diol[c]	$C_{17}H_{24}O_6$	*F. sambucinum, F. camptoceras*	*(12, 13)*
OH	OH	OH	H	OAc	15-Deacetyl-7,8-dihydroxycalonectrin	$C_{17}H_{24}O_7$	*F. culmorum*	*(14)*
H	H	H	OR^6	H	Isocrotonyltrichodermol	$C_{19}H_{26}O_4$	*T. roseum, Spicellum roseum*	*(15, 16)*
H	H	OH	OAc	OAc	3,4-Diacetoxyscirpen-15-ol[c]	$C_{19}H_{26}O_7$	*F. sambucinum*	*(12)*
H	H	OAc	OH	OAc	3,15-Diacetoxyscirpen-4-ol[c]	$C_{19}H_{26}O_7$	*F. sambucinum*	*(17, 12)*
OAc	H	OAc	OH	OH	Acuminatin[c]	$C_{19}H_{26}O_8$	*F. acuminatum*	*(18)*
							F. equiseti (*compactum*)	*(19)*
OR^7	H	OAc	OAc	OH	8-Propionylneosolaniol	$C_{22}H_{30}O_9$	*F. sporotrichioides, F. sambucinum*	*(20, 21)*
OR^8	H	OAc	H	OH	4-Deacetoxy-T-2 toxin	$C_{22}H_{32}O_7$	*F. sporotrichioides*	*(22)*
H	H	H	OR^{18}	H	Harzianum A	$C_{23}H_{28}O_6$	*Trichoderma harzianum*	*(22a)*
OAc	H	OAc	OAc	OAc	Diacetylneosolaniol[c] (TetraacetylT-2 tetraol)	$C_{23}H_{30}O_{10}$	*F. acuminatum*	*(18)*
H	H	OH	OR^9	H	(6″R)-Trichoverrol A,B	$C_{23}H_{32}O_7$	*M. verrucaria*	*(23)*
OR^{10}	H	OAc	OAc	OH	8-Isobutyrylneosolaniol	$C_{23}H_{32}O_9$	*F. sporotrichioides*	*(20)*
OR^8	H	OAc	OH	OAc	Iso-T-2 toxin[c]	$C_{24}H_{34}O_9$	*F. sporotrichioides*	*(24)*
OR^{11}	H	OAc	OAc	OH	8-n-Pentanoylneosolaniol	$C_{24}H_{34}O_9$	*F. sporotrichioides*	*(25)*
OR^{12}	H	OAc	OAc	OH	8-n-Hexanoylneosolaniol	$C_{25}H_{36}O_9$	*F. sporotrichioides*	*(25)*
H	H	OR^{13}	OR^{14}	H	(6″R)-Isotrichoverrin A,B	$C_{29}H_{40}O_9$	*M. verrucaria*	*(23)*
H	H	OR^{15}	OR^9	H	Trichoverrin C (6″R, 7″R)	$C_{29}H_{40}O_9$	*M. verrucaria*	*(23)*

Table 2. *Naturally Occurring Simple 12,13-Epoxytrichothec-9-en-8-ones That Have Been Described Since 1986*[d]

Substituents in (**9**)							
R^1	R^2	R^3	R^4	Trivial name	Formula	Organism[a, b]	Refs.
H	OH	OH	OH	7-Deoxynivalenol[e]	$C_{15}H_{20}O_6$	*F. graminearum*, *F. camptoceras*	(*26*, *13*)
H	H	H	OAc	8-Oxoisotrichodermin	$C_{17}H_{22}O_5$	*F. crookwellense*	(*9*)
OAc	OH	H	OH	7-Acetylvomitoxin	$C_{17}H_{22}O_7$	*F. camptoceras*	(*13*)
H	OAc	OAc	OH	8-Oxodiacetoxyscirpenol[c]	$C_{19}H_{24}O_8$	*F. sporotrichioides*, *F. culmorum*	(*27*, *10*)
						F. crookwellense	(*9*)

Table 3. *New Post-1986 Sources of Simple 12,13-Epoxytrichothec-9-enes*

Substituents in (**8**)								
R^1	R^2	R^3	R^4	R^5	Trivial name	Formula	Source[a, b]	Refs.
H	H	H	H	H		$C_{15}H_{22}O_2$	*F. culmorum, F. crookwellense*	(*28, 10*)
							F. graminearum	(*10*)
H	H	H	OH	H	Trichodermol	$C_{15}H_{22}O_3$	*S. cylindrospora*	(*29*)
H	H	OH	H	OH	Isoverrucarol	$C_{15}H_{22}O_4$	*F. sporotrichioides, F. oxysporum*	(*30, 31*)
H	H	OH	OH	OH	Scirpenetriol	$C_{15}H_{22}O_5$	*F. sambucinum, F. camptoceras*	(*17, 13*)
							F. acuminatum, F. equiseti	(*32, 32*)
OH	H	OH	OH	OH	T-2 tetraol	$C_{15}H_{22}O_6$	*F. acuminatum*	(*32*)
H	H	H	OAc	H	Trichodermin	$C_{17}H_{24}O_4$	*S. cylindrospora*	(*29*)
H	H	H	H	OAc	Isotrichodermin	$C_{17}H_{24}O_4$	*F. sporotrichioides, F. sambucinum*	(*20, 10*)
							F. crookwellense	(*9*)
OH	H	H	H	OAc	8-Hydroxyisotrichodermin	$C_{17}H_{24}O_5$	*F. culmorum, F. crookwellense*	(*33, 9*)
H	OH	H	H	OAc	7-Hydroxyisotrichodermin	$C_{17}H_{24}O_5$	*F. culmorum, F. crookwellense*	(*33, 9*)
H	H	OAc	H	OH	3-Deacetylcalonectrin	$C_{17}H_{24}O_5$	*F. crookwellense, F. graminearum*	(*10, 10*)
H	H	OH	H	OAc	15-Deacetylcalonectrin	$C_{17}H_{24}O_5$	*F. sporotrichioides*	(*30*)
H	H	OAc	OH	OH	15-Acetoxyscirpenediol	$C_{17}H_{24}O_6$	*F. poae, F. sambucinum*	(*34, 17*)
H	H	OH	OAc	OH	4-Acetoxyscirpenediol	$C_{17}H_{24}O_6$	*F. sambucinum, unidentified F.* sp.	(*17, 35*)
							F. camptoceras	(*13*)
OAc	H	OH	OH	OH	8-AcetylT-2 tetraol	$C_{17}H_{24}O_7$	*F. acuminatum*	(*18*)
OH	H	OAc	OH	OH	15-AcetylT-2 tetraol	$C_{17}H_{24}O_7$	*F. acuminatum*	(*18*)
OH	H	OH	OAc	OH	NT-2	$C_{17}H_{24}O_7$	*F. acuminatum*	(*18*)
H	H	OAc	OAc	H	Diacetylverrucarol[f]	$C_{19}H_{26}O_6$	*Baccharis coridifolia*	(*36*)
H	H	OAc	H	OAc	Calonectrin	$C_{19}H_{26}O_6$	*F. sporotrichioides, F. crookwellense*	(*20, 10*)
H	OH	OAc	H	OAc	7-Hydroxycalonectrin	$C_{19}H_{26}O_7$	*F. culmorum*	(*33*)

References, pp. 54–70

H	H	OAc	OAc	OH	Diacetoxyscirpenol	$C_{19}H_{26}O_7$	*F. poae, F. equiseti (compactum)*	(*34, 19*)
							F. sambucinum, F. culmorum	(*37, 10*)
							F. graminearum, F. crookwellense	(*38, 39*)
							F. avenaceum, F. oxysporum	(*38, 24*)
OAc	H	OH	OAc	OH	NT-1	$C_{19}H_{26}O_8$	*F. equiseti (compactum)*[g]	(*40*)
							F. acuminatum	(*18*)
OH	H	OAc	OAc	OH	Neosolaniol	$C_{19}H_{26}O_8$	*F. acuminatum, F. sambucinum, F. solani*	(*18, 21, 41*)
							F. crookwellense, F. tumidum	(*41a, 41b*)
OH	OH	OAc	OAc	OH		$C_{19}H_{26}O_9$	*F. camptoceras*	(*13*)
H	H	OAc	OAc	OAc	Triacetoxyscirpene	$C_{21}H_{28}O_8$	*F. sambucinum*	(*17*)
OAc	H	OAc	OAc	OH	Acetylneosolaniol	$C_{21}H_{28}O_9$	*F. acuminatum, F. equiseti (compactum)*	(*18, 19*)
							F. sambucinum	(*21*)
OR^{8}	H	OAc	OH	OH	HT-2 toxin	$C_{22}H_{32}O_8$	*F. equiseti, F. acuminatum, F. solani*	(*41, 41, 41*)
							F. sambucinum, F. oxysporum	(*21, 24*)
							F. graminearum, F. culmorum	(*41a, 41a*)
							F. moniliforme	(*41c*)
H	H	H	OR^{16}	H	Trichodermadiendiol A, B	$C_{23}H_{32}O_6$	*S. chartarum*	(*41d*)
H	H	OH	OR^{16}	H	Trichoverrol A, B	$C_{23}H_{32}O_7$	*S. chartarum, S. albipes*	(*42, 42*)
							S. kampalensis, S. microspora	(*42, 42*)
OR^{17}	H	OAc	OAc	OH	8-Butyrylneosolaniol	$C_{23}H_{32}O_9$	*F. sporotrichioides*[h], *F. sambucinum*	(*20, 21*)
OR^{8}	H	OAc	OAc	OH	T-2 toxin	$C_{24}H_{34}O_9$	*F. stilboides, F. sambucinum, F. culmorum*	(*38, 21, 38*)
							F. graminearum, F. crookwellense	(*38, 38*)
							F. solani, F. subglutinans	(*41, 41c*)
OR^{8}	H	OAc	OH	OAc	isoT-2 toxin	$C_{24}H_{34}O_9$	*F. graminearum*	(*42a*)
OR^{8}	H	OAc	OAc	OAc	AcetylT-2 toxin	$C_{26}H_{36}O_{10}$	*F. sporotrichioides, F. graminearum*	(*24, 42a*)
H	H	OH	OR^{19}	H	Roridin L2	$C_{29}H_{38}O_9$	*S. chartarum*	(*41d*)

Table 4. *New Post-1986 Sources of Simple 12,13-Epoxytrichothec-9-en-8-ones*

Substituents in (**9**)							
R^1	R^2	R^3	R^4	Trivial name	Formula	Source[a,b]	Refs.
OH	OH	H	OH	Vomitoxin (deoxynivalenol)	$C_{15}H_{20}O_6$	*F. avenaceum, F. semitectum, F. equiseti*	(*43, 43, 44*)
						F. moniliforme, F. oxysporum	(*45, 44*)
						F. acuminatum, F. crookwellense	(*41a, 41a*)
						F. subglutinans, F. poae	(*45a, 41a*)
OH	OH	OH	OH	Nivalenol	$C_{15}H_{20}O_7$	*F. crookwellense, F. poae, F. camptoceras*	(*46, 47, 13*)
						F. sporotrichioides, F. culmorum	(*41a, 41a*)
OH	OH	H	OAc	3-Acetylvomitoxin	$C_{17}H_{22}O_7$	*F. camptoceras, F. semitectum*	(*13, 43*)
OH	OAc	H	OH	15-Acetylvomitoxin	$C_{17}H_{22}O_7$	*F. sporotrichioides, F. semitectum*	(*41, 43*)
						F. avenaceum, F. crookwellense	(*41a, 41a*)
						F. culmorum, F. equiseti	(*41a, 41a*)
						F. poae	(*41a*)
OH	OH	OAc	OH	Fusarenone	$C_{17}H_{22}O_8$	*F. culmorum, F. crookwellense, F. poae*	(*43, 46, 47*)
						F. acuminatum, F. sporotrichioides	(*41a, 41a*)
H	H	OR^6	H	Trichothecin	$C_{19}H_{24}O_5$	*F. graminearum*	(*48*)
OH	OAc	H	OAc	3,15-Diacetylvomitoxin	$C_{19}H_{24}O_8$	*F. culmorum*	(*28*)
OH	OAc	OAc	OH	4,15-Diacetylnivalenol	$C_{19}H_{24}O_9$	*F. crookwellense, F. sambucinum*	(*9, 10*)
						F. camptoceras	(*13*)

Footnotes to Tables 1–4

[a] *F* = *Fusarium*, *M* = *Myrothecium*, *S* = *Stachybotrys*, *T* = *Trichothecium*.
[b] *Fusaria* are named according to the scheme used in ref. (*1*). *S. atra* and *S. chartarum* are synonyms (*2*).
[c] Known compound previously obtained by chemical modification of other naturally occurring trichothecenes.
[d] The macrolide structure assigned to the substance named isotrichothecin (*49*), from *T. roseum*, is unlikely to be correct. This substance has not been included in assessing the total number of known trichothecenes.
[e] Omitted from ref. (*1*).
[f] The derivative, 4*β*,15-diacetoxy-10,13-cyclotrichothecan-9*α*,12-diol, $C_{19}H_{28}O_7$, also isolated from *B. coridifolia* (*36*) is presumed to be an artifact.
[g] Subsequently shown (*19*) to be the isomer, acuminatin.
[h] Now characterized.

$R^6 = COCH \overset{Z}{=} CHMe$
$R^7 = COEt$
$R^8 = COCH_2CHMe_2$
$R^9 = COCH \overset{Z}{=} CHCH \overset{E}{=} CH\overset{R}{C}H(OH)\overset{R,S}{C}H(OH)Me$
$R^{10} = COCHMe_2$
$R^{11} = CO(CH_2)_3Me$
$R^{12} = CO(CH_2)_4Me$
$R^{13} = COCH \overset{E}{=} C(Me)CH_2CH_2OH$
$R^{14} = COCH \overset{E}{=} CHCH \overset{E}{=} CH\overset{R}{C}H(OH)\overset{R,S}{C}H(OH)Me$
$R^{15} = COCH_2CMe \overset{E}{=} CHCH_2OH$
$R^{16} = COCH \overset{Z}{=} CHCH \overset{E}{=} CH\overset{S}{C}H(OH)\overset{S,R}{C}H(OH)Me$
$R^{17} = CO(CH_2)_2Me$
$R^{18} = COCH \overset{Z}{=} CHCH \overset{E}{=} CHCH \overset{E}{=} CHCO_2H$
$R^{19} = COCH \overset{Z}{=} CHCH \overset{E}{=} CHCH(CH_2OH)OCH_2CH_2$—

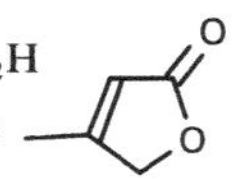

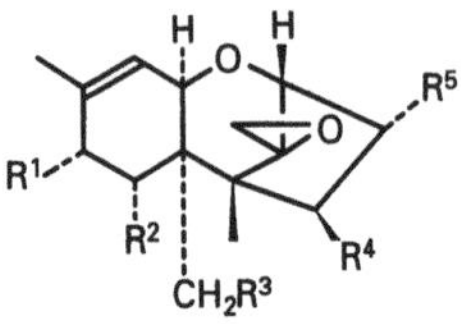

(8)

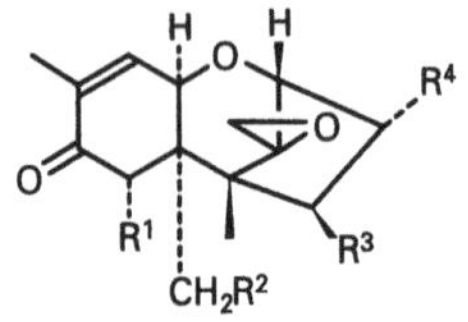

(9)

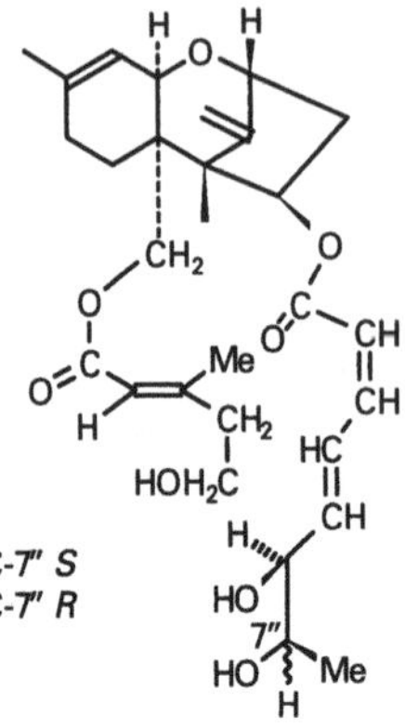

(10A) C-7″ S
(10B) C-7″ R

Many had previously been obtained by chemical modification of known naturally occurring trichothecenes. Additionally, the $C_{29}H_{40}O_8$ compounds 12,13-deoxytrichoverrin A (**10A**) and B (**10B**) have been obtained from *M. verrucaria* (*8*). New sources, reported since 1986, of simple trichothecenes known previously, are set out in Tables 3 and 4.

Several new species of *Fusarium* have been described in recent years, but only *F. crookwellense* (*50*) (Discolor Section) has been shown to produce trichothecenes. Newcomers to the list of trichothecene producers (Tables 3 and 4) are *F. stilboides* (Lateritium Section) (T-2 toxin) (*38*); *F. subglutinans* (Liseola Section) [T-2 toxin (*41c*) and (as *F. sacchari* var *subglutinans*) vomitoxin (*45a*)]; *F. tumidum* (Martiella Section) (neosolaniol) (*41b*); and *F. camptoceras* (Arthrosporiella Section) (*13*).

The species *F. acuminatum* has been amended (*51*) to include a subspecies *F. acuminatum* subsp. *armeniacum*. Trichothecene production by the latter has been investigated (*52*): no specificity was discovered, but the probability of detecting this was diminished by the inclusion of a hydrolysis step in the analytical procedure employed. The subspecies are not differentiated in Tables 1–4.

Of interest is the demise of *F. nivale* from which the name nivalenol was derived. Strains attributed to *F. nivale* which produced nivalenol and its relatives were first shown to be *F. sporotrichioides* (*1*): now, the organism is no longer considered to be a Fusarium and has been transfered to the genus *Microdochium* (*53*). *F. compactum* (*19*) is equated with *F. equiseti* (*54*).

The classification of some *Fusarium* spp. (*sambucinum, culmorum, graminearum, crookwellense*) in the Discolor Section into various chemotypes, based on differences in the structures of the trichothecenes produced, has been reported (*55*), and relates to the presence (or absence) of specific oxygenases and esterases. Further evidence has accrued (*56,57,57a*) for the division of *F. graminearum* strains into two chemotypes producing either vomitoxin- or nivalenol-related trichothecenes.

Some workers have expressed concern about the morphological diversity of strains classified as *F. sambucinum*. *F. sambucinum* sensu lato has been divided, on the basis of morphological and cultural features, into *F. sambucinum* sensu stricto, *F. torulosum*, and *F. venenotum* (*57b*). Investigation of some of these strains by three groups of workers (*57c, 57d, 57e*) showed that whereas strains classified as *F. torulosum* did not produce trichothecenes, those classified as *F. venenotum* produced only diacetoxyscirpenol, whilst *F. sambucinum* sensu stricto strains yielded mainly diacetoxyscirpenol but also T-2 toxin and neosolaniol.

In a completely different approach (*57f*), *F. sambucinum* sensu lato strains were divided into two groups. Toxic (trichothecene-producing) strains were found to yield, in addition, large amounts of a diverse range of sesquiterpene hydrocarbons but nontoxic strains gave only smaller amounts of sesquiterpene hydrocarbons with less chemical diversity.

The chemical evidence (*58,59*) for the production of diacetoxyscirpenol, T-2 toxin and trichothecolone (**9**; $R^1=R^2=R^4=H$, $R^3=OH$) by authenticated strains of *F. moniliforme* (Liseola Section) has been criticized (*60*), but appears to this reviewer to be secure. The production of both vomitoxin (*45*) and HT-2 toxin (*41c*) by this organism has recently been claimed.

The isolation of scirpenetriol together with all seven of its acetylated derivatives from a single strain of *F. sambucinum* has been recorded (*12*).

(6″*R*)-Trichoverrols [for numbering, (but not configuration) see 12,13-deoxytrichoverrin A and B (**10**)] have now been isolated in addition to the known isomers. All four trichoverrols, diastereoisomeric at positions 6″ and 7″, have thus been obtained from *M. verrucaria* (*23*). If this finding is repeated with the trichoverrins, as seems probable, it will be of significance to any discussion (*2*) of the macrolide cyclisation step in the biosynthesis of the macrocyclic trichothecenes. Although all baccharinoids and many roridins appear to be derived from trichoverrin-like precursors with 6″*S*, until recently (*41d*), isororidin E (**11**; 6′*S*, 13′*S*) was exceptional in requiring a precursor with 6″*R*. The isolation, from *S. chartarum*, of all four possible C-6′/C-13′ diastereomers (**11**) of the roridin E structure (see p. 12) has been claimed (*41d*), consistent with the existence of the four C-6″/C-7″ diastereoisomeric trichoverrols.

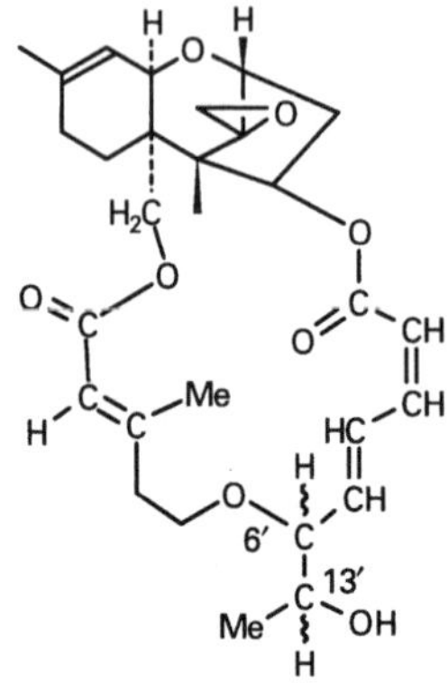

(11)

3. Unnatural Trichothecenes

Syntheses of racemic 12,13-epoxytrichothecene (**8**; $R^1=R^2=R^3=R^4=R^5=H$) (*61*), trichodermol (**8**; $R^1=R^2=R^3=R^5=H$, $R^4=OH$) (*62, 61*) and the 8-one (**11a**) (*63*) have been reported. The trichothec-9,12-diene antipode (**12**) has been prepared (*64, 65*), and a new route to 12,13-epoxytrichothecene has been applied to the synthesis of its antipode (*66*).

Semi-synthetic 12-epi analogues (**13**; R=H or Ac) of diacetoxyscirpenol have been prepared (*67, 68*).

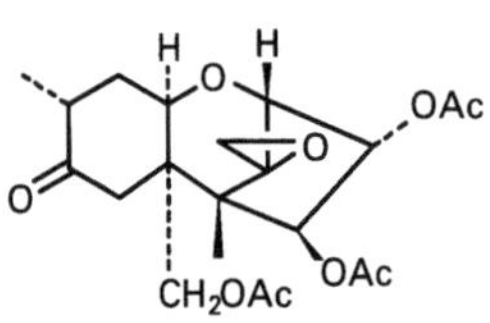

(11a)

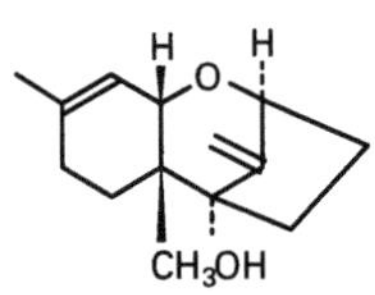

(12)

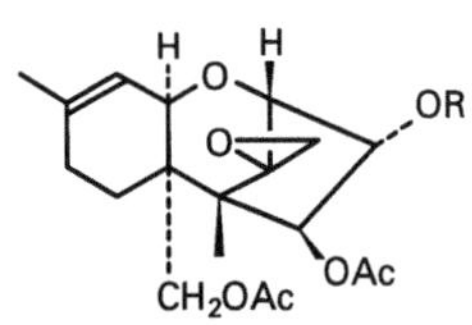

(13)

The algal product (−)-filiformin (**13a**; R = Br) (*68a*) and some analogs, including (**13a**; R = H) (*68a*) and (**13b**) (*68b*) have been synthesized.

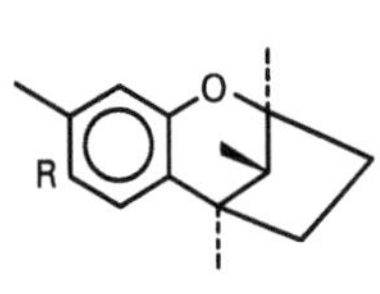

(13a)

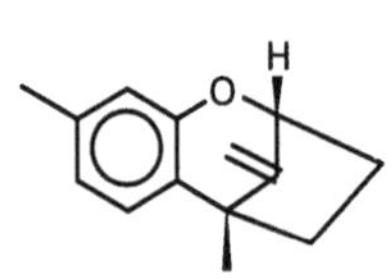

(13b)

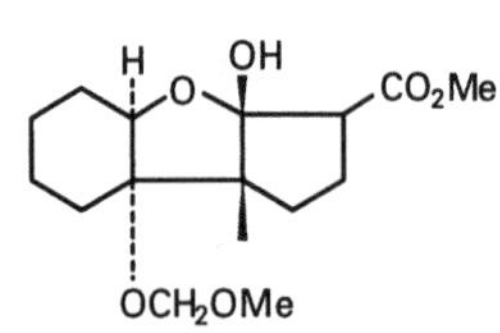

(13c)

These compounds can be regarded as ring A aromatic trichothecanes. The apotrichothecane analog (**13c**) has also been prepared (*68c*).

4. Chemistry

The classification of the simple trichothecenes into 3 groups according to the substitution pattern of ring A, as outlined in Part 1 (*1*), is maintained. The account is arranged in terms of the reactivity of functional groups.

Some further evidence is now available concerning the ring A conformation of compounds in Groups II (α-substituents at positions 8 or 10) and III (9-en-8-ones). X-Ray crystallographic structures of T-2 toxin (**8**; R^1=$OCOCH_2CHMe_2$, R^2=H, R^3=R^4=OAc, R^5=OH) (*69*) and 8-*n*-pentanoylneosolaniol (**8**; R^1=OCO$(CH_2)_3$Me, R^2=H, R^3=R^4=OAc, R^5=OH) (*25*) were consistent with the ring A half-chair conformation (**14**; 16-Me omitted for clarity) found for diacetoxyscirpenol (**8**; R^1=R^2=H, R^3=R^4=OAc, R^5=OH) (*17*). The solid state ring A conformations of these three trichothecenes were almost identical, and there was no evidence for conformational change associated with the introduction of bulky 8α-substituents. This conclusion contrasts with the chemical evidence (see Sect. 4.1.2 below).

Investigations by nmr spectroscopy showed that the ring A half-chair conformation (**14**) of T-2 toxin was retained in both $CDCl_3$ and C_6D_6 (*70*). On the ot her hand, nOe studies in $CDCl_3$ of the synthetic model 10α-benzenesulfinyl derivative (**15**) showed that ring A had adopted the skew boat conformation (**16**), with the benzenesulfinyl group equatorial (*66*).

In Group III, the ring A hemiacetal form (**17**) of nivalenol (**9**; R^1=R^2=R^3=R^4=OH), first discovered (*71*) by solid state ir spectroscopy, has now been shown on ^{13}C nmr evidence (*72*) to be present (18%) in solution in dimethyl sulfoxide-d_6. Remarkably, the equilibrium with vomitoxin (**9**; R^1=R^2=R^4=OH, R^3=H) wholly favoured the keto form in all the solvent systems examined (*72, 73*) (cf. Sect. 4.1.3.1, below). The 4,8-dione (**18**) preferred to form the 15→4 hemiacetal (**19**) (*72*) in $CDCl_3$.

Trichothecenes give uv fluorescence enhancement with 2-(diphenylacetyl)indan-1,3-dione 1-hydrazone and its derivatives (*74*). This is believed to be due to molecular association through hydrogen bonding, and is effective at 25°C. As such it has advantages over methods of detection involving a chemical reaction with the 12,13-epoxide (see Sect. 4.4.1.2).

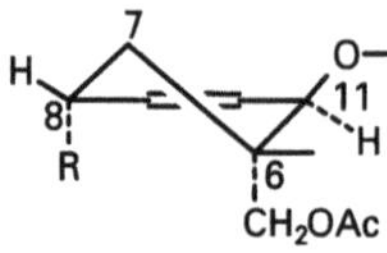

(14)

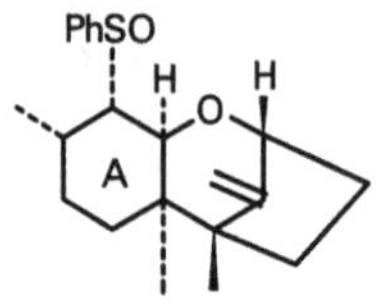

(15)

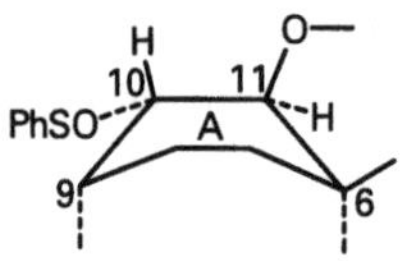

(16)

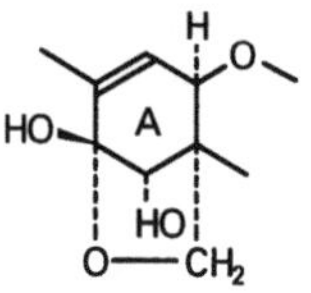

(17)

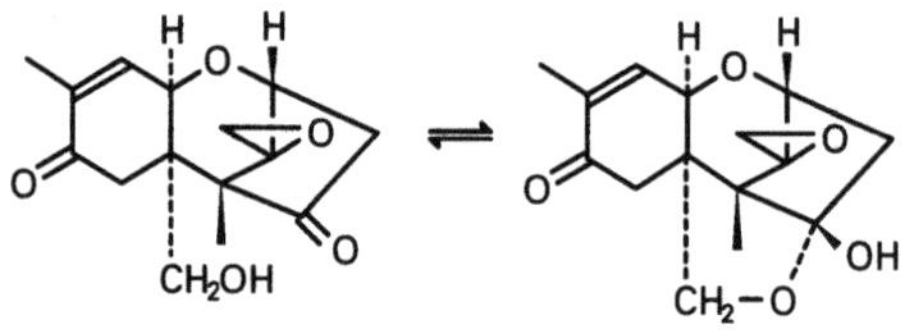

(18) **(19)**

4.1. Ethylenic Double Bonds

4.1.1. Isomerization

Isomerization of the 9(16)-ene (**20**) with trifluoroacetic acid in refluxing acetonitrile or with rhodium(III) chloride in refluxing ethanol gave the 8-ene (**21**) rather than the desired 9-ene (*75*). In so far as the isomerization of a 9(16)-ene and the dehydration of a 9α-ol [which gives a high proportion of 9-ene (*1*)] can be compared, it is clear that the two reactions do not proceed through the same type of tertiary carbocation intermediate. A number of transition metal catalysts, used in olefine isomerization, were ineffective when tried with analogs of (**20**) (*75*).

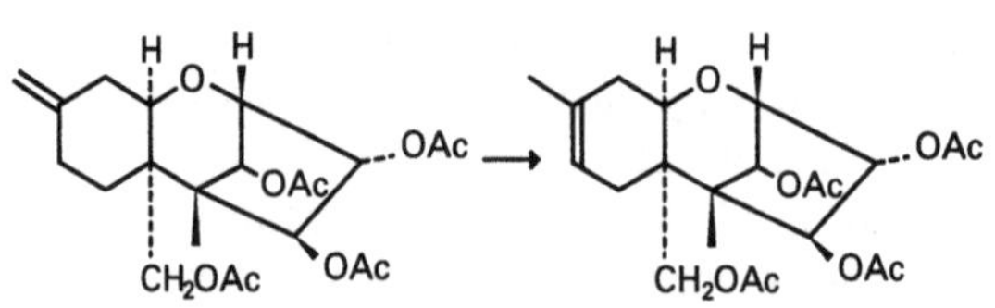

(20) **(21)**

4.1.2. Catalytic Reduction

Hydrogenation of T-2 tetraol (**8**; R^1=R^3=R^4=R^5=OH, R^2=H) (Group II) to give a mixture of 9-epimers, in contrast to the stereospecific hydrogenation of Group I trichothecenes, was attributed (*1*) to a ring A conformational change. Further evidence for this has now become available: reduction of T-2 toxin over 5% palladium on charcoal also gave a 9-epimeric mixture, and was accompanied by some C-8 deoxygenation (hydrogenolysis of the C-8 ester group) (*76*).

4.1.3. Electrophilic Addition Reactions

4.1.3.1. Epoxidation of 9-Enes

In conformity with earlier results, only the β-epoxides were obtained from T-2 toxin (*76*) and diacetylneosolaniol (**8**; R^1=R^3=R^4=R^5=OAc, R^2=H) (*77*) on treatment with 3-chloroperbenzoic acid in chloroform at room temperature. Surprisingly, this reaction was also β-stereospecific with trichothecodiol (**8**; R^1=R^4=OH, R^2=R^3=R^5=H) (*78*). With the 3-ketone (**22**), Baeyer-Villiger oxidation to the lactone (**23**) took precedence over epoxidation of the 9-ene (*76*).

With the Group III trichothecene vomitoxin (**9**; R^1=R^2=R^4=OH, R^3=H) the action of alkaline sodium hypochlorite (*79*) gave the α-epoxide, isolated as the hemiacetal (**24**) (cf. Sect. 4, above).

4.1.3.2. Epoxidation of 12-Enes

To prevent acid-catalysed rearrangement following epoxidation with 3-chloroperbenzoic acid, the use of disodium hydrogen phosphate buffer is recommended (*80*). The reaction is stereospecific giving the 12,13-epoxide. With 4β-hydroxytrichothec-9-enes, further examples of regioselectivity in the presence of a methyl group at position 15 (*62*, *61*), absence of 12-regioselectivity in the presence of a 15-hydroxyl group (*81*), and protection of the 9-ene as the bromo-ether (*80*), have been recorded. There was no regioselectivity in the absence of hydroxyl substituents at both positions *4β* and 15 (*66*). The 8-keto group in the 12-ene (**25**) presented no difficulty (*63*). Trifluoroperacetic acid selectively epoxidized the 12-ene in the deuteriated Group III trichothecene (**26**) (*82*).

4.1.3.3. Epoxidation of 7-Enes

From the ratio of the ultimate products (see Sect. 4.4.2.), epoxidation of the silyloxydiene (**27**) with 1 equivalent of 3-chloroperbenzoic acid in hexane gave a 1:1 mixture of 7β, 8β- and 7α, 8α-epoxides (*68*).

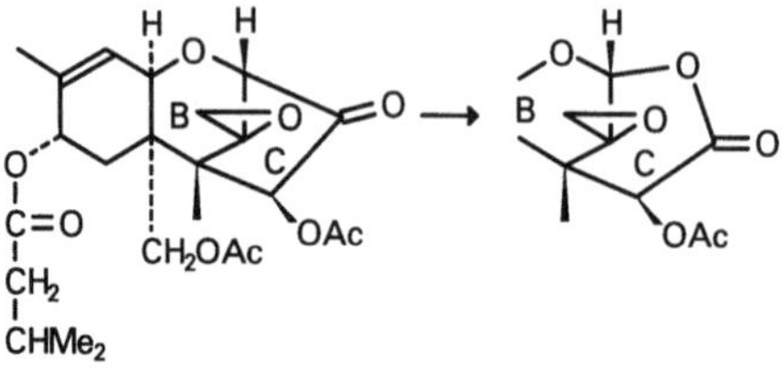

(22) (23)

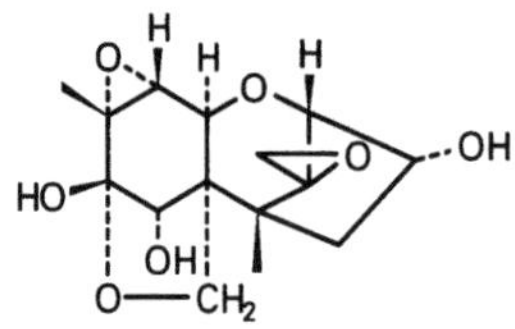

(24)

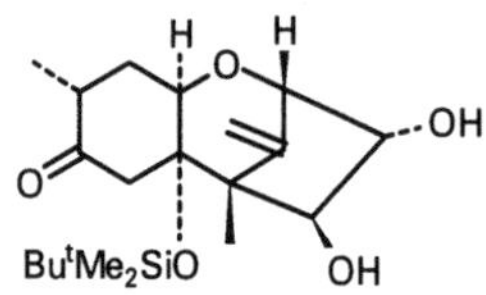

(25)

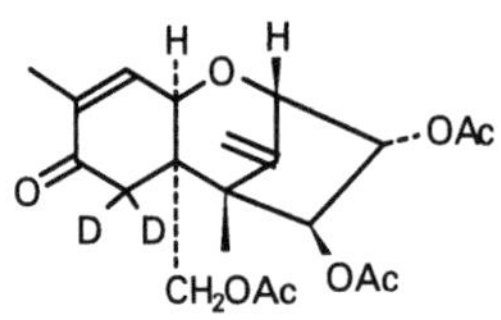

(26)

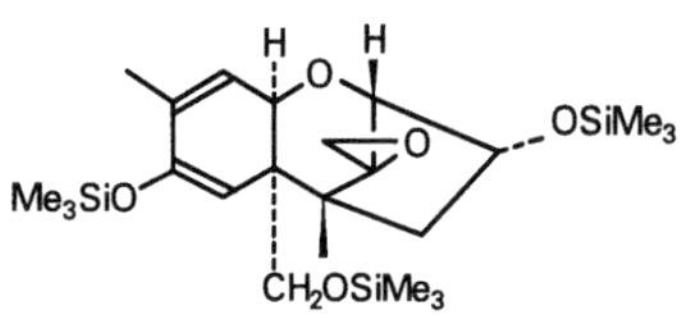

(27)

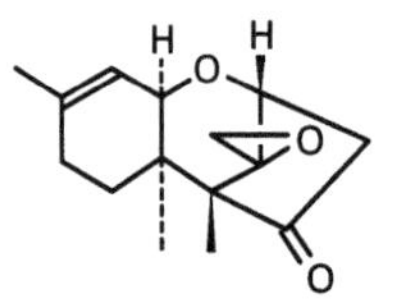

(28)

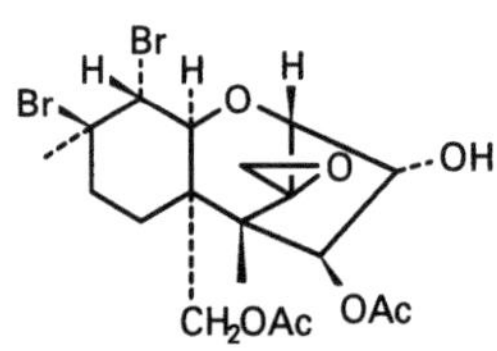

(29)

4.1.3.4. Hydroxylation

In contrast to the α-hydroxylation of trichodermone (**28**) (*1*), the 9β, 10β-diol was stated to be the product from T-2 toxin and osmium tetraoxide (*76*), although the evidence for this configurational assignment was not presented.

4.1.3.5. Bromination

The dibromide (**29**) was obtained on treating diacetoxyscirpenol with bromine and silver acetate (*83*), and also as a by-product of the allylic bromination of diacetoxyscirpenol with *N*-bromosuccinimide in dichloromethane (*77*).

4.1.3.6. Carbene Addition

Reaction of the 12-ene (**30**) with diiodomethane and the Zn-Ag couple in refluxing diethyl ether gave the cyclopropane (**31**) (*67*), which on deprotection/acetylation furnished the 12,13-methano analogue (**32**) of diacetoxyscirpenol.

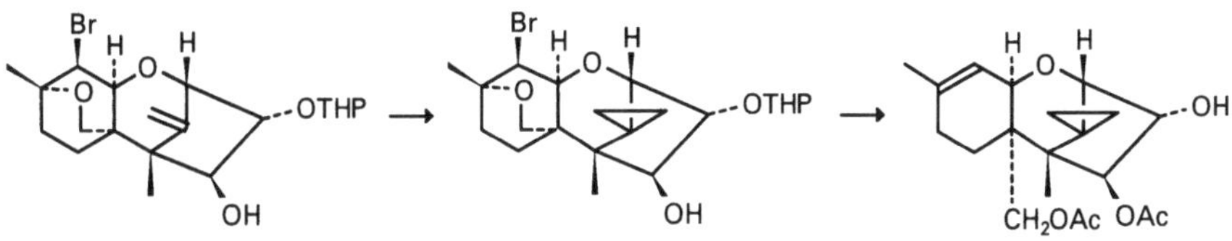

(**30**) (**31**) (**32**)

4.1.4. Reactions at the α-Carbon Atom

Further work on the C-8 allylic bromination of diacetoxyscirpenol with N-bromosuccinimide in dichloromethane or carbon tetrachloride has been reported (*83*, *77*). This reaction is in competition with that outlined in Sect. 4.1.3.5 above. Another (minor) product was the 3α, 11α-epoxytrichothecene (**33**) (*77*).

Although the 8β-ol was the main product of the oxidation of diacetoxyscirpenol and its derivatives with selenium (IV) oxide (*3*, *77*), the 8-oxo- (**34**; $R^1{=}R^2{=}H$, $R^3R^4{=}O$) and 16-oxo-(**34**; $R^1R^2{=}O$, $R^3{=}R^4{=}H$) compounds, expected by analogy with earlier work (*84*) with diacetylverrucarol (**35**; $R^1{=}R^2{=}Ac$), were also obtained (*77*). Selenium(IV) oxidation is not specific to the 8-position and oxidation at position 16 occurs concurrently (*84*, *85*).

Excellent yields of the 8-one have been obtained directly from calonectrin (**36**) using freshly prepared dipyridine chromium trioxide (*86*, *68*).

4.1.5. Ozonolysis

Further examples of the conversion of a 12-ene into the 12-norketone have been reported for compounds in which the 9-ene is protected as the bromo-ether and any OH groups are also protected (*68*, *80*).

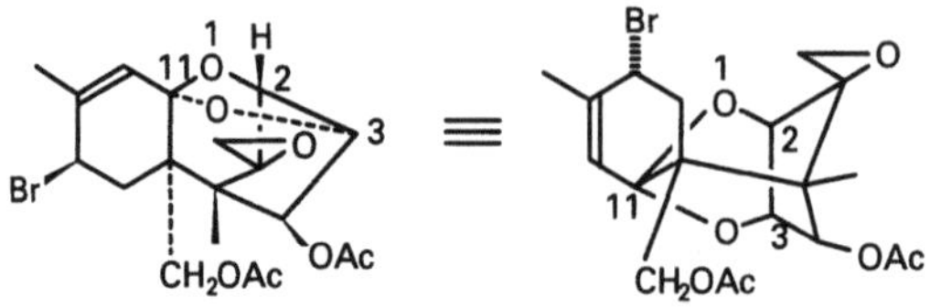

(33)

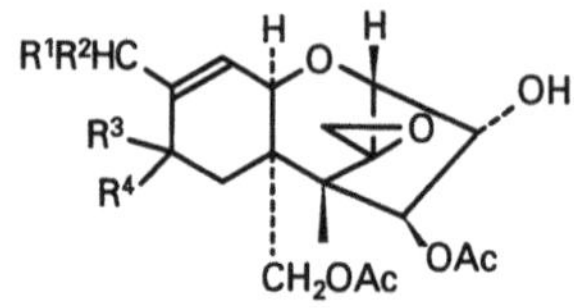

(34)

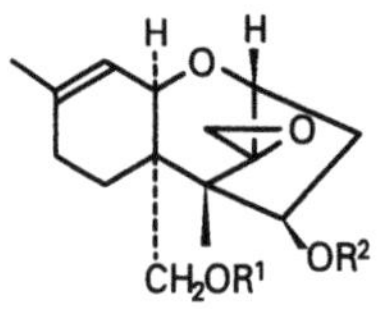

(35)

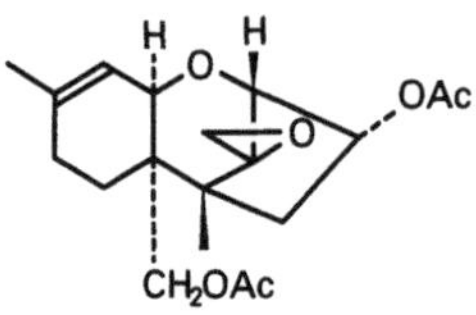

(36)

4.2. Hydroxyl Groups

4.2.1. Regioselective Esterification and Etherification, and Regeneration of the Hydroxyl Function

4.2.1.1. Verrucarol (**35**; R^1=R^2=H)

The 15-OH of verrucarol is selectively acetylated (*1*). By contrast, the 4α-OH in 4-epi-verrucarol was acetylated in preference to the 15-OH by the Steglich method and by other procedures (*87*).

Treatment of verrucarol with (2-trimethylsilylethoxycarbonyl)-imidazole and 1,8-diazabicyclo[5.4.0]undec-7-ene afforded the C-4 mono-protected derivative (**35**; R^1=H, R^2=$COO(CH_2)_2SiMe_3$) (*88*). Levulinate has been used successfully as a 15-protecting group during the attachment of a muconic acid residue at position 4β (*89*): no protecting group was needed in the 4-epi series (*87*). Selective 15-esterification with verrucarinic acid derivatives has been achieved (*1*): less satisfactory results were obtained with the more complex diester (**37**) (*88*), in the presence of bis(2-oxo-3-oxazolidinyl)phosphinic chloride (**38**), triethylamine and 4-dimethylaminopyridine as condensing agents, and use of the above-mentioned C-4 blocking group was indicated.

4.2.1.2. Scirpenetriol (**39**; R^1=R^2=R^3=H)

The ether (**39**; R^1=R^2=H, R^3=THP) underwent selective acetylation at position 15 with acetyl chloride-triethylamine in dichloromethane at

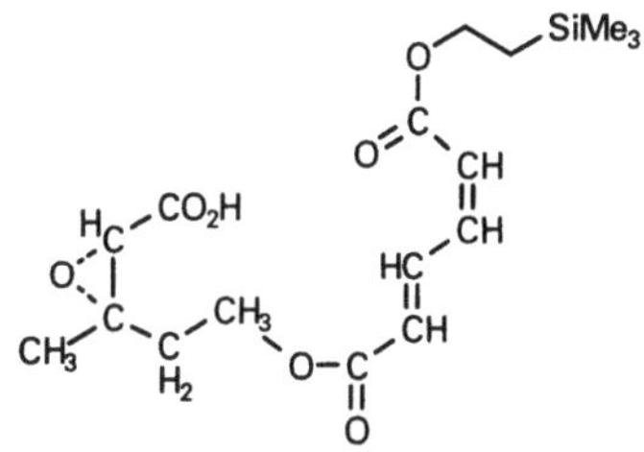

(37)

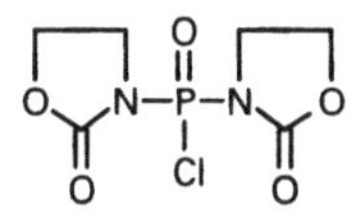

(38)

0°C (*3*), and selective esterification at the same position with levulinic acid in the presence of dicyclohexylcarbodiimide and dimethylaminopyridine (*89*).

A number of 4- and 15-esters containing the arylazido group have been prepared using standard procedures (*90*).

Attempted derivatization with sulfonating reagents of the hindered 3β-ol in the analogue (**40**) failed (*75*).

4.2.1.3. T-2 Tetraol (**41**; $R^1=R^2=R^3=R^4=H$)

The methods adopted (*91*) for the preparation of T-2 toxin-protein conjugates have been assessed (*92*) and improved. The 3α-hemiglutarate was the most efficient intermediate. The stability of T-2 toxin in buffered saline in the pH range 5–12 has been investigated (*93*). Above pH 6.7 degradation proceeded *via* sequential cleavage of the ester side chains.

Treatment of neosolaniol (**41**; $R^1=R^4=H$, $R^2=R^3=Ac$) with aqueous ammonia afforded regiospecific hydrolysis at C-15 to give the triol (**41**; $R^1=R^2=R^4=H$, $R^3=Ac$) (*76*). Selective hydrolysis of the formate residue in the triester (**41**; $R^1=HCO$, $R^2=R^3=Ac$, $R^4=H$) was achieved using triethylamine in boiling methanol (*3*). Some, but not altogether satisfactory, specificity was achieved on mild acid hydrolysis of 3α, 4β-diesters (*94*).

The 3-acetate (**41**; $R^1=H$, $R^2=R^3=R^4=Ac$) was the major product on acetylation of neosolaniol with 1 equivalent of acetic anhydride (*76*).

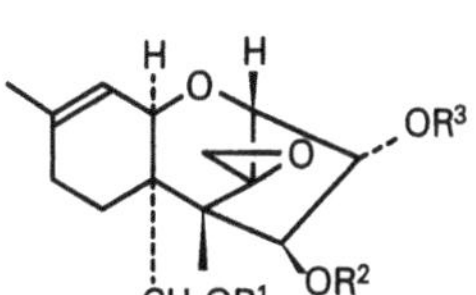

(39)

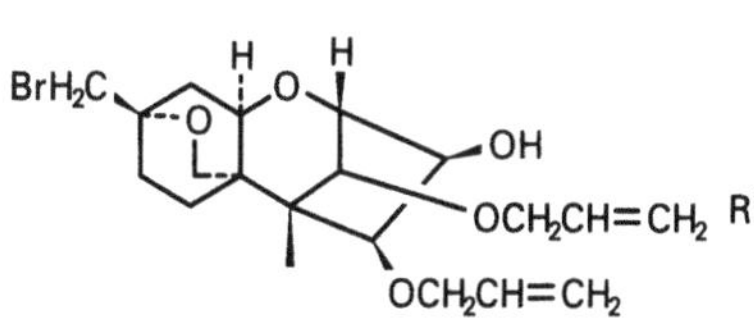

(40)

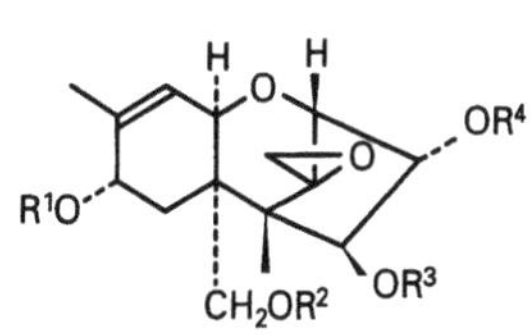

(41)

Attempts, selectively, to form the 8-Bu^tMe_2Si ether of the triol (**41**; $R^1{=}R^2{=}R^3{=}H$, $R^4{=}Bu^tMe_2Si$) failed, and only the 3,15- and 3,4-disilyl ethers were obtained in the ratio 4:1 (*95*). With 2-methoxyethoxymethyl chloride an inseparable mixture of the 8- and 15-ethers in the ratio 63:8 was obtained. Selective 8-*O*-benzylation was successful, but the benzyl group was not easily removed because of competitive reactions at the 9-ene and 12,13-epoxide. This drawback was overcome by the use of *p*-methoxybenzyl chloride: deprotection, after formation of a 4,15-macrolide, was achieved with dichlorodicyanoquinone. This route, but incorporating acetylation of the 4,15-diol, was used for a six-step conversion of T-2 toxin into neosolaniol (*95*). The reaction of HT-2 toxin (**41**; $R^1{=}COCH_2CHMe_2$, $R^2{=}Ac$, $R^3{=}R^4{=}H$) with *t*-butyldimethylsilyl chloride at room temperature for 1 day gave the 3-Bu^tMe_2Si ether; preparation of the 3,4-disilyl ether required a reaction time of 3 days (*96*). The ether (**41**; $R^1{=}R^2{=}R^3{=}H$, $R^4{=}Bu^tMe_2Si$) underwent (8,8′)-intermolecular, Lewis acid-catalyzed, acetal formation with acetone, but (8,15)-intramolecular, acetal formation with benzaldehyde (*96*).

Acid catalyzed transesterification in chloroform solution has been reported for neosolaniol with migration of an acetate residue from position 15 to position 8α (*97*). Transesterification also occurs with *cis*-4-acetoxy-3-hydroxy and -3-acetoxy-4-hydroxy compounds (*76, 75*). This process, when disadvantageous, has been circumvented by employing the allyl residue as a protecting group (*75*).

4.2.1.4. Vomitoxin (9; $R^1{=}R^2{=}R^4{=}OH$, $R^3{=}H$) and Nivalenol (9; $R^1{=}R^2{=}R^3{=}R^4{=}OH$)

Experimental details for the selective 15-acetylation of vomitoxin (*1*), and for the six-step conversion of 3-acetylvomitoxin into the 15-acetyl isomer have been published (*98*). Vomitoxin formed a 7,15-boronate with phenylboronic acid in acetone (*99*). Protection of the 3-OH of 15-acetylvomitoxin with a *t*-butyldimethylsilyl group, followed by hydrolysis, succinylation with succinic anhydride in pyridine, and removal of the protecting group with fluoride ion, allowed the formation of a 15-hemisuccinate (*99a*).

Fatty acid esters of vomitoxin at positions 3 and 15 were obtained in the ratio 3:1 by treatment with the appropriate acid chloride in the presence of pyridine (*100*). Vomitoxin 3- and 15-glucosides resulted from the standard reaction of 15- and 3-acetylvomitoxin, respectively, with acetobromoglucose in toluene in the presence of cadmium carbonate, followed by hydrolysis of the protecting acetate residues (*100*).

The reaction of nivalenol with succinic anhydride was non-selective, even in the presence of *n*-butylboronic acid, introduced as an *in situ* reagent for the protection of the 7- and 15-OH groups (*101*).

Acid hydrolysis of triacetylvomitoxin gave the 7,15-diacetyl compound in reasonable yield (*99*).

4.2.1.5. 12-Ols

In some 13-nortrichothecene analogues, *e.g.* (**42**) and (**43**), containing a secondary alcohol group at position 12, the 3α- and 4β-hydroxy groups were acetylated with acetic anhydride and pyridine with only partial acetylation of the 12-ol (*102, 75*). The diol (**44**) likewise underwent selective benzoylation at position 4 (*103*).

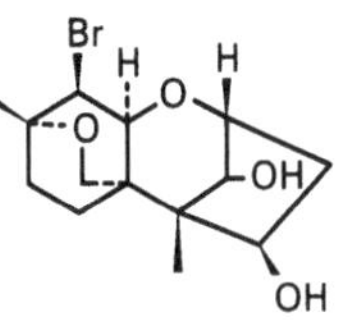

(**42**)

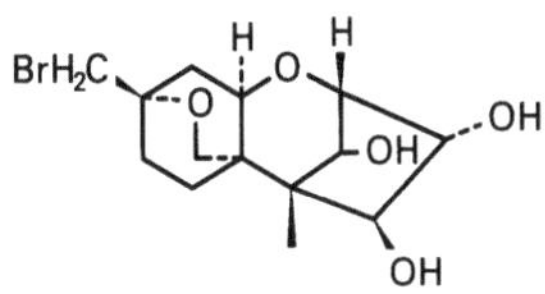

(**43**)

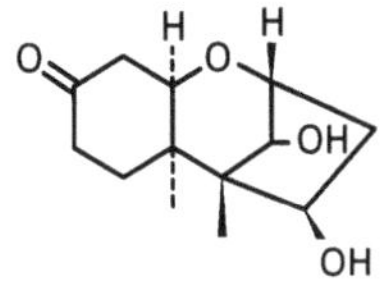

(**44**)

4.2.2. Regioselective Oxidation

4.2.2.1 8-Ols

In the oxidation of acetyl T-2 toxin with selenium dioxide in acetic acid (*104*) the yield of the 8-oxo product (**45**; R=Ac) was improved when the radical initiating agent azoisobutyronitrile was added as catalyst (*105*). The chromic oxide-pyridine reagent can only be used, *e.g.* to give (**45**; R=$SiMe_2Bu^t$), when other OH functions are protected (*106, 86, 68*). Manganese dioxide in chloroform at room temperature has been used extensively for the regiospecific oxidation of allylic 8-ols (*1*), but in the presence of a 4-ol (in acetone), the 4,8-diketone was obtained (*72*). In this example the 15-ol was unaffected, but in others (*105*), the 8,15-dioxo compound was a secondary product. Oxidation of neosolaniol (**41**; $R^1=R^4=H$, $R^2=R^3=Ac$) with pyridinium dichromate in dichloromethane afforded the 8-one (*76*).

4.2.2.2. 15-Ols

Oxidation of verrucarol with the tris(triphenylphosphine) ruthenium chloride reagent afforded the 15-aldehyde (**46**) (*107*). In the 4-epi, 11-

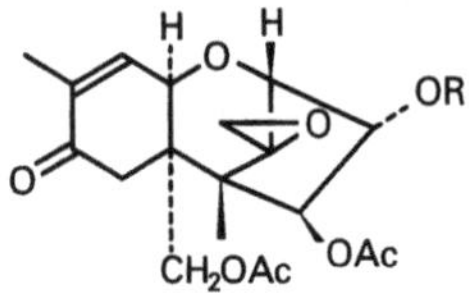

(**45**)

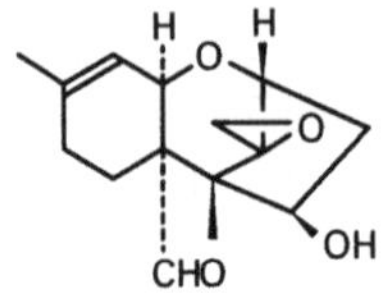

(**46**)

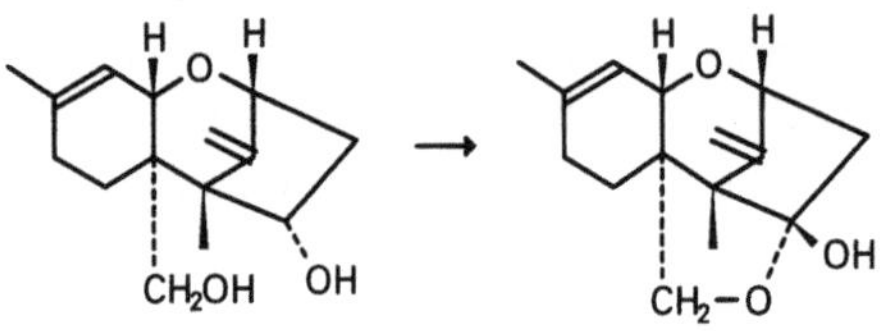

(**47**) (**48**)

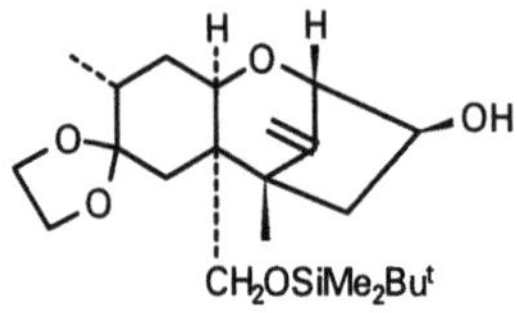

(**49**)

epiverrucarol series (*81*) the opposite result was obtained and the diol (**47**) gave the hemiacetal (**48**) on oxidation with pyridinium chlorochromate.

4.2.2.3. 3-Ols

The Swern reagent (dimethyl sulfoxide-trifluoroacetic anhydride) selectively oxidized a 3α-ol to the 3-one in the presence of a 7α-ol in 4,15-diacetylnivalenol (**9**; R^1=R^4=OH, R^2=R^4=OAc) and in 15-acetylvomitoxin (*108*); but with vomitoxin itself, no regioselectivity between positions 3α and 15 was observed (*99*) in a similar oxidation. Regioselective oxidation of the 3α-ol in 15-acetylvomitoxin was also achieved using pyridinium chlorochromate (*108*).

The chromic oxide-pyridine complex in dichloromethane cleanly oxidised the 3β-ol (**40**) to the 3-one in the presence of the allyl ether protecting groups (*75*). The 12-en-3β-ol (**49**) likewise gave the 3-one (*63*) with the chromic oxide-3,5-dimethylpyrazole complex in dichloromethane.

4.2.2.4. 4-Ols

A further example of selective oxidation of a 4β-ol in the verrucarol series with manganese dioxide has been reported (*72*). The Swern method was satisfactory when other OH groups were protected (*87*).

4.2.2.5. 12-Ols

The periodinane reagent (**50**) (*109*) has been used to obtain the 12-one (**51**) from the corresponding 12-ol (*75*).

References, pp. 54–70

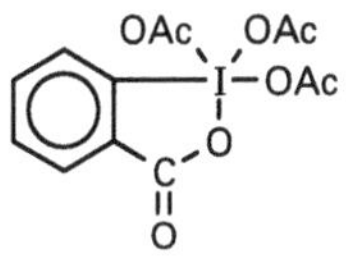

(50)

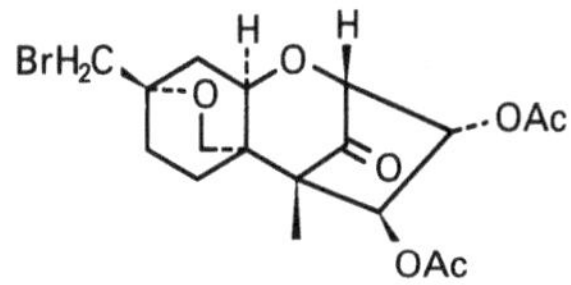

(51)

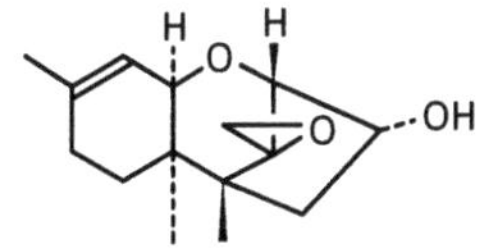

(52)

4.2.3. *Deoxygenation*

The Barton procedure, or variations thereof, has been used to convert isotrichodermol (**52**) into 12,13-epoxytrichothecene (*6*), the 15-levulinate (**39**; $R^1=C_5H_7O_2$, $R^2=H$, $R^3=THP$) into the corresponding 4-deoxy compound (*89*), T-2 toxin into the 3-deoxy derivative (*105, 76*), and HT-2 toxin into the 3,4-bis-deoxygenated product (*76*). The last was accompanied by the 3-ene resulting from mono-deoxygenation and elimination. With the 8α-ol (**53**) the reduction step also opened the 12,13-epoxide giving the 13-ol (**54**) as the major product (*106*).

Nucleophilic replacement of an 8α-OH by Br in compound (**53**) (see below) followed by reduction with tri-*n*-butyltin hydride has been used successfully (*106*).

In the early work (*110*), the 3,4-bismethanesulfonates (**55**; R=Ac or Ms) were shown to undergo regiospecific elimination *via* enol methanesulfonates to give the 3-ketones (**56**; R=Ac or Ms) on refluxing with sodium methoxide in methanol. Further examples of this reaction, by which the 3-ketones (**58**) and (**59**; R=H) were obtained from the bromoether (**57**) (*86, 68*) and HT-2 toxin bismethanesulfonate (*76*) respectively, have been described.

In the macrocyclic series (*2*), [16-^{3}H]-verrucarin A was prepared by reduction of the 16-mesyloxy derivative with tritium-labelled sodium borohydride under modified phase transfer conditions (*107*).

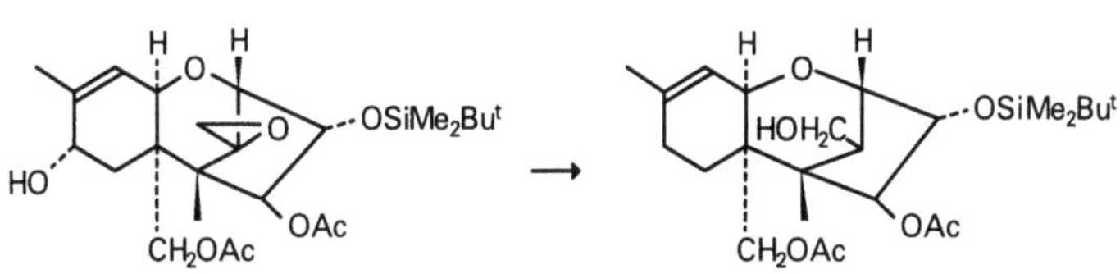

(53) (54)

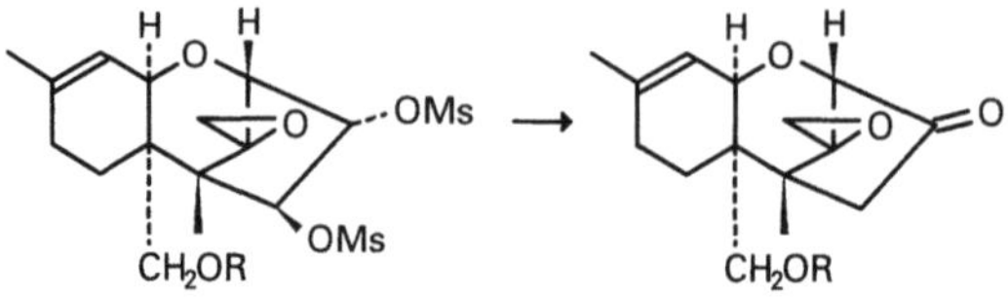

(55) (56)

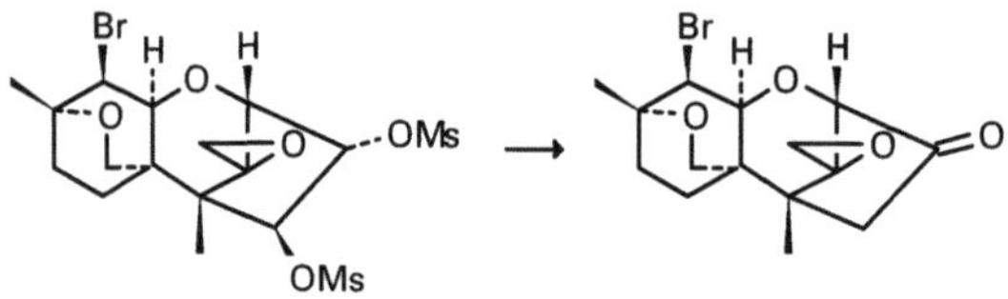

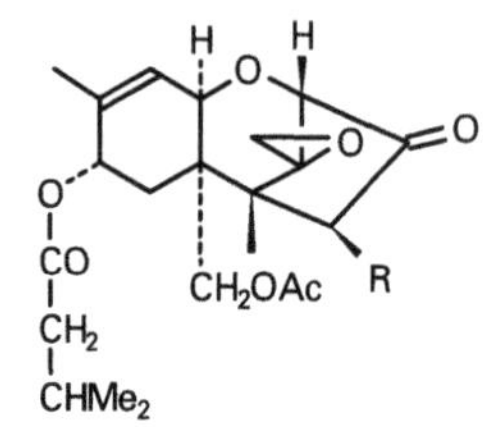

(57) (58) (59)

4.2.4. Nucleophilic Substitution

Mitsunobu's procedure (diethyl azodicarboxylate, triphenylphosphine and a carboxylic acid, followed by saponification of the resulting ester) for the epimerisation of steroidal secondary alcohols was applied satisfactorily to compounds containing an 8β-ol (*3*), but some 3α-(*3, 87*), 3β-(*75*), and 4β-OH groups (*3*) were unreactive, due, presumably, to steric hindrance.

The 4β, 8β-diol (**60**; R^1=OH, R^2=H) reacted smoothly and regioselectively under Mitsunobu conditions (isovaleric acid) to give the 8α-ester (**60**; R^1=H, R^2=$OCOCH_2CHMe_2$) (HT-2 toxin 3-THP ether) (*3*). For inversion at position 4, cesium propionate proved superior to other oxygen nucleophiles (*81*), but the formation of the ester (**61**; R^1=H, R^2=OCOEt) from the *p*-toluenesulphonate (**61**; R^1=OTs, R^2=H) was accompanied by partial elimination to form the 3-ene.

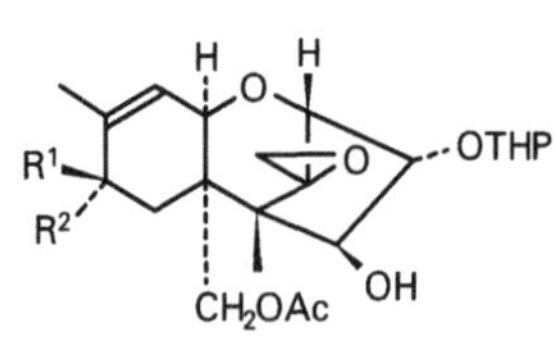

(60)

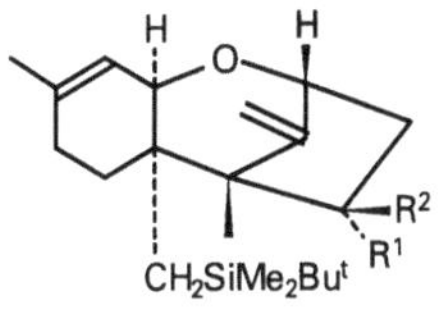

(61)

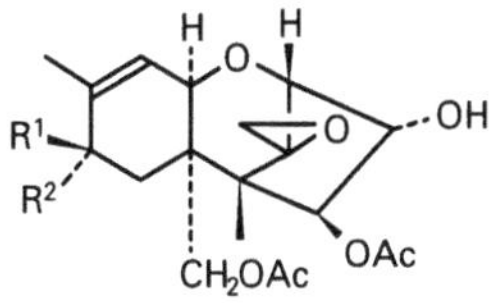

(62)

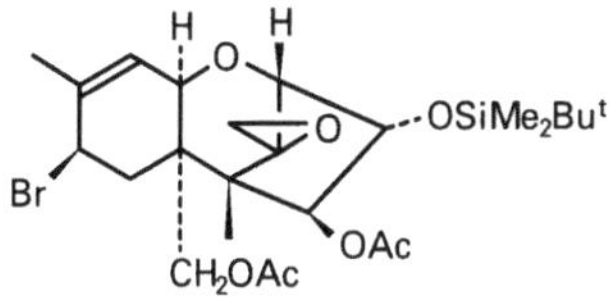

(63)

Acetolysis of the 8*β*-bromo derivative (**62**; R^1=Br, R^2=H) with sodium acetate in acetic acid gave approximately equal amounts of 8-acetylneosolaniol (**62**; R^1=H, R^2=OAc) and the 8*β*-epimer (**62**; R^1=OAc, R^2=H) (*83*), suggesting the involvement of an intermediate which is planar at C-8.

Treatment of the 8*α*-ol (**53**) with carbon tetrabromide and polymer-supported triphenylphosphine at room temperature gave the 8*β*-bromo analogue (**63**) (*106*).

The 15-fluoro derivatives of 12,13-epoxytrichothecene and isotrichodermol (**52**) and their 12-ene analogues were prepared, after protection of the 3-OH, where appropriate, by the action of tetrabutylammonium fluoride on the 15-trifluoromethanesulfonates of the corresponding 15-hydroxy compounds (*111*).

Attempts to introduce O-nucleophiles at C-3 of the scirpenetriol nucleus by the S_N2 mechanism with 3*α*-sulfonates resulted only in elimination, and the desired 3*β*-products were not obtained (*87*).

4.3. Keto Groups

4.3.1. Derivative Formation

Attempted preparation of cyclic thioacetals of the 8-one (**64**; R=$SiMe_2Bu^t$) failed (*106*).

4.3.2. Regio- and Stereo-Selective Reduction

Some 8*β*-ol is normally obtained as a minor product in the reduction of Group III 9-en-8-ones, *e.g.* (**64**; R=H) (*76*), with sodium borohydride, but the use of lithium tri-*s*-butylborohydride in tetrahydrofuran at −78°C [on the enone (**64**; R=Ac)] provided the 8*α*-ol stereospecifically (*63*). Stereospecific reduction was also obtained with diisobutylaluminium hydride (*82*).

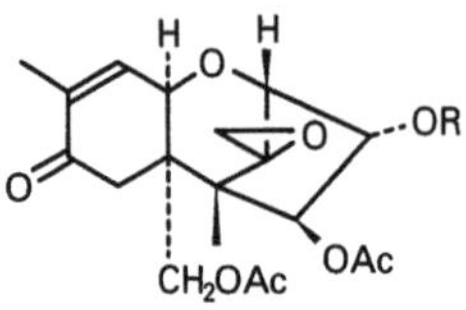

(64)

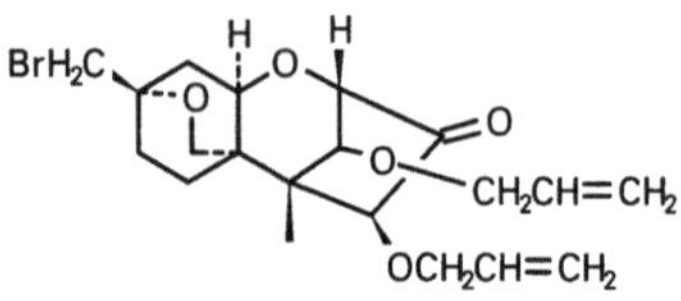

(65)

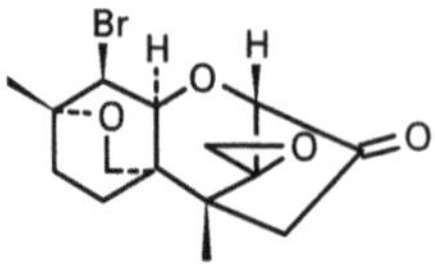

(66)

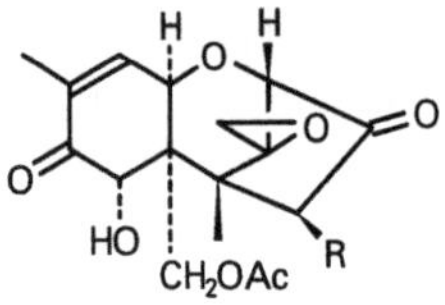

(67)

Where the β-face of ring C is masked by bulky groups, as in the 3-one (**65**), metal hydride reduction gave the 3β-ol (**40**) (*75*). After removal of the allyl protecting groups, the 3α-ol was obtained. Although reduction of the 3-one (**66**) with sodium borohydride at 0°C was stereoselective, giving the 3α-ol (*68*), nearly half of the product from a similar reduction at room temperature of the 3-one (**59**; R = OAc) had the 3β-configuration (*76*). Regio- and stereo-selective reduction of the 3-one by sodium borohydride in a number of 3,8-diones has been reported (*108*). The only isolable product from sodium borohydride reduction of the 4β-bromo-3-one (**67**; R=Br) was the corresponding 3β-ol (*108*).

4.3.3. Nucleophilic Addition Reactions

4.3.3.1. Phosphorus Ylides

Good yields were obtained of the corresponding 12-enes by treatment of the 12-ones (**68**; R=$SiMe_2F$) (*62, 61*), (**69**) (*63*), and (**70**; R^1=Et_3Si, R^2=THP) (*80*), with methylenetriphenylphosphorane. Difficulties were encountered with the 12-one (**68**; R=$SiMe_2F$) due to displacement of fluoride by *t*-butoxide (*61*).

4.3.3.2. Sulfur Ylides

Treatment of the 12-ones (**70**; R^1=R^2=Ac) (*68*), (**70**; R^1=R^2=Et_3Si) (*68*), and (**70**; R^1=Et_3Si, R^2=THP) (*67*) with dimethylsulfonium methylide gave the corresponding 12-epi, 13-epoxides (**71**). The use of Et_3Si as

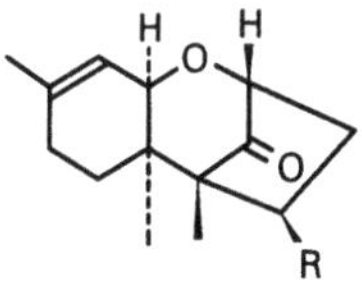

(68)

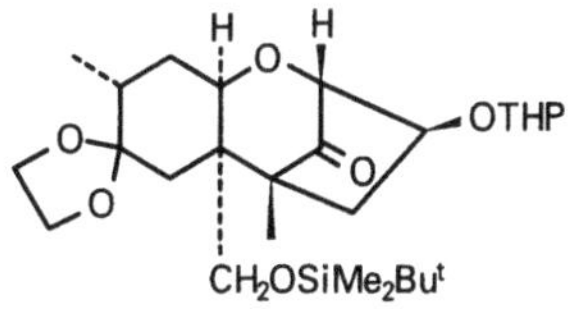

(69)

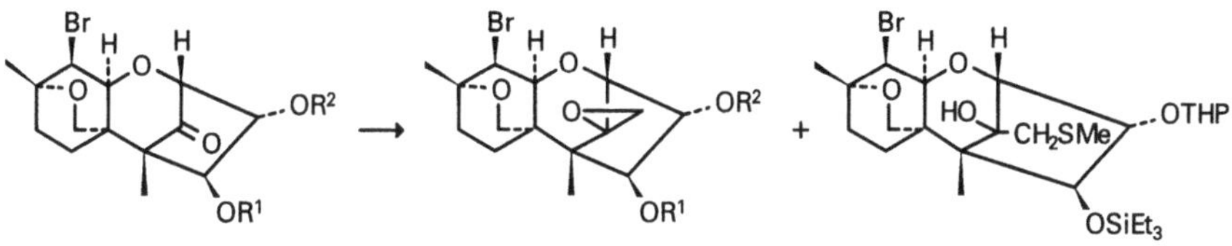

(70) (71) (72)

a protecting group gave greatly improved yields (*68*). The methanethiol adduct (**72**) was a minor product from the 12-one (**70**; R^1=$SiEt_3$, R^2=THP): separation of the products was only accomplished after removal of the protecting group (*67*).

4.3.4. *Enolization, and Reactions at the α-Carbon Atom*

Trichothecene 3-ones (but not 4-oxygenated 3-ones) readily formed 3-enol acetates under conditions normally used for acetylation of trichothecene primary and secondary alcohols (*108*). 3,4-Ketol acetates underwent epimerization and/or hydrolysis in solution in the presence of silica gel (*108*).

2-Toluene-*p*-sulfonyl-3-(*p*-nitrophenyl)oxaziridine mediated 4-hydroxylation (*112*) of the enolate of the 3-one (**73**), followed by hydride reduction gave the 3α,4β-diol (**74**) (*63*). Bromination of the 3-one (**67**; R=H) by phenyltrimethylammonium tribromide in boiling tetrahydrofuran in the presence of anhydrous potassium carbonate gave the 4β-bromo ketone (**67**; R=Br) (*108*).

Treatment of the 8-ketone (**75**) with CD_3ONa in CD_3OD introduced two deuterium atoms at position 7 (*82*).

Diacetylnivalenol (**9**; R^1=R^4=OH, R^2=R^3=OAc) was obtained (*71*) from the 7-deoxy analogue by oxidation with lead tetraacetate. In an extension (*113*) of this work oxidation of 7-deoxynivalenol to nivalenol occurred with reagents such as lead tetraacetate (preferably with the initiator azoisobutyronitrile) or hydrogen peroxide-ferrous ion-ascorbic

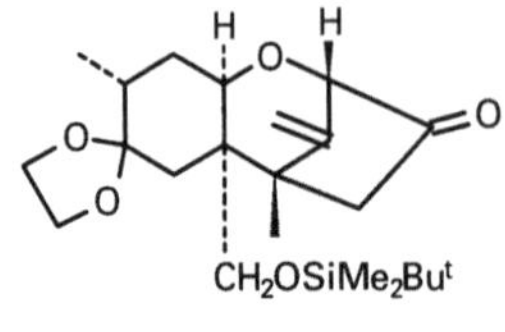

(**73**)

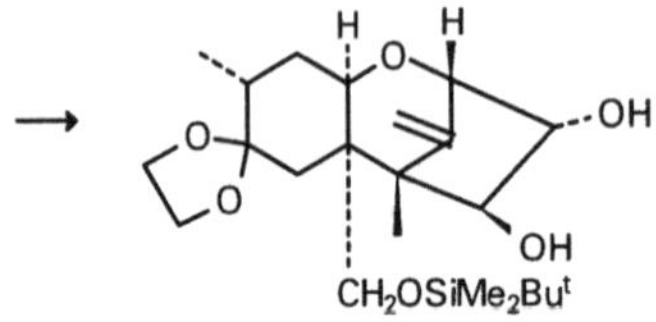

(**74**)

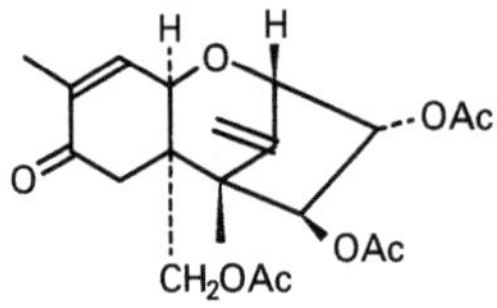

(**75**)

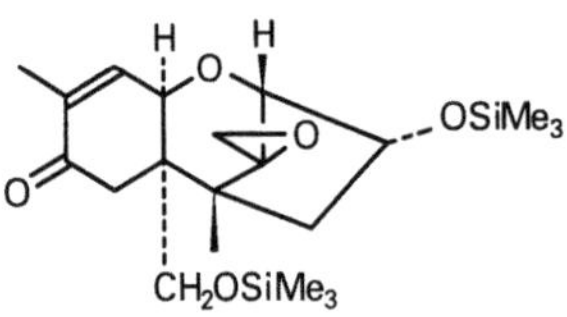

(**76**)

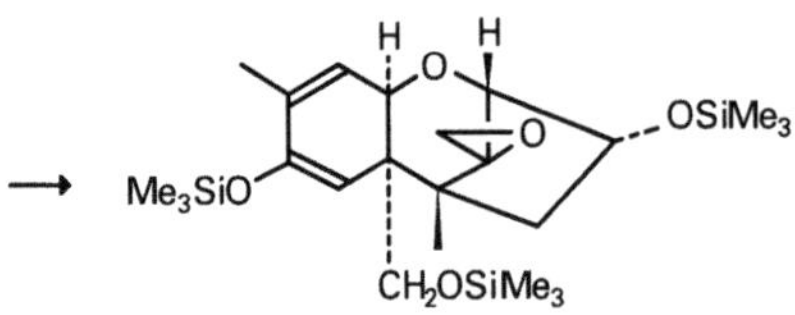

(**77**)

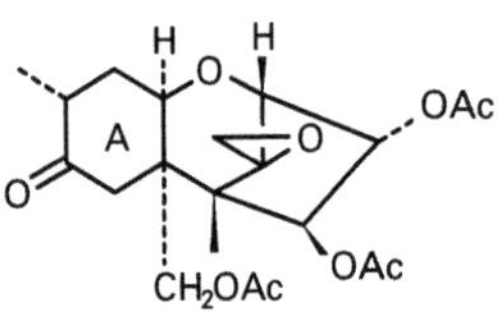

(**78**)

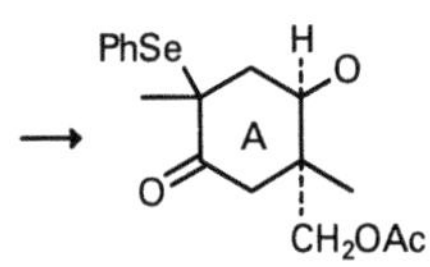

(**79**)

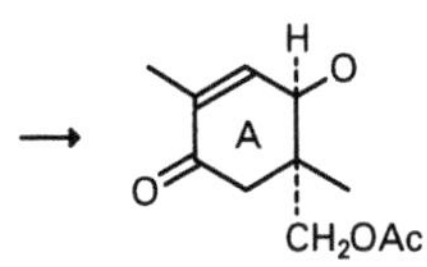

(**80**)

acid, which are known to react by a free radical pathway; methods requiring prior formation of the enol followed by attack by an electrophilic reagent, failed. Notwithstanding this conclusion, deprotonation of the 8-one (**76**) with lithium diisopropylamide and silylation of the anion with trimethylsilyl chloride at $-78\,°C$ (*86, 68*) gave the silyloxydiene (**77**) which served as an intermediate for the introduction of an oxygen substituent at position 7 (see Sect. 4.4.2).

Pyridinium toluene-*p*-sulfonate catalyzed α-selenylation of the 8-one (**78**) provided the selenide (**79**) as a regiospecifically pure mixture of epimers (*63*). Selenoxide formation and cyclo-elimination then gave the 9-en-8-one (**80**).

4.3.5. Isomerization

The 7,8-dione (**81**) has been isolated (*114*) as a minor oxidation product of the vomitoxin → isovomitoxin isomerisation [(**82**) → (**83**)] (*1*) in acetic anhydride.

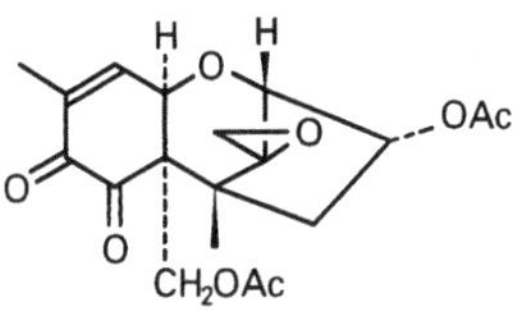

(**81**)

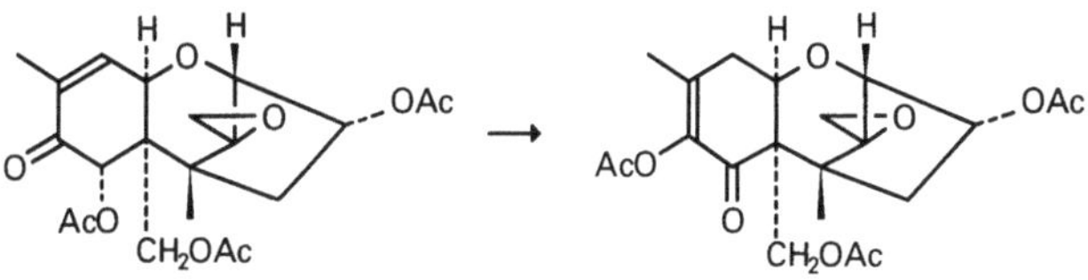

(**82**) (**83**)

4.4. Epoxide Function

4.4.1. 12,13-Epoxide

4.4.1.1. The Trichothecene → 10,13-Cyclotrichothecane Rearrangement

Ring A of 4-epi, 15-anhydroverrucarol (**84**) exists preferentially in the conformation (**84A**) (or the corresponding boat form) and the 9-ene participated in a spontaneous solvolysis of the 12,13-epoxide in aqueous media giving the 10 → 13-cyclo product (**85**) (*115*).

An analogous rearrangement of a 9,10:12,13-diepoxide is described in Sect. 4.4.2.

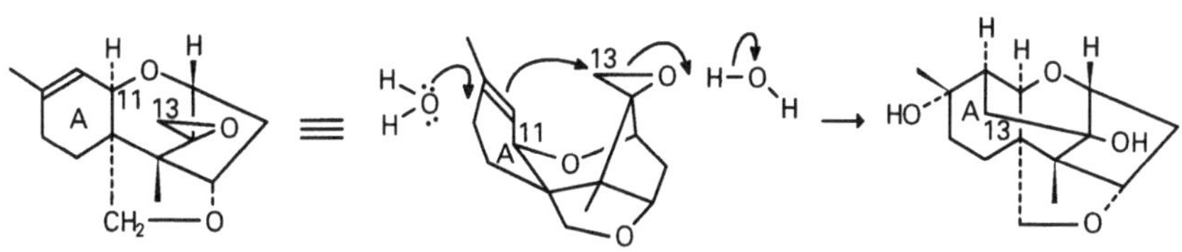

(**84**) (**84A**) (**85**)

4.4.1.2. The Trichothecene → Apotrichothecene Rearrangement

Some further examples of this rearrangement in the macrocyclic series have been reported (*2*). The apotrichothecene chemistry known when Part 1 was published (*1*) was described as being "mainly unexceptional". This chemistry was possibly unrepresentative, and some interesting reactions have recently been described, albeit in the 11-epi series. 3-Hydroxy-11-epi-apotrichothecenes readily undergo autoxidation in solution in daylight (*116, 117*) with the formation of numerous products, including the ether (**87**) and its 9β, 10β- and 9α, 10α-epoxides from the 3α-hydroxy compound (**86**); and the 10-one (**89**) from the 3β-hydroxy compound (**88**).

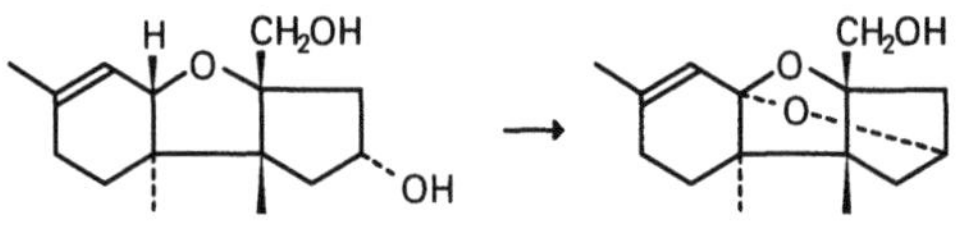

(**86**) (**87**)

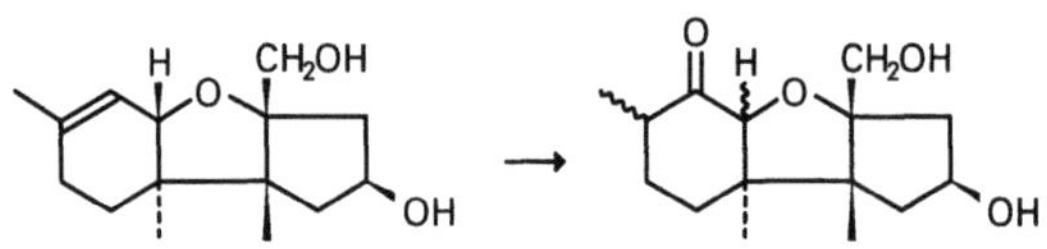

(**88**) (**89**)

If an apotrichothecene rearrangement of rings B/C of vomitoxin (**90**; 8-enol form) were to occur together with the isovomitoxin rearrangement (*1*) of ring A, the product, (**91**)=(**91A**), would be close (apart from absolute configuration) to the structure (**92**) proposed for the naturally occurring trichothecene relative gramilaurone (*118*) (see Sect. 5.2).

Several examples where the apotrichothecene rearrangement is replaced by "normal" halogenohydrin formation have been recorded (*1*) for 7α-hydroxytrichothec-9-en-8-ones. In an analogous reaction, the bromohydrin (**93**) was obtained as a by-product from the bromination of the 3-one (**67**; R=H) (*108*).

The precise chemistry of the useful alkylation reaction between 12,13-epoxytrichothecenes and 4-(*p*-nitrobenzyl)pyridine (*119*) is unknown. The severe conditions (150 °C for 30 min.) necessary in this detection procedure have been modified (90 °C for 2 min.) by the use of thorium(IV) salts as catalyst (*120*). The reagent has been compared with other reagents for the detection of trichothecenes on tlc plates (*121*).

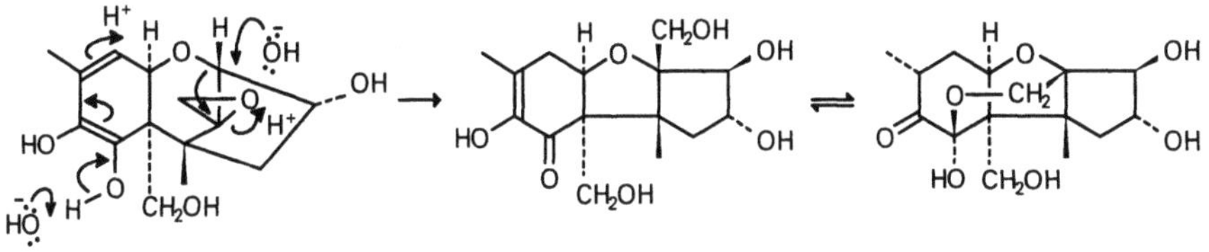

(90) (91) (91A)

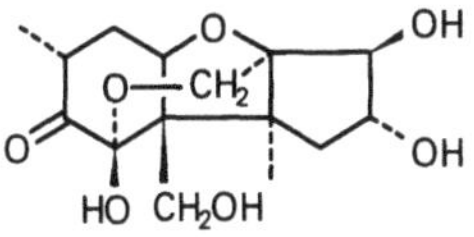

(92)

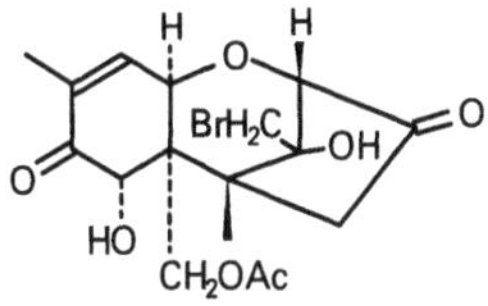

(93)

4.4.1.3. Deoxygenation

Further satisfactory deoxygenations of 12,13-epoxides to the corresponding 12-enes using the Sharpless tungsten reagent (WCl_6/*n*-BuLi) have been recorded (*35*, *82*, *122*, *111*, *80*, *68*). The failure of a number of alternative well-tried procedures is also reported (*80*). The Sharpless procedure has the advantage that any masking of a 9-ene as the bromo-ether is retained in the product (*35*).

Metabolic deoxygenation is dealt with in Sect. 6.3.

4.4.2. 7,8-Epoxide

Treatment of the silyloxydiene (**94**) with 1 equivalent of 3-chloroperbenzoic acid gave a 1:1 mixture of the 7β,13-epoxy compound (**95**) and the tristrimethylsilyl ether (**96**) (*86*, *68*). It is assumed that diastereoisomeric 7,8-oxiranes are involved, and that they undergo acid-catalyzed opening. The oxygen anion from the 7β,8β-oxirane attacks the 12,13-epoxide giving (**95**), whilst an O(8) → O(7) silyl migration in the 7α,8α-oxirane leads to the 7α-silyloxy ketone (**96**).

4.4.3. 9,10-Epoxide

In the presence of a 15-OH group the intramolecular formation of a 15 → 9 ether linkage takes precedence over the inter-molecular

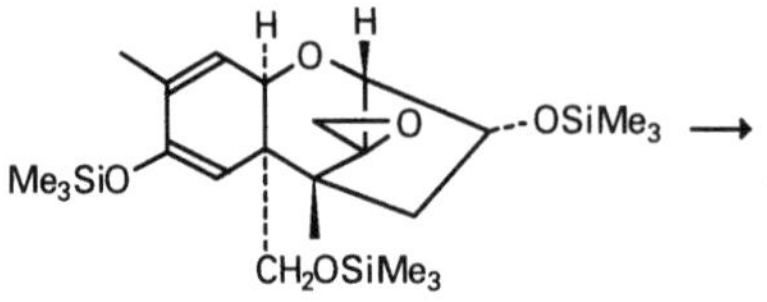

(94)

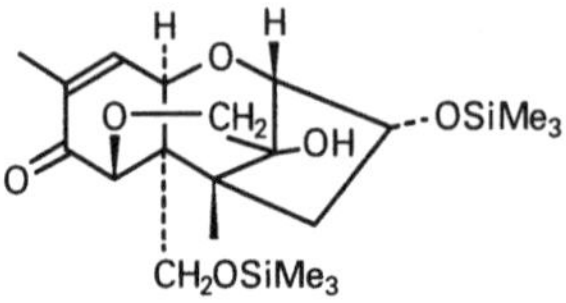

(95)

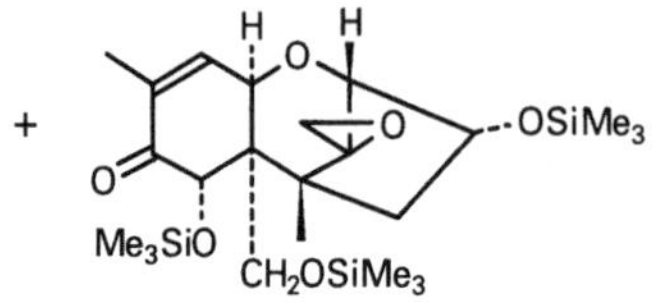

(96)

addition of a nucleophile to a 9β,10β-epoxide (*78*). The shape of the resulting rigid all-boat oxabicyclo[2.2.2]octane system precludes any attack on a 12,13-epoxide by a 10β-oxygen anion.

In the absence of a 15-oxygen substituent an 8α-OH has a profound influence on the course of events (*78*). In trichodermone 9β,10β-epoxide (Scheme 1: **98**; R^1R^2=O) and trichodermol 9β,10β-epoxide (**98**; R^1=OH, R^2=H) the normal rearrangements of the 12,13-epoxide in aqueous basic and acidic media occurred independently of reactions at the 9,10-epoxide, giving, respectively, the apotrichothecanes (**99**) and (**97**). In trichothecodiol 9β,10β-epoxide (**101**) intramolecular attack on the 9,10-epoxide led, in a basic medium, by further intramolecular attack, to the 8α,9α: 10β,13-diepoxytrichothecane (**102**), and, in an acidic medium, to trichothecolone glycol (**108**), a known rearrangement product of trichothecolone (**105**). Trichothecolone was considered to arise by way of the sequence (**101**) → (**100**) → (**103**) → (**104**) → (**105**).

The product, in a basic medium, from T-2 tetraol 9β,10β-epoxide (**107**), which contains hydroxyl functions at both positions 8α and 15, offering a choice of reaction path, was the 9α,15-ether (**106**) (*77*).

5. Biosynthesis

Most of the recent work relates to the biosynthetic pathways to diacetoxyscirpenol (**109**; R=H) or T-2 toxin (**109**; R=$OCOCH_2CHMe_2$), in *F. sambucinum* or *F. sporotrichioides*, and to 3-acetylvomitoxin (**110**; R=Ac) in *F. culmorum*. Trichodiene (**120**, Scheme 2, p. 40), previously

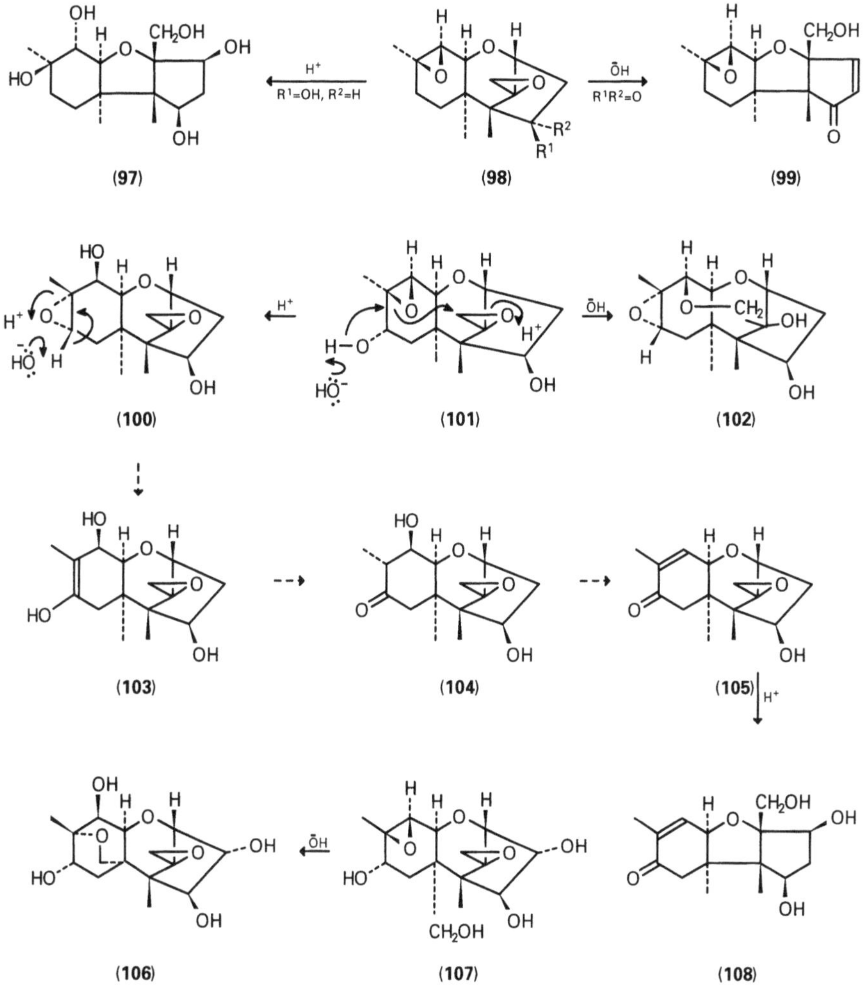

Scheme 1. Rearrangement of trichothecene 9β; 10β-epoxides in the absence and presence of an 8α-OH substituent

shown to be a precursor of 12,13-epoxytrichothecene (**144**), and trichothecin (**9**; $R^1=R^2=R^4=H$, $R^3=OCOCH\overset{Z}{=}CHMe$) in *T. roseum* (*123*), has been confirmed as a precursor of all three of the above trichothecenes (*124, 125, 126*). The known pattern of incorporation of mevalonate units into the trichothecene nucleus has been confirmed for 3-acetylvomitoxin (*127*). The biosynthesis of this trichothecene by *F. graminearum* is

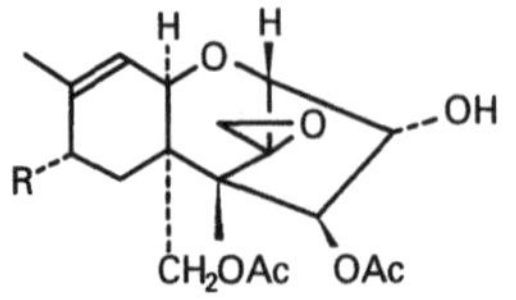

(109)

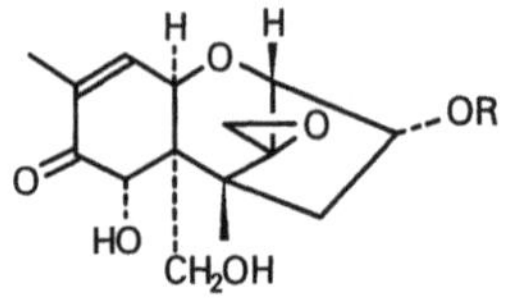

(110)

inhibited by manganese (Mn^{2+}) (*128*). It is believed that the associated higher level of lipid biosynthesis results in the depletion of acetylCoA which would otherwise be, available for trichothecene biosynthesis.

The biosynthetic pathway to the trichothecenes is conveniently divided into three sections: (i) steps between mevalonate and trichodiene; (ii) oxygenation of trichodiene and cyclisation of the resulting intermediates to 12,13-epoxytrichothecene and isotrichodermin (**146**, Scheme 2); and (iii) further oxygenation and esterification of the trichothecene nucleus.

5.1. Mevalonate to Trichodiene

This section of the pathway is now well understood, and much of the recent work has been reviewed (*129, 130*). The second review (*130*) contains interesting, hitherto unpublished, information. Further evidence for the role of nerolidyl diphosphate (**112**) as an intermediate in the conversion of farnesyl diphosphate (**111**) into trichodiene [(**113**)≡(**120**)] by the enzyme trichodiene synthase has been reported (*131, 132, 132a*). The enzyme, which can be regarded as an isomerase-cyclase, has been studied both in *F. sambucinum* (*133*) and in *F. sporotrichioides* (*133, 134*). The *F. sporotrichioides* preparation has been characterized (*134*), and has been shown to be a dimer composed of two identical subunits, *M* 45000; Mg^{2+} is a cofactor. In further work on trichodiene synthase, substrate specificity has been studied, using both the native and recombitant enzyme (*134a*), and the active site has been identified by systematic site-directed mutagenesis (*134b, 134c*). The trichodiene synthase gene (*Tox5*; *Tox*=trichothecene toxin: subsequently (*134d*) renamed *Tri5*) has been sequenced, and expressed in *Escherichia coli* (*135, 136, 137*), and, more recently, in transgenic tobacco (*Nicotiana tabacum*) (*138*). In *E. coli*, the recombitant enzyme had properties closely resembling those of the native enzyme, and yields of trichodiene of 60 μg/l were obtained: only low levels of trichodiene were obtained in tobacco tissue. By using

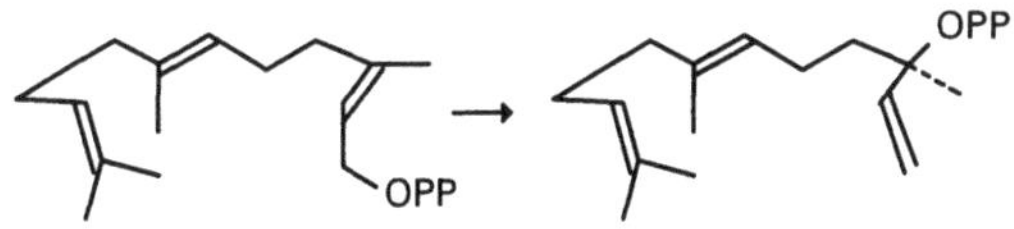

(111) (112)

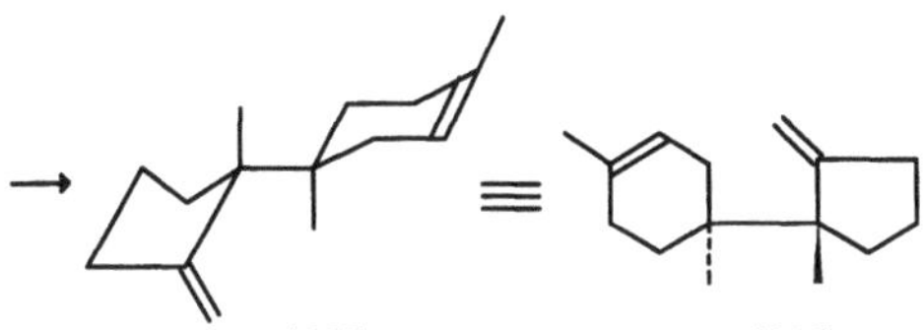

(113) (120)

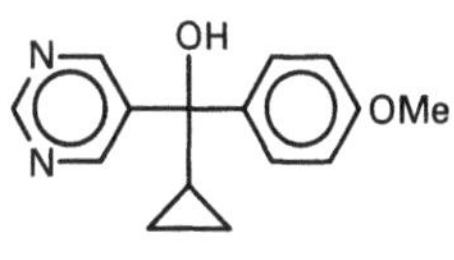

(114)

standard molecular gene disruption techniques, *Tri5*⁻ mutants have been constructed (*139, 140, 134d*).

In T-2 toxin (**109**; R=$OCOCH_2CHMe_2$) all six non-carbonyl oxygen atoms, including those at position 1 and in the 12,13-epoxide, are derived from molecular oxygen (*141*), and monooxygenase enzymes are believed to be responsible. By supplying specific cytochrome P450 monooxygenase inhibitors to the fermentations, these oxygenation reactions are suppressed, and trichodiene, the last hydrocarbon intermediate in the pathway, and in certain circumstances, post-trichodiene intermediates (see below), accumulate. Effective inhibitors with *F. sporotrichioides* fermentations were the synthetic pyrimidine ancymidol (**114**) (*124, 125*) and certain shikimate-derived flavanoids and furanocoumarins (*142*). The same result was achieved with a particular uv-induced *F. sporotrichioides* mutant strain (*143*); similar results were obtained with *F. culmorum* cultures treated with ancymidol (*144*) or with the furanocoumarin xanthotoxin (*145*). These blocked fermentations have greatly facilitated the investigation of subsequent steps in the biosynthetic pathway, as has the kinetic pulse-labelling technique (*146*) and the use of mutant strains of *F. sporotrichioides* (*30, 7*).

5.2. Trichodiene (and Its Relatives) to 12,13-Epoxytrichothecene and Isotrichodermin

Much effort has been devoted to the identification of the oxygenation steps which take place after the formation of trichodiene. Some large scale and/or blocked fermentations with *T. roseum* or *Fusarium* spp. have yielded a number of metabolic products with structures closely related to the trichothecenes. These relatives, 42 in number, are listed in Table 5, together with their sources. With the exception of gramilaurone, (see Sect. 4.4.1.2), they can be derived from trichodiene (**120**) by plausible hypothetical pathways (summarized in Scheme 2) involving, mainly, allylic hydroxylation at positions 2α and 11α (biosynthetic numbering) and rearrangement of the allylic alcohols, both steps being associated with β-epoxidation of the 12-ene. A cell-free enzyme system capable of epoxidizing the 12-ene of trichodiene has been obtained from *F. culmorum* (*165*).

Allylic rearrangement of a 2-hydroxy-12-ene to a 13-hydroxy-2-ene may be followed by a second allylic oxidation at position 3. Scheme 2 presents a generalised picture and the pathways depicted are not necessarily common to all the contributing organisms. Some hypothetical steps remain to be confirmed by suitable experiments. The variety and multiplicity of the structures points to the operation of metabolic grids rather than unique pathways (*145*, *146*), but this aspect remains to be clarified.

The diol (**115**) (or its 12-epimer) is a dead-end metabolite and its formation by *F. culmorum* (*144*) may be preceded by epoxidation of the 12-ene. Cyclization of the protonated 12-ene (**116**) generates the carbon skeleton (**117**) of sambucinic acid (**118**) which may be obtained, from *F. sambucinum*, after a further series of oxidative steps (*149*). 11α-Hydroxytrichodiene (**119**), isolated from a mutant strain of *F. sporotrichioides* (*7*), is a likely intermediate in the proven transformation of trichodiene to isotrichodiol (**124**) (*145, 166*), as are the transformation product (**123**) from *F. culmorum* (*144*) and the hypothetical intermediate (**121**) (*116*). Likewise, the 3-one (**128**), a proven transformation product in *F. culmorum* (*147*), is a likely precursor of the 9-hydroxy derivatives FS 2 (**136**) (*27*) from *F. sporotrichioides*, and FS 4 (**138**) (*17*) from *F. sambucinum*. The acetate (**139**) is a proven transformation product in *F. sambucinum* (*147*).

Central to the pathway from trichodiene to the trichothecene nucleus are the *Fusarium* metabolic products isotrichodiol (**124**) (*145*) and isotrichotriol (**130**) and their respective products of acid-catalyzed allylic rearrangement, trichodiol (**126**) and 9-epi-trichodiol (**127**), both isolated from *T. roseum* (*159*, *14*), and trichotriol (**134**) and 9-epitrichotriol (**135**),

Table 5. *Naturally Occurring Trichothecene Relatives*

Structure (Scheme 2)	Trivial name	Formula	Organism[a]	Refs.
(**148**)	FS 3	$C_{15}H_{20}O_3$	*F. sambucinum*	(*17*)
(**128**)		$C_{15}H_{22}O_2$	*F. culmorum*	(*147*)
(**147**)	FS 1	$C_{15}H_{22}O_3$	*F. sporotrichioides, F. sambucinum*	(*5, 17*)
(**138**[b])	FS 4	$C_{15}H_{22}O_3$	*F. sambucinum*	(*17*)
(**151**)	3-Deoxysambucinol	$C_{15}H_{22}O_3$	*F. culmorum, F. graminearum*	(*148, 148*)
			F. crookwellense	(*9*)
(**118**)	Sambucinic acid	$C_{15}H_{22}O_3$	*F. sambucinum*	(*149*)
(**140**)	Sambucoin	$C_{15}H_{22}O_3$	*F. sporotrichioides, F. poae*	(*20, 10*)
			F. sambucinum, F. culmorum	(*150, 151*)
			F. graminearum, F. crookwellense	(*152, 9*)
(**154**)		$C_{15}H_{22}O_3$	*F. sporotrichioides, F. culmorum*	(*116, 116*)
(**159**)	3-Dehydroapotrichodiol	$C_{15}H_{22}O_3$	*F. sporotrichioides, F. sambucinum*	(*17, 17*)
(**155**)		$C_{15}H_{22}O_4$	*F. sporotrichioides*	(*22*)
(**141**)	8β-Hydroxysambucoin	$C_{15}H_{22}O_4$	*F. sporotrichioides*	(*153*)
(**142**)	8α-Hydroxysambucoin	$C_{15}H_{22}O_4$	*F. sporotrichioides*	(*153*)
(**152**)	Sambucinol	$C_{15}H_{22}O_4$	*F. sporotrichioides, F. sambucinum*	(*20, 150*)
			F. culmorum, F. graminearum	(*151, 148*)
			F. crookwellense	(*9*)
(**161**[c])	Sporol	$C_{15}H_{22}O_4$	*F. sporotrichioides*	(*154, 155*)
(**92**[d])	Gramilaurone	$C_{15}H_{22}O_7$	*F. graminearum*	(*118*)
(**120**)	Trichodiene	$C_{15}H_{24}$	*T. roseum, F. sporotrichioides*	(*156, 125*)
			F. sambucinum, F. culmorum	(*124, 145*)
(**119**)	11α-Hydroxytrichodiene	$C_{15}H_{24}O$	*F. sporotrichioides*	(*7*)
(**123**)		$C_{15}H_{24}O_2$	*F. culmorum*	(*144*)
(**156**)	Apotrichool	$C_{15}H_{24}O_2$	*F. culmorum*	(*157*)
(**157**[e])	Apotrichodiol	$C_{15}H_{24}O_3$	*F. sporotrichioides, F. sambucinum*	(*20, 17*)
			F. culmorum, F. graminearum	(*148, 148*)
			F. crookwellense	(*9*)

Table 5 (*continued*)

Structure (Scheme 2)	Trivial name	Formula	Organism[a]	Refs.
(158[e]**)**	3-epi-Apotrichodiol	$C_{15}H_{24}O_3$	*F. sporotrichioides, F. sambucinum*	(*20, 17*)
			F. culmorum, F. graminearum	(*148, 148*)
			F. crookwellense	(*9*)
(149)		$C_{15}H_{24}O_3$	*F. sporotrichioides*	(*158*)
(136[b]**)**	FS 2	$C_{15}H_{24}O_3$	*F. sporotrichioides*	(*27*)
(137)	3-epi-FS 2	$C_{15}H_{24}O_3$	*F. sporotrichioides*	(*17*)
(126[f, g]**)**	Trichodiol[h]	$C_{15}H_{24}O_3$	*T. roseum*	(*159*)
(127)	9-epi-Trichodiol[i]	$C_{15}H_{24}O_3$	*T. roseum*	(*14*)
(124[f]**)**	Isotrichodiol	$C_{15}H_{24}O_3$	*F. culmorum*	(*145*)
(122[j]**)**		$C_{15}H_{24}O_3$	*F. sporotrichioides*	(*27, 160*)
(134[f, k]**)**	Trichotriol	$C_{15}H_{24}O_4$	*F. sporotrichioides, F. culmorum*	(*27, 160*)
(135)	9-epi-Trichotriol[l]	$C_{15}H_{24}O_4$	*F. sporotrichioides, F. culmorum*	(*6, 160*)
(130[f]**)**	Isotrichotriol	$C_{15}H_{24}O_4$	*F. sporotrichioides, F. culmorum*	(*7, 14*)
(125)	8α-Hydroxyisotrichodiol	$C_{15}H_{24}O_4$	*F. culmorum*	(*160*)
(131)		$C_{15}H_{24}O_5$	*F. sporotrichioides*	(*7*)
(132)		$C_{15}H_{24}O_5$	*F. sporotrichioides*	(*7*)
(133)		$C_{15}H_{24}O_5$	*F. sporotrichioides*	(*7*)
(115[m]**)**		$C_{15}H_{26}O_2$	*F. culmorum*	(*144*)
(139[b]**)**	AcetylFS 4	$C_{17}H_{24}O_4$	*F. sambucinum*	(*147*)
(150)		$C_{17}H_{26}O_4$	*F. sporotrichioides*	(*158*)
(162)		$C_{17}H_{26}O_5$	*F. sporotrichioides*	(*22*)
(153)	Diacetylsambucinol	$C_{19}H_{26}O_6$	*F. sporotrichioides*	(*22*)
(163)		$C_{19}H_{28}O_6$	*F. sporotrichioides*	(*22*)
(164)		$C_{19}H_{28}O_6$	*F. sporotrichioides*	(*22*)

Footnotes to Table 5

[a] *F* = *Fusarium*, *T* = *Trichothecium*

[b] Originally formulated (*27, 17, 147*) as the 9α-hydroxy epimer, but changed (*160*) on spectroscopic evidence.

[c] An earlier structure (*155*) has been renamed neosporol and has been synthesized (*161*), as has the revised structure (*161*) (*162*).

[d] As published. The analogous structure (**91A**) could be derived from vomitoxin (see Sect. 4.4.1.2.).

[e] Initially formulated as an apotrichothecene (*148, 163*) but subsequently shown (*164, 116*) to belong to the 11-epi series (11β-H).

[f] Some groups of workers, e.g. (*144, 145*), consistently (and mistakenly) write this compound as the 12-epimer.

[g] Originally (*159*) formulated without assignment of configuration at position 9 but commonly written as the 9α-hydroxy compound. It was subsequently shown to be the 9β-hydroxy epimer (*160*).

[h] Trichodiol A (**160**) is considered to be an artifact (*159*).

[i] Originally (*14*) called 9β-trichodiol.

[j] The metabolite was believed (*27*) to be trichodiol. This conclusion has been criticized (*160*), but the metabolite does not have the newly assigned (*160*) 12-ene structure: It is most probably (**122**).

[k] Initially (*27*) written as the 9α-hydroxy epimer by analogy with the commonly written structure for trichodiol.

[l] Originally (*6*) called 9β-trichotriol.

[m] Or C-12 epimer.

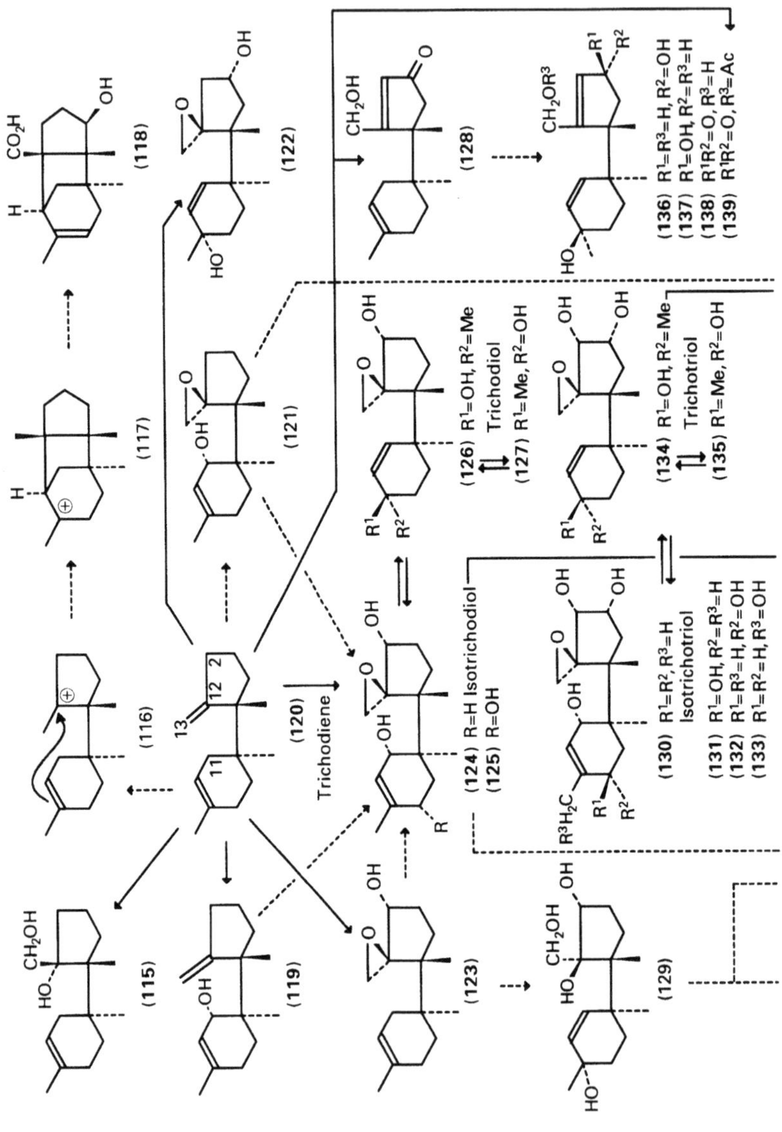
(115)
(116)
(117)
(118)
(119)
(120) Trichodiene
(121)
(122)
(123)
(124) R=H Isotrichodiol
(125) R=OH
(126) R1=OH, R2=Me Trichodiol
(127) R1=Me, R2=OH
(128)
(129)
(130) R1=R2, R3=H Isotrichotriol
(131) R1=OH, R2=R3=H
(132) R1=R3=H, R2=OH
(133) R1=R2=H, R3=OH
(134) R1=OH, R2=Me Trichotriol
(135) R1=Me, R2=OH
(136) R1=R3=H, R2=OH
(137) R1=OH, R2=R3=H
(138) R1R2=O, R3=H
(139) R1R2=O, R3=Ac

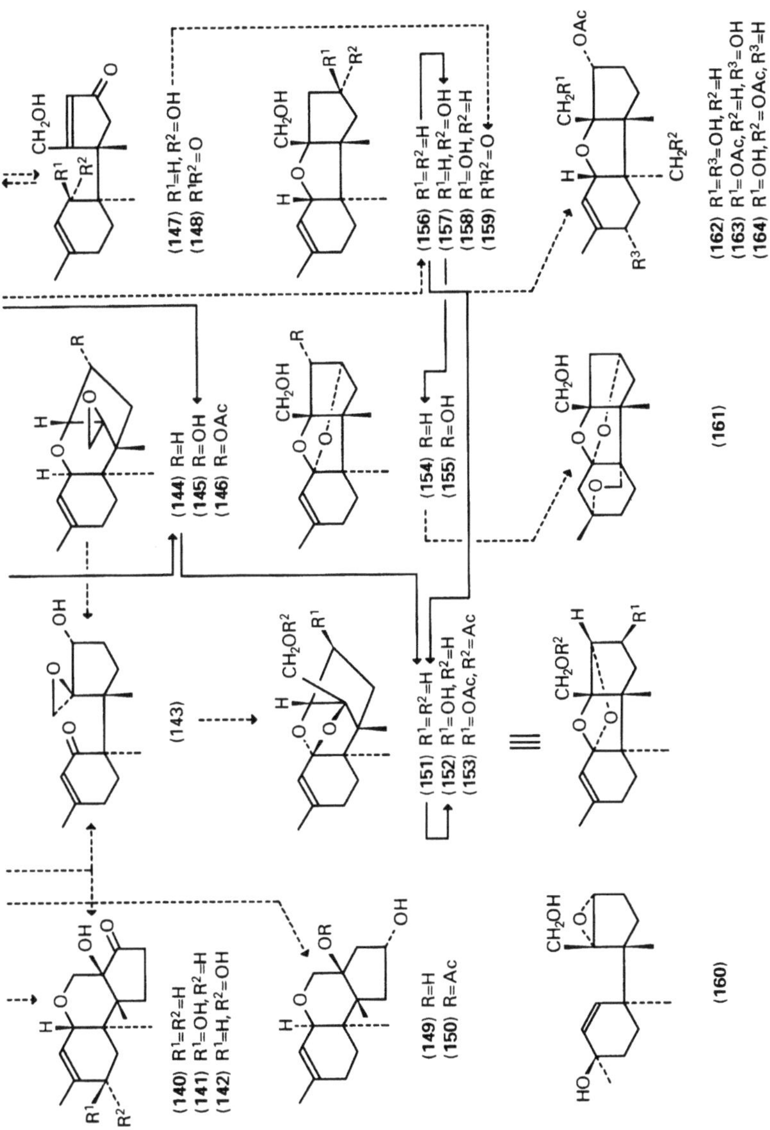

Scheme 2. Proven (—) and hypothetical (···) pathways from trichodiene (**120**) to the trichothecene relatives listed in Table 5. ( ) Isolated trichothecene relative. ( ) Hypothetical intermediate

both obtained from *F. sporotrichioides* (*27, 6*). The conversion of trichodiol and trichotriol into 12,13-epoxytrichothecene (**144**) and isotrichodermol (**145**), respectively, proceeds spontaneously at acid pH by mechanism (A) (Scheme 3) (*27*). For this reason it was at first assumed that trichodiol was an intermediate in the formation of 12,13-epoxytrichothecene, although there was no direct evidence for this. More recently it has been suggested, in a thesis summary (*14*), and subsequently in the full paper (*160*), that trichodiol and trichotriol are non-enzymic metabolites and that isotrichodiol and isotrichotriol are the true intermediates on the pathway to the trichothecene nucleus in *F. culmorum*. Isotrichodiol, trichodiol, and 9-epitrichodiol were interconverted in acid solution *via* an intermediate allylic carbocation and a similar interconversion operated in the trichotriol series. Slow acid-catalysed cyclization of isotrichodiol to 12,13-epoxytrichothecene was demonstrated, but the rapid *in vivo* incorporation pointed to an enzyme-mediated process in the fungus (*166*).

Trichotriol, 9-epitrichotriol and isotrichotriol were all converted into T-2 toxin by mutant strains of *F. sporotrichioides* but trichodiol was not (*6*). These results indicate that in the biosynthesis of T-2 toxin, oxygenation at C-3 occurs before the cyclization step. This sequence also applies to the biosynthesis of 3-acetylvomitoxin in *F. culmorum* (*166*) and was confirmed by feeding experiments with 12,13-epoxytrichothecene and isotrichodermin (see Sect. 5.3). Since isotrichodiol is converted into isotrichodermin in *F. culmorum* (*145*), it must first be oxidized to isotrichotriol.

It is uncertain whether 8α-hydroxyisotrichodiol (**125**) is a dead-end metabolite or whether it is involved in the biosynthesis of 8α-hydroxytrichothecenes (*160*).

Two routes to sambucoin (**140**) in *F. sambucinum* have been proposed (*150*), involving intramolecular nucleophilic attack either by the 11-OH of isotrichodiol at C-13 or by the 13-OH of the tetraol (**129**) at C-11 with allylic displacement of the 9-OH. The second route permits formation of either the 11β-H in sambucoin or the 11α-H in the 11-episambucoins (**149–150**). In *F. sporotrichioides* further hydroxylation of sambucoin at C-8 may follow (*153*) to give the diols (**141**) and (**142**).

The *F. sambucinum* metabolites FS 1 (**147**) and FS 4 (**138**) bear the same relationship to each other as isotrichodiol and trichodiol. Alternative hypothetical routes, by way of FS 1 (mechanism B, Scheme 3) or the hypothetical intermediate (**121**) (mechanism C) (*116*), lead to the 11-epiapotrichothecene, apotrichool, (**156**). Apotrichool was converted by *F. culmorum* into apotrichodiol (**157**) (*147*), which in turn was oxidized photochemically (*116*) to the 11-acetal (**154**), a likely precursor of sporol

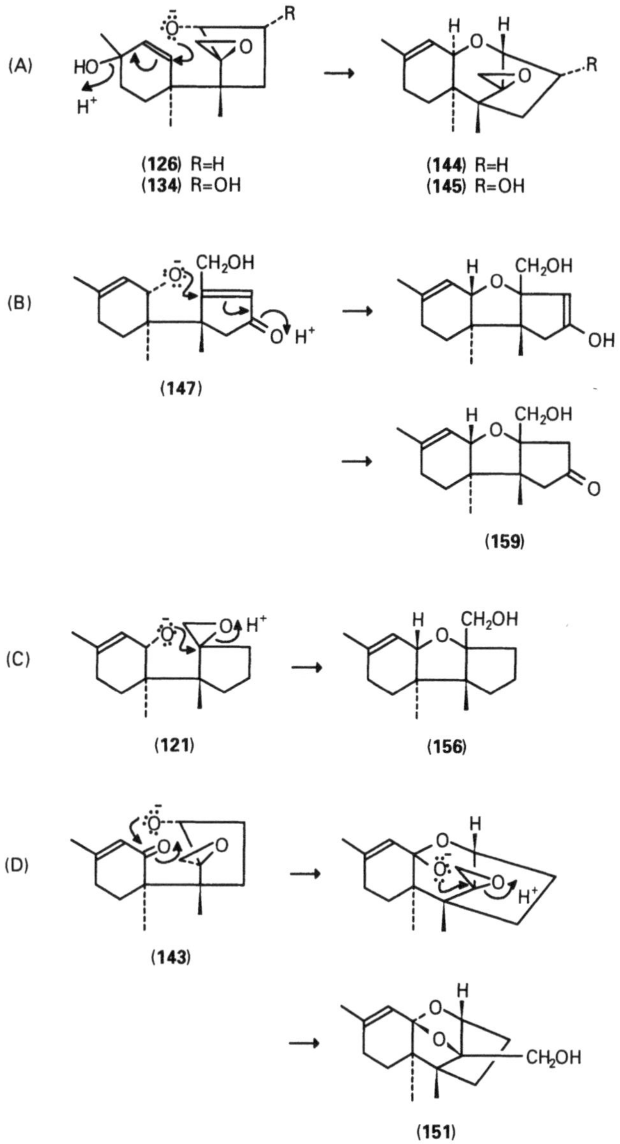

Scheme 3. Possible cyclization mechanisms in the biosynthesis of 12,13-epoxytrichothecene and its relatives

(**161**) (*22*). Further hydroxylation of apotrichool could give the derivatives (**162**)–(**164**).

The 11-ketone (**143**) is a likely intermediate between isotrichodiol and 3-deoxysambucinol (**151**), ketalization occurring according to mechanism D with opening of the 12,13-epoxide. 3-Deoxysambucinol is a known precursor of sambucinol (**152**) (*167*) in *F. culmorum*. The precise mechanisms whereby 12,13-epoxytrichothecene (*157, 167*) and, particularly, apotrichool (**156**) (*157*) could be converted into sambucinol are uncertain, although the 11-ketone (**143**) might be involved in the former conversion. The evidence for the formation of sambucinol from 12,13-epoxytrichothecene has been criticized (*116*), and may require further investigation. Isotrichodermin was not a precursor of sambucinol in *F. culmorum* (*167*).

When bazzanene (**165**), the 6-epimer of trichodiene, was fed to *F. culmorum*, the 6-epiapotrichothecenes (**167**; R=H or Ac) were isolated (*168*), implying that 11-hydroxylation occurs in the *β*-position, *cis* to the 15-methyl group, in the hypothetical intermediate (**166**), *cf.* structure (**121**).

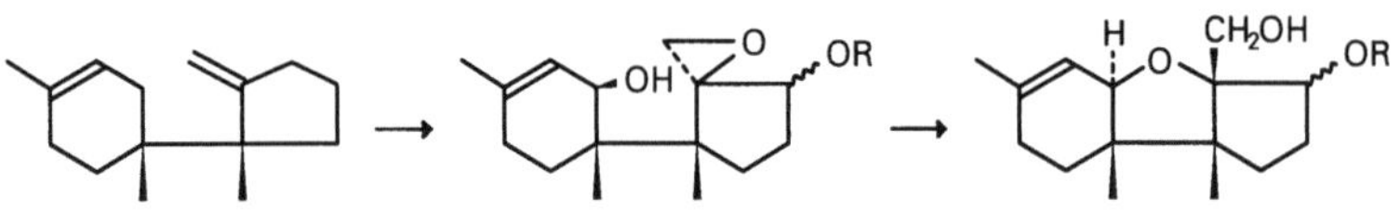

(165) (166) (167)

5.3. Further Oxygenation and Esterification of the Trichothecene Nucleus

Although isotrichodermol (**145**) [or its acetate, isotrichodermin (**146**)] is a precursor both of 3-acetylvomitoxin (**110**; R=Ac) in *F. culmorum* (*157*) and of T-2 toxin (**109**; R=$OCOCH_2CHMe_2$) in *F. sporotrichioides* (*6*), 12,13-epoxytrichothecene is not (*167, 6*), thus reversing (for 3-acetylvomitoxin) an earlier statement (*157*) to the contrary. 12,13-Epoxytrichothecene was not converted into isotrichodermin by *F. culmorum* (*167*), and neither trichodermol (**168**; R^1=H, R^2=OH) nor 15-hydroxy-12,13-epoxytrichothecene (**168**; R^1=OH, R^2=H) were precursors of T-2 toxin in *F. sporotrichioides* (*6*). These results reinforce previous conclusions (Sect. 5.2) that in trichothecene biosynthesis 3α-hydroxylation precedes the cyclization step. Nevertheless, 3-hydroxylations of the 11-epiapotrichothecene (3α) (*147*) and sambucinol (3*β*) (*167*) nuclei occur after cyclization.

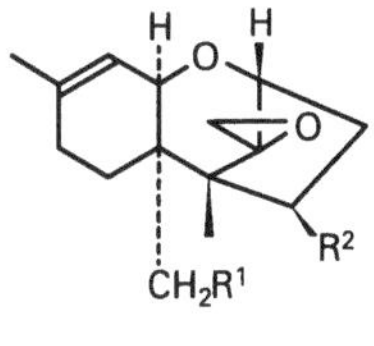

(168)

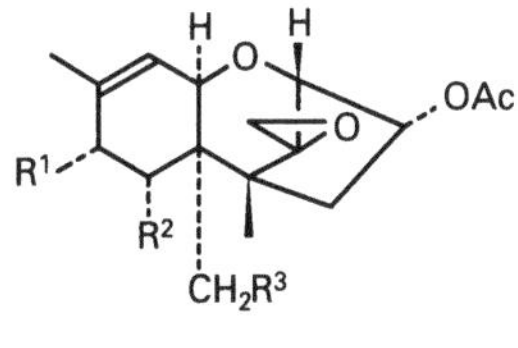

(169)

The steps in the conversion of isotrichodermin into 3-acetylvomitoxin follow logically from the pattern of secondary metabolites isolated from *F. graminearum* (*169*). Kinetic pulse-labelling techniques revealed 15-deacetylcalonectrin (**169**; $R^1=R^2=H$, $R^3=OH$) to be the first intermediate in this section of the pathway (*33*). Calonectrin (**169**; $R^1=R^2=H$, $R^3=OAc$), and 7-hydroxy-(**169**; $R^1=R^3=H$, $R^2=OH$) and 8-hydroxyisotrichodermin (**169**; $R^1=OH$, $R^2=R^3=H$) were all shown to be precursors of 3-acetylvomitoxin (**110**; R=Ac) (*33*), as were 7,8-dihydroxycalonectrin (**169**; $R^1=R^2=OH$, $R^3=OAc$) and 15-deacetyl-7,8-dihydroxycalonectrin (**169**; $R^1=R^2=R^3=OH$) (*14, 170*). Oxidation of the 8-OH to the 8-one was shown to be the final step (*170*). 15-Deacetylcalonectrin had previously (*30*) been isolated from a mutant strain of *F. sporotrichioides* in which T-2 toxin production was blocked.

The isovalerate moiety present in T-2 toxin is derived by decarboxylative transamination from L-leucine (*171, 172*). T-2 toxin is thus a product of mixed biogenesis involving two primary metabolic pathways. Leucine limitation enhances the production of neosolaniol (**109**; R=OH).

Some strains of *F. sambucinum* (teleomorph: *Gibberella pulicaris*) are particularly amenable to genetic manipulation (*173, 174, 175*), and the genes controlling the production of trichothecenes have been studied. Most wild type strains produce only diacetoxyscirpenol, but a small number produce trichothecenes with an additional oxygen substituent at C-8. Results from crosses differing in this trait have led to the identification of a specific locus, *Tri1*, for C-8 hydroxylation (*176, 177*). Other loci determine acetylation of the C-8 oxygen (*176*) and the yield of trichothecene. Although it was believed initially (*176*) that these loci were unlinked, recent work (*175*) has shown that some of the trichothecene biosynthetic pathway genes in *G. pulicaris* may be linked to *Tri5* (see Sect. 5.1). Pathway genes are also clustered in *F. sporotrichioides* (*178*). Some of this work has been the subject of a review (*179*).

More recent work on the trichothecene biosynthesis pathway genes in *Fusarium sporotrichioides* has shown that *Tri3* encodes the transacetylase that converts 15-deacetylcalonectrin (**169**; $R^1=R^2=H$. $R^3=OH$) to calonectrin (**169**; $R^1=R^2=H$, $R^3=OAc$) (*179a*), whilst *Tri4* encodes a cytochrome P450 monooxygenase that converts trichodiene to an, as yet, uncharacterized oxygenated derivative (*179b*). The *Tri3–5* genes are clustered with *Tri6* which encodes a protein involved in the regulation of trichothecene biosynthesis (*179c*).

5.4. Inhibitors of Trichothecene Biosynthesis

A series of enantiomeric aza analogues of the α-terpinyl and bisabolyl cations have been prepared as potential inhibitors of trichodiene synthase (*180*), and have been compared with trimethylamine as standard. Inorganic pyrophosphate was required for effective competitive inhibition.

The inhibition of trichothecene biosynthesis in a *Fusarium* sp. in the presence of 0.1*M*-sodium hydrogen carbonate (*180a*) is attributed to a pH-related inhibition of mevalonate kinase (*180b*).

A number of analogues of 12,13-epoxytrichothecene and isotrichodermin, some containing a 15-fluoro substituent, have been synthesised as potential inhibitors of the later stages of trichothecene biosynthesis in *F. culmorum* (*111*). Some were effective in suppressing the production of all trichothecenes, but others, interestingly, whilst inhibiting the production of 3-acetylvomitoxin, enhanced the production of isotrichodermin and 15-deacetylcalonectrin (*181*).

Of a series of farnesyl diphosphate analogs, the 10-fluoro-compound (**169a**) was the most effective inhibitor of trichodiene synthase (*134a*).

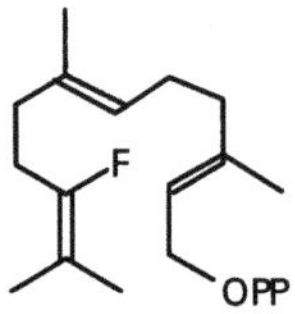

(**169a**)

6. Metabolism

In Part 1 (*1*) metabolism was only briefly mentioned in the Section on the chemical deoxygenation of 12,13-epoxides. The subject is dealt with more fully in this account which incorporates some work published in 1984–1986. Valuable reviews of work on T-2 toxin are available (*182, 183*).

References, pp. 54–70

In general, in mammalian systems 12,13-deoxygenation, which causes complete detoxification of the molecule, is accompanied, or is more often preceded by, hydroxylation (when possible) and cleavage of the ester side chains. The products are then involved in glucuronide formation. All these products are more polar than the parent trichothecene and, if the 12,13-epoxide is retained, may retain some, much diminished, toxicity. Both animals and plants produce the same metabolites. None of the many biologically inactive chemically derived trichothecene rearrangement products (see Sect. 4.4) have ever been detected.

Published investigations have been concerned mainly with the potential grain contaminants, vomitoxin (**110**; R=H, p. 34), T-2 toxin (**109**; R=$OCOCH_2CHMe_2$), and diacetoxyscirpenol (**109**; R=H).

6.1. Cleavage of Ester Groups

Whilst some 3,4-transesterification has been reported in cows (*184*), the metabolites isolated from the excreta of farm animals [and also a monkey (*184a*)] to which T-2 toxin had been administered in various ways include all the possible products of ester cleavage, namely, HT-2 toxin, T-2 triol, neosolaniol, 4- and 15-deacetylneosolaniol and T-2 tetraol (*185, 186, 187, 188*) (for structures see Sect. 4.2.1.3). Some, or all, of these compounds are also produced in a diverse group of essentially *in vitro* systems, including perfused rat liver (*189*) and intestine (*190*), monkey (*191*) and guinea pig (*192, 193*) skin, mouse and monkey liver homogenates (*194*), rat brain homogenate (*195*), hepatic microsomal fractions (*196, 197, 198*), human and rat blood (*199*), rumen fluid (micro-organisms) (*200, 201, 202*), soil and freshwater bacteria (*203*), leaves and twigs of shrubs of *Baccharis* sp. (*204*), and a considerable number of cell lines (*205, 206*).

15-Acetoxyscirpenediol and scirpenetriol were obtained from body fluids and excreta following the oral administration of diacetoxyscirpenol to cattle (*207*), pigs (*208, 207*), and rats (*209*). The same metabolites resulted from the action of rumen micro-organisms (*200*), *Baccharis* sp. cuttings (*204*), and a carboxyl esterase isoenzyme isolated from mouse liver microsomes (*210*). 15-Acetoxyscirpenediol was converted into scirpenetriol by three insect sp. (*211*).

A cell-free extract from a *Fusarium* sp. efficiently converted 3-acetylvomitoxin into vomitoxin and triacetoxyscirpene into diacetoxyscirpenol (*212*).

6.2. Hydroxylation

Oxidation at the 3′-position of the 8-isovaleryl ester side chain of T-2 toxin administered to farm animals to give the ester (**109**; R=$OCOCH_2C(OH)Me_2$) was first reported in 1982 (*213*). This oxidation accompanies the cleavage of ester groups elsewhere in the molecule and has subsequently been confirmed in other farm animals (*214,215*) and in rats (*188*). It also occurs in homogenates of mouse and monkey liver (*194*), and in *Baccharis* sp. cuttings (*204*). In hepatic microsomal fractions of several species (*216*) the oxidation was catalyzed by the cytochrome P-450-dependent monooxygenase system (*198*). When T-2 toxin metabolism by rat liver homogenates is modified by the presence of esterase inhibitors, the metabolism is completely shifted to hydroxylation at the 3′-position (*217*).

Oxidation by rat liver homogenates at the 4′-position of T-2 toxin, giving the ester (**109**; R=$OCOCH_2CH(Me)CH_2OH$) (diastereo-isomers), has been reported (*218*). On further incubation this ester is deacetylated to 4′-hydroxy HT-2 toxin.

Some 7-hydroxylation of the trichothecene nucleus occurred in cows (*219*) and in perfused rat liver (*220*). Vomitoxin was unaffected by rat liver microsomal preparations (*221*).

6.3. 12,13-Deoxygenation (see also Sect. 4.4.1.3)

12,13-De-epoxy-derivatives of the following compounds have been isolated from, or detected in, body fluids and/or excreta of animals to which the appropriate parent trichothecene mycotoxin had been administered orally: vomitoxin (*222,223,224,225*); T-2 toxin (*226*) and its degradation products 3′-hydroxy HT-2 toxin (*227, 188*), 3′-hydroxy T-2 triol (*227*), and T-2 tetraol (*227,228,188*); diacetoxyscirpenol and 15-acetoxyscirpenediol (*209*); and nivalenol (*229*).

The ability of gut or bovine rumen microorganisms to yield de-epoxy metabolites *in vitro* has been demonstrated with vomitoxin (*230,231,232,200,224,233*); T-2 toxin (*200,234,235*); and diacetoxyscirpenol (*200,234,235*).

De-epoxy derivatives of T-2 toxin hydrolysis and hydroxylation products have been detected in the tissues of intravascularly dosed swine (*236*).

6.4. Glucuronide Formation

3α-(1′-β-D-Glucopyranosiduronyl) conjugates of both diacetoxyscirpenol (*237*) and T-2 toxin (*238*) have been prepared by *in vitro* procedures, and the T-2 toxin conjugate has been detected in swine urine (*215*). However, most of the glucuronide conjugates isolated or detected *in vivo* have been derived from the T-2 toxin degradation products 3′-hydroxy T-2 toxin, HT-2 toxin, 3′-hydroxy HT-2 toxin and T-2 tetraol (*215, 189, 193*). Glucuronide conjugates of 15-acetoxyscirpenediol and scirpenetriol (from diacetoxyscirpenol) (*220*) and T-2 toxin degradation products (*220, 239*) have been detected in isolated perfused rat livers.

The 3α-(1′-β-D-glucopyranosyl) conjugate of vomitoxin has been obtained from *Zea mays* suspension cultures (*240*), and the excretion of glucuronide conjugates of both vomitoxin and de-epoxyvomitoxin has been studied in sheep (*241*).

7. Spectroscopy

7.1. Nuclear Magnetic Resonance Spectra

Some unpublished data for derivatives and analogues of diacetoxyscirpenol, particularly the 10,13-cyclotrichothecane and aporearrangement products, which were included in Table 10 of Part 1 (*1*) have now been reported in full (*242*). The spectral assignments for the trichothecenes and certain of their relatives (Scheme 2 and Table 5) have been reviewed (*243*). Differing assignments (*164*) for apotrichodiol (**157**) (^{1}H: 14-Me and 15-Me; ^{13}C: C-2 and C-4) require further investigation.

7.1.1. ^{1}H-Spectra

^{1}H Spectral data for 36 trichothecenes, oxygenated at position 3 and either intermediates in the biosynthesis of 3-acetylvomitoxin or analogues of T-2 toxin, have been collated (*99*). The observed chemical shifts and coupling constants are consistent with the values given in Tables 5–9 of Part 1 (*1*). The spectra of T-2 toxin and some derivatives and analogues have been analyzed more fully (*97, 70, 93*). Chemical shifts and coupling constants not previously recorded (*1*) include δ_H *ca.* 4.7 (ABq, $J = 1.0$ Hz) for the 13-H_2 group of a 12-ene.

With modern instrumentation, long-range couplings (0.5–1.6 Hz) are obtained more readily. In addition to the expected allylic (16-Me–10-H) and homoallylic (16-Me–11-H) couplings in the trichothecene molecule, four-bond (W) couplings (7α-H–11α-H; 7β-H–15-H) have been reported, and are of diagnostic value, for example, in determining the configuration at position 11, or in the identification of the 15-Me signal, and hence the C-15 signal in the ^{13}C spectrum.

The ring A spectra of the apotrichothecenes are very similar to those of the trichothecenes with, for example, $J_{10,11}$=5 Hz. The naturally occurring 11-epiapotrichothecenes (Table 5 and Scheme 2), are distinguished by having $J_{10,11}$=1.7 Hz. The ring C spectra, on the other hand, show characteristic features. Apotrichothecenes derived by rearrangement of 12,13-epoxytrichothecenes have a 2β-substituent, and the ABXY and ABX spectra resulting from 3- or 4-mono- and 3,4-di-substitution, respectively, are not always amenable to first order interpretation (*244*). Similar considerations apply to the ring C ABCDX spectra of the naturally occurring 11-epiapotrichothecenes with substituents at positions 2α or 3. Modern instrumentation and computer programs have done much to eliminate these problems.

Whereas the rigidity of ring C in the trichothecene molecule determines that only two dihedral angles, 0° and 120°, are possible between hydrogens at positions 3 and 4, ring C in the apotrichothecenes is more flexible and a wide range of dihedral angles and coupling constants is observed (*244*). In 2β, 4β-disubstituted apotrichothecenes, steric overcrowding of the β-face is relieved by adoption of the envelope conformation (**170**) in which dihedral angles $\phi_{2\alpha 3\beta}$ and $\phi_{3\beta 4\alpha}$ approach 180° with $J_{2\alpha 3\beta}$ and $J_{3\beta 4\alpha}$ equal to 10–12 Hz (*244*, *78*). This conformation is also adopted, both in the solid state and in solution (*116*), by the 3α-substituted apotrichodiol (**157**). Overcrowding of the β-face in the 3β-epimer (**158**) produces a more planar conformation with dihedral angles closer to 0° and 120°. Apotrichothecenes with conformation (**170**) can show a 2β, 4β W-coupling of ~1.5 Hz. Exceptionally, ring C in the 3,11-epoxy-compound (**154**) has, perforce, the alternative envelope conformation (**171**) (*22*) in which 2α, 4α W-coupling (2.0 Hz) is observed (*116*). In this conformation, the dihedral angles $\phi_{2\alpha 3\beta}$ and $\phi_{3\beta 4\alpha}$ are close to 90° and the values of $J_{2\alpha 3\beta}$ and $J_{3\beta 4\alpha}$ are small (*116*, *245*). Remarkably, this conformation is also found in the apotrichothecene (**172**), derived from the macrolide verrucarin A: $J_{2\alpha 3\beta}=J_{3\beta 4\alpha}=0$ and H-2 and H-4 consequently give doublets (*244*, *245*). Ring C *gem* couplings are in the range 12–16 Hz.

Apotrichothecenes resulting from the rearrangement of trichothecenes, and most 11-epiapotrichothecenes, have a 13-OH substituent which causes deshielding of the hydrogens in the 14-Me group. 14-Me chemical shifts range from 0.92 in apotrichodiol (**157**), through 1.07 in the 3β-epimer (**158**), to 1.35 in the 2β-chloroapotrichothecane

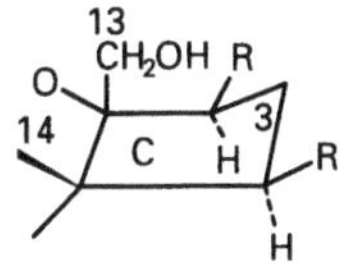
(170)

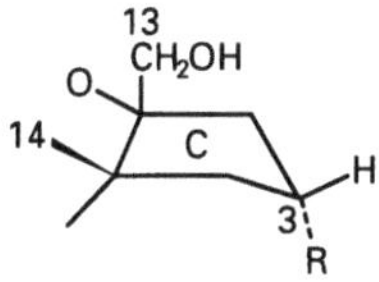
(171)

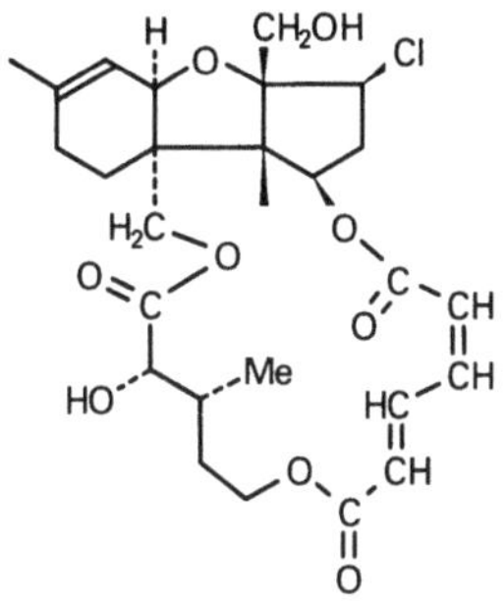
(172)

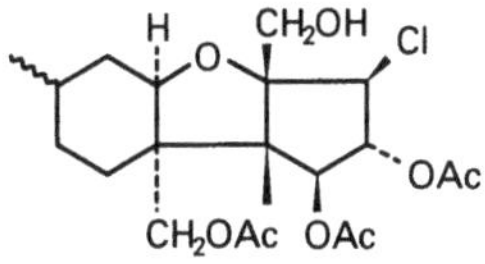
(173)

(**173**) derived from diacetoxyscirpenol (*244*). These figures should be compared with those for trichothecenes in Part 1, Table 9 (*1*).

7.1.2. ^{13}*C-Spectra*

Some erroneous literature assignments have been corrected and a general strategy for the assignment of signals in trichothecene ^{13}C-spectra has been outlined (*242*). With the ^{1}H resonance assigned, a ^{1}H–^{13}C correlation spectrum then identifies the hydrogen-bearing carbons. Finally, quaternary carbon resonances are assigned from heteronuclear nOe experiments. For C-5 and C-6 this approach was considered to be more reliable than the conventional use of long range couplings (*246*), but this view has been criticized (*70*).

A summary of ^{13}C assignments based on the spectra of 30 trichothecenes has been presented (*243*). The ^{13}C chemical shifts for some other commonly occurring trichothecenes lie outside the limits quoted in this paper. The effect of progressive hydroxylation of the trichothecene skeleton on the ^{13}C-spectra was set out in Part 1 (*1*), which also included data for some 8-ones and a 7-one. Table 6 records recent data for representative 3- and 4-ones (**174** and **175**); a 7,8-dione (**176**); 7,8- and 9,10-epoxides (**177** and **178**); and a 9 → 15 ether (**179**).

Table 6. *Chemical Shifts (δ) of Carbon Nuclei in the* ^{13}C *NMR Spectra of Some Representative Trichothecenes and Derivatives*

Position/Structure	(**174**)	(**175**)	(**176**)	(**177**)	(**178**)	(**179**)
Ref.	(*76*)	(*76*)	(*114*)	(*242*)	(*76*)	(*76*)
C-2	79.3	75.5	78.9	79.1	77.7	78.3
C-3	211.7	73.2	70.6	36.7	77.1	77.6
C-4	48.1	209.1	40.1	74.1	83.7	77.6
C-5	42.5	53.8	45.3	48.2	47.0	45.8
C-6	45.6	45.5	59.2	41.7	42.6	41.7
C-7	27.0	25.9	182.7	59.2	25.8	28.3
C-8	68.3	67.2	194.0	50.7	68.9	67.6[a]
C-9	136.9	136.6	141.1	137.7	57.8	73.6
C-10	123.6	122.7	141.5	123.0	58.9	52.0
C-11	68.3	67.6	70.4	70.1	67.6	70.1[a]
C-12	64.1	60.8	64.7	65.7	63.7	65.9
C-13	48.8	48.3	50.3	47.6	48.6	46.4
C-14	11.6	6.6	13.3	6.7	6.5	5.6
C-15	64.3	62.9	64.7	16.3	65.5	63.4
C-16	20.6	20.0	15.3	21.9	18.7	20.0

[a] Assignments may be interchanged

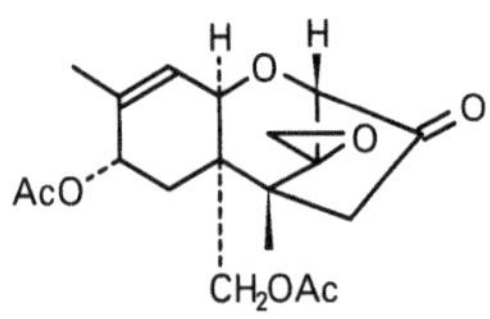

(**174**)

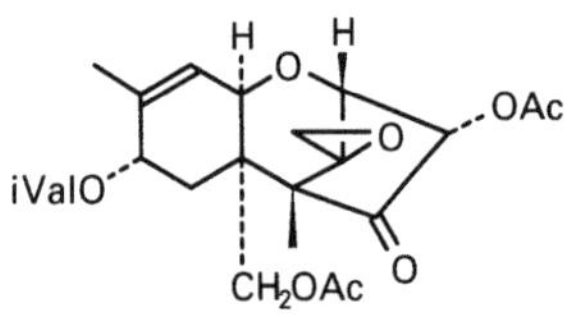

(**175**)

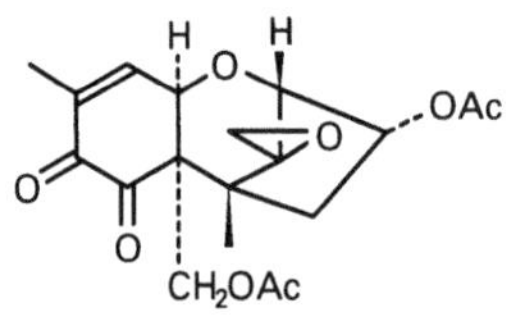

(**176**)

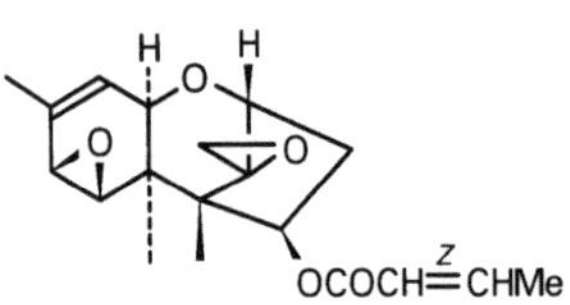

(**177**)

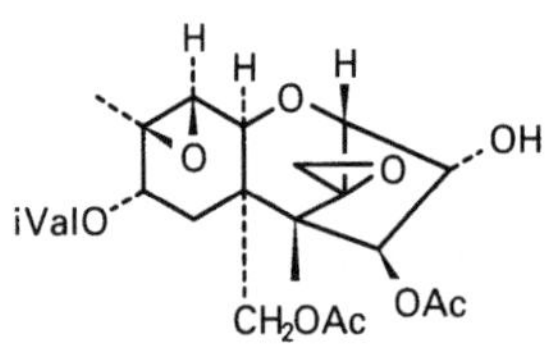

(**178**)

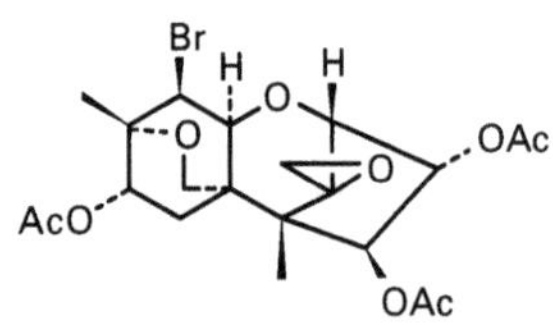

(**179**)

The ^{13}C-spectra for some 11-epiapotrichothecenes have been tabulated (*243*).

7.2. Mass Spectra

Recent developments in technique in the fields of chromatography and mass spectrometry (ms) have been applied to the detection and estimation of trichothecenes: applications relating to the macrocyclic trichothecenes have already been reviewed (*2*).

The separation and identification of T-2 toxin and HT-2 toxin by thin layer chromatography and fast atom bombardment (FAB) ms have been described (*247*). Gas chromatography has been combined with both Fourier transform infrared spectroscopy and ms for the analysis of trichothecenes in a *Fusarium* liquid culture extract (*247a*).

Earlier gas chromatography-ms (gcms) procedures for the detection of a wide range of trichothecenes in body fluids (*248*), using both electron impact (EI) and negative ion chemical ionisation (NICI) with selective ion monitoring, after either hydrolysis or heptafluorobutyrate derivatization as appropriate, have been extended to the analysis of foodstuffs and environmental samples such as pollen (*249*). These procedures have been adapted to include tandem ms (*250, 251*). The literature on the heptafluorobutyrylation of trichothecenes has been reviewed and the use of polymer-bound 4-dimethylaminopyridine as catalyst is recommended (*252*). The greater volatility of the trifluoroacetyl esters is advantageous in certain cases (*253*): These esters have been used (*254*) with the inexpensive ion trap detector [chemical ionization (CI) mode with methanol as reagent gas] for the determination of trichothecenes in cereals.

The CI conditions for the production of ammonium adduct ions from ten commonly occurring trichothecenes have been optimized, and the ions resulting from both collisionally activated dissociation (CAD) and collisionally activated reaction (CAR) have been analyzed in a tandem (triple stage quadrupole) ms system (*250, 255, 256, 257*). More extensive studies have been carried out with diacetoxyscirpenol, T-2 toxin, and vomitoxin (*257*).

A computer-assisted search of an EI and CI trichothecene gcms database has been described (*258*), and retention index standards for trichothecene fluoroacetates (EI-gcms spectra) have been recorded (*259*).

The application of tandem ms as an alternative to gcms for the qualitative and quantitative analysis of trichothecenes in liquid and solid cultures, contaminated grain samples, and biological fluids has been assessed (*260, 261*). In a combination of both major developments, CI tandem ms has been used (*262*), but detection limits in wheat and human plasma samples were better by two orders of magnitude when the technique was combined with gc. The use of dimethyl ether as a reagent gas failed to achieve any reliable classification of the CI mass spectra of a group of 23 structurally diverse trichothecenes, including two macrocycles (*262a*).

The use of hplc dispenses with the time-consuming derivatization step necessary with gc, but coupling to the spectrometer is more difficult.

The frit-FAB technique has been used to identify a number of trichothecenes from their high resolution spectra (*263*). The popular Thermospray interface has been used for the determination of diacetoxyscirpenol, T-2 toxin, and vomitoxin in spiked grain samples (*264*); for the detection of metabolites of these trichothecenes in the body fluids of animals (*265*); and to obtain the spectra of a number of related trichothecenes (*266*). Under certain conditions specific fragmentation patterns were obtained (*264*), but under others (*264*, *266*) the spectra exhibited only an abundant ammonium adduct ion without significant fragmentation. In connection with this work, retention index standards were developed (*267*) for improving the reliability of hplc-ms identifications.

Supercritical fluid chromatography (SFC) is more readily interfaced with the mass spectrometer than is conventional hplc, and NICI spectra for T-2 toxin and vomitoxin have been obtained using a heated frit restrictor (*268*). SFC retention indices for these and related trichothecenes have also been determined (*269*).

References

1. Part 1: GROVE, J.F.: Non-Macrocyclic Trichothecenes. Nat. Prod. Rep., **5**, 187 (1988).

2. GROVE, J.F.: Macrocyclic Trichothecenes. Nat. Prod. Rep., **10**, 429 (1993).

3. WANI, M.C., D.H. RECTOR, and C.E. COOK: Synthesis of HT-2 Toxin, Neosolaniol, T-2 Toxin, 3′-Hydroxy T-2 Toxin, and Sporotrichiol from Anguidine by Routes Involving Hydroxyl Inversion/Esterification. J. Org. Chem., **52**, 3468 (1987).

4. GODTFREDSEN, W.O., J.F. GROVE, and CH. TAMM: Zur Nomenklatur einer neueren Klasse von Sesquiterpenen. Helv. Chim. Acta, **50**, 1666 (1967).

5. CORLEY, D.G., G.E. ROTTINGHAUS, J.K. TRACY, and M.S. TEMPESTA: New Trichothecene Mycotoxins of *Fusarium sporotrichioides* (MC-72083). Tetrahedron Lett., **27**, 4133 (1986).

6. MCCORMICK, S.P., S.L. TAYLOR, R.D. PLATTNER, and M.N. BEREMAND: Bioconversion of Possible T-2 Toxin Precursors by a Mutant Strain of *Fusarium sporotrichioides* NRRL 3299. Appl. Environ. Microbiol., **56**, 702 (1990).

7. MCCORMICK, S.P., S.L. TAYLOR, R.D. PLATTNER, and M.N. BEREMAND: New Modified Trichothecenes Accumulated in Solid Culture by Mutant Strains of *Fusarium sporotrichioides*. Appl. Environ. Microbiol., **55**, 2195 (1989).

8. JARVIS, B.B., J.O. MIDIWO, and M.-D. GUO: 12,13-Deoxytrichoverrins from *Myrothecium verrucaria*. J. Nat. Prod., **52**, 663 (1989).

9. LAUREN, D.R., A. ASHLEY, B.A. BLACKWELL, R. GREENHALGH, J.D. MILLER, and G.A. NEISH: Trichothecenes Produced by *Fusarium crookwellense* DAOM 193611. J. Agric. Food Chem., **35**, 884 (1987).

10. LAUREN, D.R., S.T. SAYER, and M.E. DI MENNA: Trichothecene Production by *Fusarium* Species Isolated from Grain and Pasture Throughout New Zealand. Mycopathologia, **120**, 167 (1992).

11. KONONENKO, G.P., N.A. SOBOLEVA, and A.N. LEONOV: 3,7,8,15-Tetrahydroxy-12,13-epoxytrichothecene-9-en in a Culture of *Fusarium graminearum*. Khim. Prir. Soedin., 267 (1990); Chem. Nat. Compd. (Engl. Transl.), **26**, 219 (1990).
12. RICHARDSON, K.E., G.E. TONEY, C.A. HANEY, and P.B. HAMILTON: Occurrence of Scirpentriol and Its Seven Acetylated Derivatives in Culture Extracts of *Fusarium sambucinum* NRRL 13495. J. Food Prot., **52**, 871 (1989).
13. LUO, Y., F. LIN, J. YANG, J. ZHANG, Y. YE, Y. LI, Y. JIANG, N. ZHANG, and Z. WANG: Analysis and Identification of *Fusarium* Mycotoxins in Corn Culture of *Fusarium camptoceras*. Chem. Abstr., **119**, 153704 (1993).
14. HESKETH, A.R.: Metabolic Studies on the Transformation of Trichodiene to Trichothecene Mycotoxins. Mycotoxin Res., **8**, 52 (1992).
15. PLATTNER, R.D., M.B. AL-HETTI, D. WEISLEDER, and J.B. SINCLAIR: A New Trichothecene from *Trichothecium roseum*. J. Chem. Res. (S), 311 (1988).
16. LANGLEY, P., A. SHUTTLEWORTH, P.J. SIDEBOTTOM, S.K. WRIGLEY, and P.J. FISHER: A Trichothecene from *Spicellum roseum*. Mycol. Res., **94**, 705 (1990).
17. SANSON, D.R., D.G. CORLEY, C.L. BARNES, S. SEARLES, E.O. SCHLEMPER, M.S. TEMPESTA, and G.E. ROTTINGHAUS: New Mycotoxins from *Fusarium sambucinum*. J. Org. Chem., **54**, 4313 (1989).
18. VISCONTI, A., C.J. MIROCHA, A. LOGRIECO, A. BOTTALICO, and M. SOLFRIZZO: Mycotoxins Produced by *Fusarium acuminatum*. Isolation and Characterization of Acuminatin: A New Trichothecene. J. Agric. Food Chem., **37**, 1348 (1989).
19. GREENHALGH, R., J.D. MILLER, and A. VISCONTI: Toxigenic Potential of *Fusarium compactum* R8287 and R8293. J. Agric. Food Chem., **39**, 809 (1991).
20. GREENHALGH, R., B.A. BLACKWELL, M. SAVARD, J.D. MILLER, and A. TAYLOR: Secondary Metabolites Produced by *Fusarium sporotrichioides* DAOM 165006 in Liquid Culture. J. Agric. Food Chem., **36**, 216 (1988).
21. DESJARDINS, A.E., and R.D. PLATTNER: Trichothecene Toxin Production by Strains of *Gibberella pulicaris* (*Fusarium sambucinum*) in Liquid Culture and in Potato Tubers. J. Agric. Food Chem., **37**, 388 (1989).
22. GREENHALGH, R., D.A. FIELDER, B.A. BLACKWELL, J.D. MILLER, J.-P. CHARLAND, and J.W. APSIMON: Some Minor Secondary Metabolites of *Fusarium sporotrichioides* DAOM 165006. J. Agric. Food Chem., **38**, 1978 (1990).
22a. CORLEY. D.G., M. MILLER-WIDEMAN, and R.C. DURLEY: Isolation and Structure of Harzianum A: A New Trichothecene from *Trichoderma harzianum*. J. Nat. Prod., **57**, 422 (1994).
23. JARVIS, B.B., T. DESILVA, J.B. MCALPINE, S.J. SWANSON, and D.N. WHITTERN: New Trichoverroids from *Myrothecium verrucaria* Isolated by High Speed Countercurrent Chromatography. J. Nat. Prod., **55**, 1441 (1992).
24. MIROCHA, C.J., H.K. ABBAS, T. KOMMEDAHL, and B.B. JARVIS: Mycotoxin Production by *Fusarium oxysporum* and *Fusarium sporotrichioides* Isolated from *Baccharis* spp. from Brazil. Appl. Environ. Microbiol., **55**, 254 (1989).
25. BEKELE, E., A.A. ROTTINGHAUS, G.E. ROTTINGHAUS, H.H. CASPER, D.M. FORT, C.L. BARNES, and M.S. TEMPESTA: Two New Trichothecenes from *Fusarium sporotrichioides*. J. Nat. Prod., **54**, 1303 (1991).
26. VESONDER, R.F., A. CIEGLER, A.H. JENSEN, W.K. ROHWEDDER, and D. WEISLEDER: Co-Identity of the Refusal and Emetic Principle from *Fusarium*-Infected Corn. Appl. Environ. Microbiol., **31**, 280 (1976).
27. CORLEY, D.G., G.E. ROTTINGHAUS, and M.S. TEMPESTA: Toxic Trichothecenes from *Fusarium sporotrichioides* (MC-72083). J. Org. Chem., **52**, 4405 (1987).

28. BALDWIN, N.C.P., B.W. BYCROFT, P.M. DEWICK, D.C. MARSH, and J. GILBERT: Trichothecene Mycotoxins from *Fusarium culmorum* Cultures. Z. Naturforsch., **C42**, 1043 (1987).

29. AYER, W.A., and S. MIAO: Secondary Metabolites of the Aspen Fungus *Stachybotrys cylindrospora.* Canad. J. Chem., **71**, 487 (1993).

30. PLATTNER, R.D., L.W. TJARKS, and M.N. BEREMAND: Trichothecenes Accumulated in Liquid Culture of a Mutant of *Fusarium sporotrichioides* NRRL 3299. Appl. Environ. Microbiol., **55**, 2190 (1989).

31. KIM, K.-H., Y.-W. LEE, C.J. MIROCHA, and R.W. PAWLOSKY: Isoverrucarol Production by *Fusarium oxysporum* CJS-12 Isolated from Corn. Appl. Environ. Microbiol., **56**, 260 (1990).

32. MCLACHLAN, A., K.J. SHAW, A.D. HOCKING, J.I. PITT, and T.H.L. NGUYEN: Production of Trichothecene Mycotoxins by Australian *Fusarium* Species. Food Addit. Contam., **9**, 631 (1992).

33. ZAMIR, L.O., K.A. DEVOR, and F. SAURIOL: Biosynthesis of the Trichothecene 3-Acetyldeoxynivalenol. Identification of the Oxygenation Steps After Isotrichodermin. J. Biol. Chem., **266**, 14992 (1991).

34. EVIDENTE, A., G. RANDAZZO, A. VISCONTI, and A. BOTTALICO: Isolation of 15-Acetoxyscirpendiol from a Culture of *Fusarium poae* on Corn. Mycotoxin Res., **5**, 30 (1989).

35. COLVIN, E.W., and S. CAMERON: Selective Chemical Transformations of the Trichothecene 4β-Acetoxyscirpene-3α,15-diol. Heterocycles, **25**, 133 (1987).

36. HABERMEHL, G.: Isolation and Structure of New Toxins from Plants. Pure Appl. Chem., **61**, 377 (1989).

37. RICHARDSON, K.E., and P.B. HAMILTON: Preparation of 4,15-Diacetoxyscirpenol from Cultures of *Fusarium sambucinum* NRRL 13495. Appl. Environ. Microbiol., **53**, 460 (1987).

38. HUSSEIN, H.H., M. BAXTER, I.G. ANDREW, and R.A. FRANICH: Mycotoxin Production by *Fusarium* Species Isolated from New Zealand Maize Fields. Mycopathologia, **113**, 35 (1991).

39. VESONDER, R.F., P. GOLINSKI, R. PLATTNER, and D.L. ZEITKIEWICZ: Mycotoxin Formation by Different Geographic Isolates of *Fusarium crookwellense.* Mycopathologia, **113**, 11 (1991).

40. COLE, R.J., J.W. DORNER, J. GILBERT, D.N. MORTIMER, C. CREWS, J.C. MITCHELL, R.M. WINDINGSTAD, P.E. NELSON, and H.G. CUTLER: Isolation and Identification of Trichothecenes from *Fusarium compactum* Suspected in the Aetiology of a Major Intoxication of Sandhill Cranes. J. Agric. Food Chem., **36**, 1163 (1988).

41. BOSCH, U., and C.J. MIROCHA: Toxin Production by *Fusarium* Species from Sugar Beets and Natural Occurrence of Zearalenone in Beets and Beet Fibers. Appl. Environ. Microbiol., **58**, 3233 (1992).

41a. ABRAMSON, D., R.M. CLEAR, and D.M. SMITH: Trichothecene Production by *Fusarium* spp. Isolated from Manitoba Grain. Can. J. Plant Pathol., **15**, 147 (1993).

41b. ALTOMARE, C., A. RITIENI, G. PERRONE, V. FOGLIANO, L. MANNINA, and A. LOGRIECO: Production of Neosolaniol by *Fusarium tumidum.* Mycopathologia, **130**, 179 (1995).

41c. EL-MAGHRABY, O.M.O., I.A. EL-KADY, and S. SOLIMAN: Mycoflora and *Fusarium* Toxins of Three Types of Corn Grains in Egypt with Special Reference to Production of Trichothecene Toxins. Microbiol. Res., **150**, 225 (1995).

41d. JARVIS, B.B., J. SALEMME, and A. MORAIS: *Stachybotrys* Toxins, 1. Nat. Toxins, **3**, 10 (1995).

42. El-Maghraby, O.M.O., G.A. Bean, B.B. Jarvis, and M.B. Aboul-Nasr: Macrocyclic Trichothecenes Produced by *Stachybotrys* Isolated from Egypt and Eastern Europe. Mycopathologia, **113**, 109 (1991).
42a. Soares da Silva, N., and C. Kemmelmeir: Identification of Mycotoxins Produced by *Fusarium graminearum* Isolate Grown on Maize (*Zea mays* L). Chem. Abstr., **122**, 284356 (1995).
43. Bosch, U., C.J. Mirocha, H.K. Abbas, and M. di Menna: Toxicity and Toxin Production by *Fusarium* Isolates from New Zealand. Mycopathologia, **108**, 73 (1989).
44. Abbas, H.K., and U. Bosch: Evaluation of Trichothecene and Nontrichothecene Mycotoxins Produced by *Fusarium* in Soybeans. Mycotoxin Res., **6**, 13 (1990).
45. Ramakrishna, Y., R.V. Bhat, and V. Ravindranath: Production of Deoxynivalenol by *Fusarium* Isolates from Sorghum Cultivars Associated with Specific Plant Diseases and Samples of Wheat Associated with a Human Mycotoxicosis Outbreak. Appl. Environ. Microbiol., **55**, 2619 (1989).
45a. Styriak, I., E. Conkova, and J. Bohm: Occurrence of *Fusarium sacchari* var. *subglutinans* and Its Mycotoxin Production Ability in Broiler Feed. Folia Microbiol., **39**, 579 (1994).
46. Golinski, P., R.F. Vesonder, D. Latus-Zietkiewicz, and J. Perkowski: Formation of Fusarenone X, Nivalenol, Zearalenone, α-*trans*-Zearalenol, β-*trans*-Zearalenol, and Fusarin C by *Fusarium crookwellense*. Appl. Environ. Microbiol., **54**, 2147 (1988).
47. Sugiura, Y., K. Fukasaku, T. Tanaka, Y. Matsui, and Y. Ueno: *Fusarium poae* and *Fusarium crookwellense*, Fungi Responsible for the Natural Occurrence of Nivalenol in Hokkaido. Appl. Environ. Microbiol., **59**, 3334 (1993).
48. Combrinck, S., W.C.A. Gelderblom, H.S.C. Spies, B.V. Burger, P.G. Thiel, and W.F.O. Marasas: Isolation and Characterization of Trichothecin from Corn Cultures of *Fusarium graminearum* MRC 1125. Appl. Environ. Microbiol., **54**, 1700 (1988).
49. Flesch, P., and I. Voigt-Scheuermann: Isolation and Identification of Iso-Trichothecin from Cultures of the Fungus *Trichothecium roseum*. Wein Wiss., **48**, 15 (1993).
50. Burgess, L.W., P.E. Nelson, and T.A. Toussoun: Characterization, Geographic Distribution and Ecology of *Fusarium crookwellense* sp. nov. Trans. Brit. Mycol. Soc., **79**, 497 (1982).
51. Burgess, L.W., G.A. Forbes, C. Windels, P.E. Nelson, and W.F.O. Marasas: Characterization and Distribution of *Fusarium acuminatum* subsp. *armeniacum* subsp. nov. Mycologia, **85**, 119 (1993).
52. Wing, N., D.R. Lauren, W.L. Bryden, and L.W. Burgess: Toxicity and Trichothecene Production by *Fusarium acuminatum* subsp. *acuminatum* and *Fusarium acuminatum* subsp. *armeniacum*. Nat. Toxins, **1**, 229 (1993).
53. Gams, W.: Taxonomy and Nomenclature of *Microdochium nivale* (*Fusarium nivale*). In: *Fusarium* Mycotoxins, Taxonomy and Pathogenicity (J. Chelkowski, ed.), p. 195. Amsterdam: Elsevier, 1989.
54. Booth, C.: The Genus *Fusarium*, p. 193. Kew, U.K.: Commonwealth Mycological Institute, 1971.
55. Miller, J.D., R. Greenhalgh, Y.-Z. Wang, and M. Lu: Trichothecene Chemotypes of Three *Fusarium* Species. Mycologia, **83**, 121 (1991).
56. Blaney, B.J., and R.L. Dodman: Production of the Mycotoxins Zearalenone, 4-Deoxynivalenol, and Nivalenol by Isolates of *Fusarium graminearum* Groups 1 and 2 from Cereals in Queensland. Aust. J. Agr. Res., **39**, 21 (1988).

57. SYDENHAM, E.W., W.F.O. MARASAS, P.G. THIEL, G.S. SHEPHARD, and J.J. NIEUWENHUIS: Production of Mycotoxins by Selected *Fusarium graminearum* and *F. crookwellense* Isolates. Food Addit. Contam., **8**, 31 (1991).
57a. SZECSI, A., and T. BARTOK: Trichothecene Chemotypes of *Fusarium graminearum* Isolated from Corn in Hungary. Mycotox. Res., **11**, 85 (1995).
57b. NIRENBERG, H.: Morphological Differentiation of *Fusarium sambucinum* Fuckel s. st., *F. torulosum* (Berk et Curt.) Nirenberg comb. nov. and *F. venenatum* Nirenberg sp. nov. Mycopathologia, **129**, 131 (1995).
57c. ALTOMARE, C., A. LOGRIECO, A. BOTTALICO, G. MULE, A. MORETTI, and A. EVIDENTE: Production of Type A Trichothecenes and Enniatin B by *Fusarium sambucinum* Fuckel sensu lato. Mycopathologia, **129**, 177 (1995).
57d. THRANE, U., and U. HANSEN: Chemical and Physiological Characterization of Taxa in the *Fusarium sambucinum* Complex. Mycopathologia, **129**, 183 (1995).
57e. SCHMIDT, R., P. ZAJKOWSKI, and J. WINK: Toxicity of *Fusarium sambucinum* Fuckel sensu lato to Brine Shrimp. Mycopathologia, **129**, 173 (1995).
57f. JELEN, H.H., C.J. MIROCHA, E. WASOWICZ, and E. KAMINSKI: Production of Volatile Sesquiterpenes by *Fusarium sambucinum* Strains with Different Abilities to Synthesize Trichothecenes. Appl. Environ. Microbiol., **61**, 3815 (1995).
58. GHOSAL, S., K. BISWAS, R.S. SRIVASTAVA, D.K. CHAKRABARTI, and K.C.B. CHAUDHARY: Toxic Substances Produced by *Fusarium*: Occurrence of Zearalenone, Diacetoxyscirpenol, and T-2 Toxin in Moldy Corn Infected with *Fusarium moniliforme* Sheld. J. Pharm. Sci., **67**, 1768 (1978).
59. CHAKRABARTI, D.K., and S. GHOSAL: Occurrence of Free and Conjugated 12,13-Epoxytrichothec-9-enes and Zearalenone in Banana Fruits Infected with *Fusarium moniliforme*. Appl. Environ. Microbiol., **51**, 217 (1986).
60. MIROCHA, C.J., H.K. ABBAS, and R.F. VESONDER: Absence of Trichothecenes in Toxigenic Isolates of *Fusarium moniliforme*. Appl. Environ. Microbiol., **56**, 520 (1990).
61. PEARSON, A.J., and M.K. O'BRIEN: Trichothecene Synthesis Using Organoiron Complexes: Diastereoselective Total Syntheses of (±)-Trichodiene, (±)-12,13-Epoxytrichothec-9-ene, and (±)-Trichodermol. J. Org. Chem., **54**, 4663 (1989).
62. O'BRIEN, M.K., A.J. PEARSON, A.A. PINKERTON, W. SCHMIDT, and K. WILLMAN: A Total Synthesis of (±)-Trichodermol. J. Amer. Chem. Soc., **111**, 1499 (1989).
63. COLVIN, E.W., M.J. EGAN, and F.W. KERR: Synthesis of the Trichothecene Mycotoxin, T-2 Tetraol. Chem. Commun., 1200 (1990).
64. GILBERT, J.C., and R.D. SELLIAH: Highly Convergent Enantioselective Route to Trichothecenes. Tetrahedron Lett., **33**, 6259 (1992).
65. GILBERT, J.C., and R.D. SELLIAH: Enantioselective Synthesis of an *ent*-Trichothecene. Tetrahedron, **50**, 1651 (1994).
66. HUA, D.H., S. VENKATARAMAN, R. CHAN-YU-KING, and J.V. PAUKSTELIS: Enantioselective Total Synthesis of (+)-12,13-Epoxytrichothec-9-ene and Its Antipode. J. Amer. Chem. Soc., **110**, 4741 (1988).
67. ROUSH, W.R., and S. RUSSO-RODRIGUEZ: Trichothecene Degradation Studies, 3: Synthesis of 12,13-Deoxy-12,13-methanoanguidine and 12-Epianguidine, Two Optically Active Analogues of the Epoxytrichothecene Mycotoxin Anguidine. J. Org. Chem., **52**, 603 (1987).
68. CAMERON, S., and E.W. COLVIN: Trichothecene Mycotoxin Interconversions: Partial Syntheses of Calonectrin and Deoxynivalenol, and of a Trichothecene *epi*-Epoxide, 3α, 4β, 15-Triacetoxy-12,13-*epi*-epoxytrichothec-9-ene. J. Chem. Soc., Perkin Trans. 1, 887 (1989).

68a. NEMOTO, H., J. MIYATA, H. HAKAMATA. M. NAGAMOCHI, and K. FUKUMOTO: A Novel and Efficient Route to Chiral A-Ring Aromatic Trichothecanes—The First Enantiocontrolled Total Synthesis of (—)-Debromofiliformin and (—)-Filiformin. Tetrahedron, **51**, 5511 (1995).

68b. NEMOTO, H., J. MIYATA, and K. FUKUMOTO: Cyclobutane Strategy for the Synthesis of A-Ring Aromatic Trichothecanes. Heterocycles, **42**, 165 (1996).

68c. ISHIHARA, J., R. NONAKA, Y. TERASAWA, K. TADANO, and S. OGAWA: Synthetic Studies on the Trichothecene Family from D-Glucose. Tetrahedron Assymm., **5**, 2217 (1994).

69. GILARDI, R., C. GEORGE, and J.L. FLIPPEN-ANDERSON: The Structure of T-2 Toxin. Acta Crystallogr., **C46**, 645 (1990).

70. MESILAAKSO, M., and E. RAHKAMAA: ^{1}H and ^{13}C NMR Study of the Trichothecenes T-2 Toxin and T-2 Triol. J. Prakt. Chem., **334**, 53 (1992).

71. GROVE, J.F.: Phytotoxic Compounds Produced by *Fusarium equiseti*, Part V: Transformation Products of 4β,15-Diacetoxy-3α,7α-dihydroxy-12,13-epoxytrichothec- 9-en-8-one and the Structures of Nivalenol and Fusarenone. J. Chem. Soc., C, 375 (1970).

72. JARVIS, B.B., D.B. MAZZOCCHI, H.L. AMMON, E.P. MAZZOLA, J.L. FLIPPEN-ANDERSON, R.D. GILARDI, and C.F. GEORGE: Conformational Effects in Trichothecenes: Structures of 15-Hydroxy C4 and C8 Ketones. J. Org. Chem., **55**, 3660 (1990).

73. GREENHALGH, R., A.W. HANSON, J.D. MILLER, and A. TAYLOR: Production and X-ray Crystal Structure of 3α-Acetoxy-7α, 15-dihydroxy-12,13-epoxytrichothec-9-en-8-one. J. Agric. Food Chem., **32**, 945 (1984).

74. NOVAK, T.J., and K. QUINN-DOGGETT: 2-(Diphenylacetyl)-1,3-indandione 1-hydrazone (DIPAIN) Derivatives for Detection of Trichothecene Mycotoxins. Anal. Lett., **24**, 913 (1991).

75. ZIEGLER, F.E., and S.B. SOBOLOV: Synthesis of a Highly Functionalized Carbon Ring Skeleton for the Trichothecene Anguidine. J. Amer. Chem. Soc., **112**, 2749 (1990).

76. ANDERSON, D.W., R.M. BLACK, C.G. LEE, C. POTTAGE, R.L. RICKARD, M.S. SANDFORD, T.D. WEBBER, and N.E. WILLIAMS: Structure-Activity Studies of Trichothecenes: Cytotoxicity of Analogues and Reaction Products Derived from T-2 Toxin and Neosolaniol. J. Med. Chem., **32**, 555 (1989).

77. GROVE, J.F.: Phytotoxic Compounds Produced by *Fusarium equiseti*, Part 10: The Preparation and Rearrangement of Diacetylneosolaniol 9β,10β-Epoxide. J. Chem. Soc. Perkin Trans. 1, 1199 (1990).

78. GROVE, J.F.: Phytotoxic Compounds Produced by *Fusarium equiseti*, Part 9: Reactions of Some 9β,10β:12,13-Diepoxytrichothecanes. J. Chem. Soc. Perkin Trans. 1, 115 (1990).

79. BURROWS, E.P., and L.L. SZAFRANIEC: Hypochlorite-Promoted Transformations of Trichothecenes 3: Deoxynivalenol. J. Nat. Prod., **50**, 1108 (1987).

80. ROUSH, W.R., and S. RUSSO-RODRIGUEZ: Trichothecene Degradation Studies, 2: Synthesis of [13-^{14}C]Anguidine. J. Org. Chem., **52**, 598 (1987).

81. TROST, B.M., P.G. MCDOUGAL, and K.J. HALLER: A Tandem Cycloaddition-Ene Strategy for the Synthesis of (±)-Verrucarol and (±)-4,11-Diepi-12,13-deoxyverrucarol. J. Amer. Chem. Soc., **106**, 383 (1984).

82. KRAUS, G.A., and P.J. THOMAS: Synthesis of 7,7,8-Trideuteriated Trichothecenes. J. Org. Chem., **53**, 1395 (1988).

83. DILLEN, J.L.M., C,P. GORST-ALLMAN, and P.S. STEYN: Trichothecene Chemistry: Conversion of Diacetoxyscirpenol into Neosolaniol Monoacetate and Its Epimer. S. Afr. Tydskr. Chem., **39**, 111 (1986).

84. MULLER, B., R. ACHINI, and CH. TAMM: Biosynthese der Verrucarine und Roridine, Teil 3: Der Einbau von (3*R*)-[5-^{14}C]-, [2-^{14}C]- und an C(2) stereospezifisch tritiiertem Mevalonat in Verrucarol. Helv. Chim. Acta, **58**, 471 (1975).
85. JARVIS, B.B., J.O. MIDIWO, and E.P. MAZZOLA: Antileukemic Compounds Derived by Chemical Modification of Macrocyclic Trichothecenes, 2: Derivatives of Roridins A and H and Verrucarins A and J. J. Med. Chem., **27**, 239 (1984).
86. COLVIN, E.W., and S. CAMERON: Partial Syntheses of the Trichothecene Mycotoxins, Calonectrin and Deoxynivalenol. Tetrahedron Lett., **29**, 493 (1988).
87. JEKER, N., and CH. TAMM: Synthesis of New Unnatural Macrocyclic Trichothecenes: 4-Epiverrucarin. A. Helv. Chim. Acta, **71**, 1904 (1988).
88. ROUSH, W.R., and T.A. BLIZZARD: Synthesis of Verrucarin B. J. Org. Chem., **49**, 4332 (1984).
89. JEKER, N., and CH. TAMM: Synthesis of New Unnatural Macrocyclic Trichothecenes: 3-Isoverrucarin A [(1″ – *O*) (3 → 4) *abeo*-Verrucarin A], Verrucinol, and Verrucene. Helv. Chim. Acta, **71**, 1895 (1988).
90. RICHARDSON, S.K., A. JEGANATHAN, R.S. MANI, B.E. HALEY, D.S. WATT, and L.R. TRUSAL: Synthesis and Biological Activity of C-4 and C-15 Aryl Azide Derivatives of Anguidine. Tetrahedron, **43**, 2925 (1987).
91. CHU, F.S., S. GROSSMAN, R.-D. WEI, and C.J. MIROCHA: Production of Antibody Against T-2 Toxin. Appl. Env. Microbiol., **37**, 104 (1979).
92. OHTANI, K., O. KAWAMURA, and Y. UENO: Improved Preparation of T-2 Toxin-Protein Conjugates. Toxicon, **26**, 1107 (1988).
93. DUFFY, M.J., and R.S. REID: Measurement of the Stability of T-2 Toxin in Aqueous Solution. Chem. Res. Toxicol., **6**, 524 (1993).
94. SAVARD, M.E., and R. GREENHALGH: Synthesis and NMR Analysis of New Trichothecenes. J. Nat. Prod., **50**, 953 (1987).
95. ANDERSON, D.W., R.M. BLACK, D.A. LEIGH, and J.F. STODDART: Novel 4,15-Polyether Analogues of Macrocyclic Trichothecenes. Tetrahedron Lett., **28**, 2653 (1987).
96. ANDERSON, D.W., R.M. BLACK, D.A. LEIGH, and J.F. STODDART: Novel 3,4- and 8,15-Polyether Analogues of Macrocyclic Trichothecenes. Tetrahedron Lett., **28**, 2657 (1987).
97. MESILAAKSO, M., M. MOILANEN, and E. RAHKAMAA: ^{1}H and ^{13}C NMR Analysis of Some Trichothecenes. Arch. Environ. Contam. Toxicol., **18**, 365 (1989).
98. GROVE, J.F., A.J. MCALEES, and A. TAYLOR: Preparation of 10-g Quantities of 15-O-Acetyl-4-deoxynivalenol. J. Org. Chem., **53**, 3860 (1988).
99. SAVARD, M.E., B.A. BLACKWELL, and R. GREENHALGH: A ^{1}H NMR Study of Derivatives of 3-Hydroxy-12,13-epoxytrichothec-9-enes. Canad. J. Chem., **65**, 2254 (1987).
99a. SINHA, R.C., M.E. SAVARD, and R. LAU: Production of Monoclonal Antibodies for the Specific Detection of Deoxynivalenol and 15-Acetyldeoxynivalenol by ELISA. J. Agric. Food Chem., **43**, 1740 (1995).
100. SAVARD, M.E.: Deoxynivalenol Fatty Acid and Glucoside Conjugates. J. Agric. Food Chem., **39**, 570 (1991).
101. LAUREN, D.R., W.A. SMITH, and A.L. WILKINS: Preparation, Purification, and NMR Spectra of Some Mono- and Dihemisuccinates of the Trichothecene Mycotoxin Nivalenol. J. Agric. Food Chem., **42**, 828 (1994).
102. ROUSH, W.R., and T.E. D'AMBRA: Total Synthesis of (±)-Verrucarol. J. Amer. Chem. Soc., **105**, 1058 (1983).
103. STILL, W.C., and M.-Y. TSAI: Total Synthesis of (±)-Trichodermol. J. Amer. Chem. Soc., **102**, 3654 (1980).

104. BAMBURG, J.R., N.V. RIGGS, and F.M. STRONG: The Structures of Toxins from Two Strains of *Fusarium tricinctum*. Tetrahedron, **24**, 3329 (1968).
105. EHRLICH, K.C., and K.W. DAIGLE: Protein Synthesis Inhibition by 8-Oxo-12,13-epoxytrichothecenes. Biochim. Biophys. Acta, **923**, 206 (1987).
106. ANDERSON, D.W., R.M. BLACK, D.A. LEIGH, J.F. STODDART, and N.E. WILLIAMS: The Facile Conversion of T-2 Toxin and Neosolaniol into Anguidine. Tetrahedron Lett., **28**, 2661 (1987).
107. YAGAN, B., and B.B. JARVIS: Synthesis of Tritium Labelled Verrucarol and Verrucarin A. J. Lab. Comp. Radiopharm., **27**, 675 (1989).
108. GROVE, J.F.: Phytotoxic Compounds Produced by *Fusarium equiseti*, Part II: Regioselective Reactions with Derivatives of the Trichothecene Mycotoxins, Nivalenol and Vomitoxin. J. Nat. Prod., **57**, 1491 (1994).
109. DESS, D.B., and J.C. MARTIN: Readily Accessible 12-I-5 Oxidant for the Conversion of Primary and Secondary Alcohols to Aldehydes and Ketones. J. Org. Chem., **48**, 4155 (1983).
110. SIGG, H.P., R. MAULI, F. FLURY, and D. HAUSER: Die Konstitution von Diacetoxyscirpenol. Helv. Chim. Acta, **48**, 962 (1965).
111. ZAMIR, L.O., A. NIKOLAKAKIS, and F. SAURIOL: Target-Oriented Inhibitors of the Late Stages of Trichothecene Biosynthesis, 1: Design, Syntheses, and Proof of Structures of Putative Inhibitors. J. Agric. Food Chem., **40**, 676 (1992).
112. DAVIS, F.A., L.C. VISHWAKARMA, J.M. BILLMERS, and J. FINN: Synthesis of α-Hydroxy Carbonyl Compounds (Acyloins): Direct Oxidation of Enolates Using 2-Sulfenyloxaziridines. J. Org. Chem., **49**, 3241 (1984).
113. EHRLICH, K.C.: Preparation of the *Fusarium* Toxin, Nivalenol, by Oxidation of the Putative Biosynthetic Precursor, 7-Deoxynivalenol. Mycopathologia, **107**, 111 (1989).
114. KING, R.R., and R. GREENHALGH: Structural Elucidation of a Novel Deoxynivalenol Analogue. J. Org. Chem., **52**, 1605 (1987).
115. JARVIS, B.B., M.E. ALVAREZ, G. WANG, and H.L. AMMON: Solvolytic Cyclization of 4,15-Anhydroverrucarol. A Facile Trichothecene → 10,13-CycloTrichothecene Rearrangement. J. Org. Chem., **54**, 4493 (1989).
116. GREENHALGH, R., D.A. FIELDER, L.A. MORRISON, J.-P. CHARLAND, B.A. BLACKWELL, M.E. SAVARD, and J.W. APSIMON: Secondary Metabolites of *Fusarium* Species: ApoTrichothecene Derivatives. J. Agric. Food Chem., **37**, 699 (1989).
117. GREENHALGH, R., D.A. FIELDER, L.A. MORRISON, J.-P. CHARLAND, B.A. BLACKWELL, J.D. MILLER, M.E. SAVARD, and J.W. APSIMON: ApoTrichothecenes – Minor Metabolites of the *Fusarium* Species. Bioactive Mols., **10**, 223 (1989).
118. KONONENKO, G.P., A.R. BEKKER, A.N. LEONOV, and N.A. SOBOLEVA: Gramilaurone, a Novel Natural Sesquiterpenoid from *Fusarium graminearum* Schw. Tetrahedron Lett., **32**, 1893 (1991).
119. TAKITANI, S., Y. ASABE, T. KATO, M. SUZUKI, and Y. UENO: Spectrodensitometric Determination of Trichothecene Mycotoxins with 4-(p-Nitrobenzyl)pyridine on Silica Gel Thin-Layer Chromatograms. J. Chromatogr., **172**, 335 (1979).
120. NOVAK, T.J., and K.A. QUINN: Catalytic NBP Spot Test for the Detection of Trichothecene Mycotoxin T-2. Anal. Lett., **19**, 2001 (1986).
121. RAMAKRISHNA, Y., and R.V. BHAT: Comparison of Different Spray Reagents for Identification of Trichothecenes. Curr. Sci., **56**, 524 (1987).
122. CAMERON, S., and E.W. COLVIN: Chemical Deoxygenation of the Trichothecenes Diacetoxyscirpenol and Deoxynivalenol. J. Chem. Soc., Perkin Trans. 1, 365 (1989).
123. MACHIDA, Y., and S. NOZOE: Biosynthesis of Trichothecin. Tetrahedron Lett., 1969 (1972).

124. VANMIDDLESWORTH, F., A.E. DESJARDINS, S.L. TAYLOR, and R.D. PLATTNER: Trichodiene Accumulation by Ancymidol Treatment of *Gibberella pulicaris*. Chem. Commun., 1156 (1986).

125. DESJARDINS, A.E., R.D. PLATTNER, and M.N. BEREMAND: Ancymidol Blocks Trichothecene Biosynthesis and Leads to Accumulation of Trichodiene in *Fusarium sporotrichioides* and *Gibberella pulicaris*. Appl. Environ. Microbol., **53**, 1860 (1987).

126. ZAMIR, L.O., M.J. GAUTHIER, K.A. DEVOR, Y. NADEAU, and F. SAURIOL: Trichodiene Is a Precursor to Trichothecenes. Chem. Commun., 598 (1989).

127. ZAMIR, L.O., Y. NADEAU, C.-D. NGUYEN, K. DEVOR, and F. SAURIOL: Mechanism of 3-Acetyldeoxynivalenol Biosynthesis. Chem. Commun., 127 (1987).

128. VASAVADA, A.B., and D.P.H. HSIEH: Manganese Inhibition of 3-Acetyldeoxynivalenol Biosynthesis in *Fusarium graminearum* R 2118. Appl. Microbiol. Biotechnol., **33**, 335 (1990).

129. CANE, D.E.: Stereochemical Studies of Natural Products Biosynthesis. Pure Appl. Chem., **61**, 493 (1989).

130. CANE, D.E.: Enzymatic Formation of Sesquiterpenes. Chem. Rev., **90**, 1089 (1990).

131. CANE, D.E., and H.-J. HA: Trichodiene Biosynthesis and the Role of Nerolidyl Pyrophosphate in the Enzymatic Cyclization of Farnesyl Pyrophosphate. J. Amer. Chem. Soc., **110**, 6865 (1988).

132. CANE, D.E., J.L. PAWLAK, R.M. HORAK, and T.M. HOHN: Studies of the Cryptic Allylic Pyrophosphate Isomerase Activity of Trichodiene Synthase Using the Anomalous Substrate 6,7-Dihydrofarnesyl Pyrophosphate. Biochemistry, **29**, 5476 (1990).

132a. CANE, D.E., and G. YANG: Trichodiene Synthase. Stereochemical Studies of the Cryptic Allylic Diphosphate Isomerase Activity Using an Anomalous Substrate. J. Org. Chem., **59**, 5794 (1994).

133. HOHN, T.M., and M.N. BEREMAND: Regulation of Trichodiene Synthase in *Fusarium sporotrichioides* and *Gibberella pulicaris* (*Fusarium sambucinum*). Appl. Environ. Microbiol., **55**, 1500 (1989).

134. HOHN, T.M., and F. VANMIDDLESWORTH: Purification and Characterization of the Sesquiterpene Cyclase Trichodiene Synthetase from *Fusarium sporotrichioides*. Arch. Biochem. Biophys., **251**, 756 (1986).

134a. CANE, D.E., G. YANG, Q. XUE, and J.H. SHIM: Trichodiene Synthase. Substrate Specificity and Inhibition. Biochemistry, **34**, 2471 (1995).

134b. CANE, D.E., J.H. SHIM, Q. XUE, B.C. FITZSIMONS, and T.M. HOHN: Trichodiene Synthase. Identification of Active Site Residues by Site-Directed Mutagenesis. Biochemistry, **34**, 2480 (1995).

134c. CANE, D.E., and Q. XUE: Trichodiene Synthase. Enzymatic Formation of Multiple Sesquiterpenes by Alteration of the Cyclase Active Site. J. Am. Chem. Soc., **118**, 1563 (1996).

134d. PROCTOR, R.H., T.M. HOHN, and S.P. MCCORMICK: Reduced Virulence of *Gibberella zeae* Caused by Disruption of a Trichothecene Toxin Biosynthetic Gene. Mol. Plant-Microbe Interact., **8**, 593 (1995).

135. HOHN, T.M., and R.D. PLATTNER: Expression of the Trichodiene Synthase Gene of *Fusarium sporotrichioides* in *Escherichia coli* Results in Sesquiterpene Production. Arch. Biochem. Biophys., **275**, 92 (1989).

136. HOHN, T.M., and P.D. BEREMAND: Isolation and Nucleotide Sequence of a Sesquiterpene Cyclase Gene from Trichothecene-Producing Fungus *Fusarium sporotrichioides*. Gene, **79**, 131 (1989).

137. CANE, D.E., Z. WU, J.S. OLIVER, and T.M. HOHN: Overproduction of Soluble Trichodiene Synthase from *Fusarium sporotrichioides* in *Escherichia coli*. Arch. Biochem. Biophys., **300**, 416 (1993).

138. HOHN, T.M., and J.B. OHLROGGE: Expression of a Fungal Sesquiterpene Cyclase Gene in Transgenic Tobacco. Plant Physiol., **97**, 460 (1991).

139. DESJARDINS, A.E., T.M. HOHN, and S.P. MCCORMICK: Effect of Gene Disruption of Trichodiene Synthase on the Virulence of *Gibberella pulicaris*. Mol. Plant-Microbe Interact., **5**, 214 (1992).

140. HOHN, T.M., and A.E. DESJARDINS: Isolation and Gene Disruption of the *Tox5* Gene Encoding Trichodiene Synthase in *Gibberella pulicaris*. Mol. Plant-Microbe Interact., **5**, 249 (1992).

141. DESJARDINS, A.E., R.D. PLATTNER, and F. VANMIDDLESWORTH: Trichothecene Biosynthesis in *Fusarium sporotrichioides*: Origin of the Oxygen Atoms of T-2 Toxin. Appl. Environ. Microbiol., **51**, 493 (1986).

142. DESJARDINS, A.E., R.D. PLATTNER, and G.F. SPENCER: Inhibition of Trichothecene Toxin Biosynthesis by Naturally Occurring Shikimate Aromatics. Phytochem., **27**, 767 (1988).

143. BEREMAND, M.N.: Isolation and Characterization of Mutants Blocked in T-2 Toxin Biosynthesis. Appl. Environ. Microbiol., **53**, 1855 (1987).

144. ZAMIR, L.O., K.A. DEVOR, N. MORIN, and F. SAURIOL: Biosynthesis of Trichothecenes: Oxygenation Steps Post-Trichodiene. Chem. Commun., 1033 (1991).

145. HESKETH, A.R., L. GLEDHILL, D.C. MARSH, B.W. BYCROFT, P.M. DEWICK, and J. GILBERT: Isotrichodiol: A Post-Trichodiene Intermediate in the Biosynthesis of Trichothecene Mycotoxins. Chem. Commun., 1184 (1990).

146. ZAMIR, L.O., and K.A. DEVOR: Kinetic Pulse-Labeling Study of *Fusarium culmorum*. Biosynthetic Intermediates and Dead-End Metabolites. J. Biol. Chem., **262**, 15348 (1987).

147. ZAMIR, L.O., K.A. DEVOR, A. NIKOLAKAKIS, Y. NADEAU, and F. SAURIOL: Structures of New Metabolites from *Fusarium* species: An Apotrichothecene and Oxygenated Trichodienes. Tetrahedron Lett., **33**, 5181 (1992).

148. GREENHALGH, R., B.A. BLACKWELL, J.R.J. PARE, J.D. MILLER, D. LEVANDIER, R.-M. MEIER, A. TAYLOR, and J.W. APSIMON: Isolation and Characterization by Mass Spectrometry and NMR Spectroscopy of Secondary Metabolites of Some *Fusarium* Species. In: Mycotoxins and Phycotoxins (P.S. Steyn and R. Vleggaar, eds.), p. 137. Amsterdam: Elsevier, 1986.

149. ROESSLEIN, L., CH. TAMM, W. ZURCHER, A. REISEN, and M. ZEHNDER: Sambucinic Acid, a New Metabolite of *Fusarium sambucinum*. Helv. Chim. Acta, **71**, 588 (1988).

150. MOHR, P., CH. TAMM, W. ZURCHER, and M. ZEHNDER: Sambucinol and Sambucoin, Two New Metabolites of *Fusarium sambucinum* Possessing Modified Trichothecene Structures. Helv. Chim. Acta, **67**, 406 (1984).

151. GREENHALGH, R., D. LEVANDIER, W. ADAMS, J.D. MILLER, B.A. BLACKWELL, A.J. MCALEES, and A. TAYLOR: Production and Characterization of Deoxynivalenol and Other Secondary Metabolites of *Fusarium culmorum* (CMI 14764, HLX 1503). J. Agric. Food Chem., **34**, 98 (1986).

152. GREENHALGH, R., R.-M. MEIER, B.A. BLACKWELL, J.D. MILLER, A. TAYLOR, and J.W. APSIMON: Minor Metabolites of *Fusarium roseum* (ATCC 28114). J. Agric. Food Chem., **34**, 115 (1986).

153. CORLEY, D.G., G.E. ROTTINGHAUS, and M.S. TEMPESTA: Secondary Metabolites from *Fusarium*. Two New Modified Trichothecenes from *Fusarium sporotrichioides* MC-72083. J. Nat. Prod., **50**, 897 (1987).

154. ZIEGLER, F.E., A. NANGIA, and M.S. TEMPESTA: Sporol: A Structure Revision. Tetrahedron Lett., **29**, 1665 (1988).
155. CORLEY, D.G., G.E. ROTTINGHAUS, and M.S. TEMPESTA: Novel Trichothecenes from *Fusarium sporotrichioides*. Tetrahedron Lett., **27**, 427 (1986).
156. NOZOE, S., and Y. MACHIDA: Structure of Trichodiene. Tetrahedron Lett., 2671 (1970).
157. ZAMIR, L.O.: Biosynthesis of 3-Acetyldeoxynivalenol and Sambucinol. Tetrahedron, **45**, 2277 (1989).
158. FORT, D.M., C.L. BARNES, M.S. TEMPESTA, H.H. CASPER, E. BEKELE, A.A. ROTTINGHAUS, and G.E. ROTTINGHAUS: Two New Modified Trichothecenes from *Fusarium sporotrichioides*. J. Nat. Prod., **56**, 1890 (1993).
159. NOZOE, S., and Y. MACHIDA: The Structures of Trichodiol and Trichodiene. Tetrahedron, **28**, 5105 (1972).
160. HESKETH, A.R., B.W. BYCROFT, P.M. DEWICK, and J. GILBERT: Revision of the Stereochemistry in Trichodiol, Trichotriol and Related Compounds, and Concerning Their Role in the Biosynthesis of Trichothecene Mycotoxins. Phytochem., **32**, 105 (1993).
161. ZIEGLER, F.E., A. NANGIA, and G. SCHULTE: The Synthesis of Neosporol: A Trichothecene in Search of a Natural Product. Tetrahedron Lett., **29**, 1669 (1988).
162. ZIEGLER, F.E., C.A. METCALF, and G. SCHULTE: Confirmation by Total Synthesis of the Revised Structure of Sporol: An Application of Cyclic Thionocarboate-Initiated Radical Cyclization. Tetrahedron Lett., **33**, 3117 (1992).
163. APSIMON, J.W., B.A. BLACKWELL, R. GREENHALGH, R.-M. MEIER, D. MILLER, J.R.J. PARE, and A. TAYLOR: Secondary Metabolites Produced by Some *Fusarium* Species. In: Mycotoxins and Phycotoxins (P.S. Steyn and R. Vleggaar, eds.), p. 125. Amsterdam: Elsevier, 1986.
164. ZAMIR, L.O., K.A. DEVOR, Y. NADEAU, and F. SAURIOL: Structure Determination and Biosynthesis of a Novel Metabolite of *Fusarium culmorum*, Apotrichodiol. J. Biol. Chem., **262**, 15354 (1987).
165. GLEDHILL, L., A.R. HESKETH, B.W. BYCROFT, P.M. DEWICK, and J. GILBERT: Biosynthesis of Trichothecene Mycotoxins: Cell-Free Epoxidation of a Trichodiene Derivative. FEMS Microbiol. Lett., **81**, 241 (1991).
166. HESKETH, A.R., L. GLEDHILL, D.C. MARSH, B.W. BYCROFT, P.M. DEWICK, and J. GILBERT: Biosynthesis of Trichothecene Mycotoxins: Identification of Isotrichodiol as a Post-Trichodiene Intermediate. Phytochem., **30**, 2237 (1991).
167. ZAMIR, L.O., K.A. DEVOR, A. NIKOLAKAKIS, and F. SAURIOL: Biosynthesis of *Fusarium culmorum* Trichothecenes. J. Biol. Chem., **265**, 6713 (1990).
168. SAVARD, M.E., B.A. BLACKWELL, and R. GREENHALGH: The Role of ^{13}C-Labeled Trichodiene and Bazzanene in the Secondary Metabolism of *Fusarium culmorum*. J. Nat. Prod., **52**, 1267 (1989).
169. GREENHALGH, R., R.-M. MEIER, B.A. BLACKWELL, J.D. MILLER, A. TAYLOR, and J.W. APSIMON: Minor Metabolites of *Fusarium roseum* (ATCC 28114). J. Agric. Food Chem., **32**, 1261 (1984).
170. HESKETH, A.R., L. GLEDHILL, B.W. BYCROFT, P.M. DEWICK, and J. GILBERT: Potential Inhibitors of Trichothecene Biosynthesis in *Fusarium culmorum*: Epoxidation of a Trichodiene Derivative. Phytochem., **32**, 93 (1993).
171. BEREMAND, M.N., F. VANMIDDLESWORTH, S. TAYLOR, R.D. PLATTNER, and D. WEISLEDER: Leucine Auxotrophy Specifically Alters the Pattern of Trichothecene Production in a T-2 Toxin-Producing Strain of *Fusarium sporotrichioides*. Appl. Environ. Microbiol., **54**, 2759 (1988).

172. VanMiddlesworth, F., M.N. Beremand, T.A. Isbell, and D. Weisleder: T-2 Toxin Biosynthesis: Origin of the Isovalerate Side Chain. J. Org. Chem., **55**, 1237 (1990).

173. Desjardins, A.E., and M. Beremand: A Genetic System for Trichothecene Toxin Production in *Gibberella pulicaris* (*Fusarium sambucinum*). Phytopathol., **77**, 678 (1987).

174. Beremand, M.N.: Genetic and Mutational Tools for Investigating the Genetics and Molecular Biology of Trichothecene Production in *Gibberella pulicaris* (*Fusarium sambucinum*). Mycopathologia, **107**, 67 (1989).

175. Hohn, T.M., A.E. Desjardins, and S.P. McCormick: Analysis of *Tox5* Gene Expression in *Gibberella pulicaris* Strains with Different Trichothecene Production Phenotypes. Appl. Environ. Microbiol., **59**, 2359 (1993).

176. Beremand, M.N., and A.E. Desjardins: Trichothecene Biosynthesis in *Gibberella pulicaris*: Inheritance of C-8 Hydroxylation. J. Ind. Microbiol., **3**, 167 (1988).

177. Beremand, M.N., A.E. Desjardins, T.M. Hohn, and F.L. VanMiddlesworth: Survey of *Fusarium sambucinum* (*Gibberella pulicaris*) for Mating Type, Trichothecene Production, and Other Selected Traits. Phytopathol., **81**, 1452 (1991).

178. Hohn, T.M., S.P. McCormick, and A.E. Desjardins: Evidence for a Gene Cluster Involving Trichothecene-Pathway Biosynthetic Genes in *Fusarium sporotrichioides*. Curr. Genet., **24**, 291 (1993).

179. Desjardins, A.E., T.M. Hohn, and S.P. McCormick: Trichothecene Biosynthesis in *Fusarium* Species: Chemistry, Genetics, and Significance. Microbiol. Rev., **57**, 595 (1993).

179a. McCormick, S.P., T.M. Hohn, and A.E. Desjardins: Isolation and Characterization of *Tri3*, a Gene Encoding 15-*O*-Acetyltransferase from *Fusarium sporotrichioides*. Appl. Environ. Microbiol., **62**, 353 (1996).

179b. Hohn, T.M., A.E. Desjardins, and S.P. McCormick: The *Tri4* Gene of *Fusarium sporotrichioides* Encodes a Cytochrome P450 Monooxygenase Involved in Trichothecene Biosynthesis. Mol. Gen. Genet., **248**, 95 (1995).

179c. Proctor, R.H., T.M. Hohn, S.P. McCormick, and A.E. Desjardins: *Tri6* Encodes an Unusual Zinc Finger Protein Involved in Regulation of Trichothecene Biosynthesis in *Fusarium sporotrichioides*. Appl. Environ. Microbiol., **61**, 1923 (1995).

180. Cane, D.E., G. Yang, R.M. Coates, H.-J. Pyun, and T.M. Hohn: Trichodiene Synthase. Synergistic Inhibition by Inorganic Pyrophosphate and Aza Analogs of the Bisabolyl Cation. J. Org. Chem., **57**, 3454 (1992).

180a. Roinestad, K.S., T.J. Montville, and J.D. Rosen: Inhibition of Trichothecene Biosynthesis in *Fusarium tricinctum* by Sodium Bicarbonate. J. Agric. Food Chem., **41**, 2344 (1993).

180b. Roinestad, K.S., T.J. Montville, and J.D. Rosen: Mechanism for Sodium Bicarbonate Inhibition of Trichothecene Biosynthesis in *Fursarium tricinctum*. J. Agric. Food Chem., **42**, 2025 (1994).

181. Zamir, L.O., B. Rotter, K.A. Devor, and F. Vairinhos: Target-Oriented Inhibitors of the Late Stages of Trichothecene Biosynthesis, 2: In vivo Inhibitors and Chick Embryotoxicity Bioassay. J. Agric. Food Chem., **40**, 681 (1992).

182. Mirocha, C.J.: Metabolism and Residue of Trichothecene Toxins in Animal and Plant Systems. In: Mycotoxins and Phycotoxins (P.S. Steyn and R. Vleggar, eds.), p. 409. Amsterdam: Elsevier, 1986.

183. Yagen, B., and M. Bialer: Metabolism and Pharmacokinetics of T-2 Toxin and Related Trichothecenes. Drug Metab. Rev., **25**, 281 (1993).

184. VISCONTI, A., L.M. TREEFUL, and C.J. MIROCHA: Identification of Iso-TC-1 as a New T-2 Toxin Metabolite in Cow Urine. Biomed. Mass Spectrom., **12**, 689 (1985).

184a. NASEEM, S.M., J.G. PACE, and R.W. WANNEMACHER: A High-Performance Liquid Chromatographic Method for Determining [^{3}H]T-2 and Its Metabolites in Biological Fluids of the Cynomolgus Monkey. J. Anal. Toxicol., **19**, 151 (1995).

185. PACE, J.G., M.R. WATTS, E.P. BURROWS, R.E. DINTERMAN, C. MATSON, E.C. HAUSER, and R.W. WANNEMACHER: Fate and Distribution of ^{3}H-Labeled T-2 Mycotoxin in Guinea Pigs. Toxicol. Appl. Pharmacol., **80**, 377 (1985).

186. SINTOV, A., M. BIALER, and B. YAGEN: Pharmacokinetics of T-2 Toxin and Its Metabolite HT-2 Toxin After Intravenous Administration in Dogs. Drug Metab. Dispos., **14**, 250 (1986).

187. SINTOV, A., M. BIALER, and B. YAGEN: Pharmacokinetics of T-2 Tetraol, a Urinary Metabolite of the Trichothecene Mycotoxin, T-2 Toxin, in Dog. Xenobiotica, **17**, 941 (1987).

188. PFEIFFER, R.L., S.P. SWANSON, and W.B. BUCK: Metabolism of T-2 Toxin in Rats: Effects of Dose, Route, and Time. J. Agric. Food Chem., **36**, 1227 (1988).

189. PACE, J.G.: Metabolism and Clearance of T-2 Mycotoxin in Perfused Rat Livers. Fund. Appl. Toxicol., **7**, 424 (1986).

190. CONRADY-LORCK, S., M. GAREIS, X.C. FENG, W. AMSELGRUBER, W. FORTH, and B. FICHTL: Metabolism of T-2 Toxin in Vascularly Autoperfused Jejunal Loops of Rats. Toxicol. Appl. Pharmacol., **94**, 23 (1988).

191. KEMPPAINEN, B.W., R.T. RILEY, J.G. PACE, F.J. HOERR, and J. JOYAVE: Evaluation of Monkey Skin as a Model for *in vitro* Percutaneous Penetration and Metabolism of [^{3}H] T-2 Toxin in Human Skin. Fund. Appl. Toxicol., **7**, 367 (1986).

192. KEMPPAINEN, B.W., R.T. RILEY, S. BILES-THURLOW, and R.B. RUSSELL: Comparison of Penetration and Metabolism of [^{3}H] Diacetoxyscirpenol, [^{3}H] Verrucarin A and [^{3}H] T-2 Toxin in Skin. Food Chem. Toxicol., **25**, 379 (1987).

193. KEMPPAINEN, B.W., J.G. PACE, and R.T. RILEY: Comparison of *in vivo* and *in vitro* Percutaneous Absorption of T-2 Toxin in Guinea Pigs. Toxicon, **25**, 1153 (1987).

194. YOSHIZAWA, T., T. SAKAMOTO, and K. OKAMOTO: *In vitro* Formation of 3′-HydroxyT-2 and 3′-HydroxyHT-2 Toxins from T-2 Toxin by Liver Homogenates from Mice and Monkeys. Appl. Environ. Microbiol., **47**, 130 (1984).

195. YAGEN, B., F. BERGMANN, S. BAREL, and A. SINTOV: Metabolism of T-2 Toxin by Rat Brain Homogenate. Biochem. Pharmacol., **42**, 949 (1991).

196. KNUPP, C.A., S.P. SWANSON, and W.B. BUCK: *In vitro* Metabolism of T-2 Toxin by Rat Liver Microsomes. J. Agric. Food Chem., **34**, 865 (1986).

197. JOHNSEN, H., E. ODDEN, O. LIE, B.A. JOHNSEN, and F. FONNUM: Metabolism of T-2 Toxin by Rat Liver Carboxylesterase. Biochem. Pharmacol., **35**, 1469 (1986).

198. KOBAYASHI, J., T. HORIKOSHI, J.-C. RYU, F. TASHIRO, K. ISHII, and Y. UENO: The Cytochrome P-450-Dependent Hydroxylation of T-2 Toxin in Various Animal Species. Food Chem. Toxicol., **25**, 539 (1987).

199. JOHNSEN, H., E. ODDEN, B.A. JOHNSEN, and F. FONNUM: Metabolism of T-2 Toxin by Blood Cell Carboxylesterases. Biochem. Pharmacol., **37**, 3193 (1988).

200. SWANSON, S.P., J. NICOLETTI, H.D. ROOD, W.B. BUCK, L.-M. COTE, and T. YOSHIZAWA: Metabolism of Three Trichothecene Mycotoxins, T-2 Toxin, Diacetoxyscirpenol and Deoxynivalenol, by Bovine Rumen Microorganisms. J. Chromatogr., **414**, 335 (1987).

201. MUNGER, C.E., G.W. IVIE, R.J. CHRISTOPHER, B.D. HAMMOCK, and T.D. PHILLIPS: Acetylation/Deacetylation Reactions of T-2, AcetylT-2, HT-2, and AcetylHT-2 Toxins in Bovine Rumen Fluid *in vitro*. J. Agric. Food Chem., **35**, 354 (1987).

202. WESTLAKE, W., R.I. MACKIE, and M.F. DUTTON: T-2 Toxin Metabolism by Ruminal Bacteria and Its Effect on Their Growth. Appl. Environ. Microbiol., **53**, 587 (1987).

203. BEETON, S., and A.T. BULL: Biotransformation and Detoxification of T-2 Toxin by Soil and Freshwater Bacteria. Appl. Environ. Microbiol., **55**, 190 (1989).

204. MIROCHA, C.J., H.K. ABBAS, L. TREEFUL, and G. BEAN: T-2 Toxin and Diacetoxyscirpenol Metabolism by *Baccharis* spp. Appl. Environ. Microbiol., **54**, 2277 (1988).

205. TRUSAL, L.R.: Metabolism of T-2 Mycotoxin by Cultured Cells. Toxicon, **24**, 597 (1986).

206. PORCHER, J.-M., C. DAHEL, C. LAFARGE-FRAYSSINET, F.S. CHU, and C. FRAYSSINET: Uptake and Metabolism of T-2 Toxin in Relation to Its Cytotoxicity in Lymphoid Cells. Food Chem. Toxicol., **26**, 587 (1988).

207. COPPOCK, R.W., S.P. SWANSON, H.B. GELBERG, G.D. KORITZ, W.B. BUCK, and W.E. HOFFMANN: Pharmacokinetics of Diacetoxyscirpenol in Cattle and Swine: Effects of Halothane. Am. J. Vet. Res., **48**, 691 (1987).

208. BAUER, J., W. BOLLWAHN, M. GAREIS, B. GEDEK, and K. HEINRITZI: Kinetic Profiles of Diacetoxyscirpenol and Two of Its Metabolites in Blood Serum of Pigs. Appl. Environ. Microbiol., **49**, 842 (1985).

209. SAKAMOTO, T., S.P. SWANSON, T. YOSHIZAWA, and W.B. BUCK: Structure of New Metabolites of Diacetoxyscirpenol in the Excreta of Orally Administered Rats. J. Agric. Food Chem., **34**, 698 (1986).

210. WU, S.-E., and M.A. MARLETTA: Carboxylesterase Isoenzyme Specific Deacylation of Diacetoxyscirpenol (Anguidine). Chem. Res. Toxicol., **1**, 69 (1988).

211. DOWD, P.F., and F. VANMIDDLESWORTH: *In vitro* Metabolism of the Trichothecene 4-Monoacetoxyscirpenol by Fungus and Non-Fungus-Feeding Insects. Experientia, **45**, 393 (1989).

212. UDELL, M.N., and P.M. DEWICK: Metabolic Conversions of Trichothecene Mycotoxins: De-esterification Reactions Using Cell-Free Extracts of *Fusarium.* Z. Naturforsch., **C44**, 660 (1989).

213. YOSHIZAWA, T., T. SAKAMOTO, Y. AYANO, and C.J. MIROCHA: 3′-HydroxyT-2 and 3′-HydroxyHT-2 Toxins: New Metabolites of T-2 Toxin, a Trichothecene Mycotoxin, in Animals. Agric. Biol. Chem., **46**, 2613 (1982).

214. VISCONTI, A., and C.J. MIROCHA: Identification of Various T-2 Toxin Metabolites in Chicken Excreta and Tissues. Appl. Environ. Microbiol., **49**, 1246 (1985).

215. CORLEY, R.A., S.P. SWANSON, and W.B. BUCK: Glucuronide Conjugates of T-2 Toxin and Metabolites in Swine Bile and Urine. J. Agric. Food Chem., **33**, 1085 (1985).

216. KNUPP, C.A., S.P. SWANSON, and W.B. BUCK: Comparative *in vitro* Metabolism of T-2 Toxin by Hepatic Microsomes Prepared from Phenobarbital-Induced or Control Rats, Mice, Rabbits and Chickens. Food Chem. Toxicol., **25**, 859 (1987).

217. WEI, R.-D., and F.S. CHU: Modification of *in vitro* Metabolism of T-2 Toxin by Esterase Inhibitors. Appl. Environ. Microbiol., **50**, 115 (1985).

218. KNUPP, C.A., D.G. CORLEY, M.S. TEMPESTA, and S.P. SWANSON: Isolation and Characterization of 4′-Hydroxy T-2 Toxin, a New Metabolite of the Trichothecene Mycotoxin T-2. Drug Metab. Dispos., **15**, 816 (1987).

219. PAWLOSKY, R.J., and C.J. MIROCHA: Structure of a Metabolic Derivative of T-2 Toxin (TC-6) Based on Mass Spectrometry. J. Agric. Food Chem., **32**, 1420 (1984).

220. GAREIS, M., A. HASHEM, J. BAUER, and B. GEDEK: Identification of Glucuronide Metabolites of T-2 Toxin and Diacetoxyscirpenol in the Bile of Isolated Perfused Rat Liver. Toxicol. Appl. Pharmacol., **84**, 168 (1986).

221. COTE, L.-M., W. BUCK, and E. JEFFERY: Lack of Hepatic Microsomal Metabolism of Deoxynivalenol and Its Metabolite, DOM-1. Food Chem. Toxicol., **25**, 291 (1987).

222. YOSHIZAWA, T., H. TAKEDA, and T. OHI: Structure of a Novel Metabolite from Deoxynivalenol, a Trichothecene Mycotoxin, in Animals. Agric. Biol. Chem., **47**, 2133 (1983).
223. YOSHIZAWA, T., L.-M. COTE, S.P. SWANSON, and W.B. BUCK: Confirmation of DOM-1, a Deepoxidation Metabolite of Deoxynivalenol, in Biological Fluids of Lactating Cows. Agric. Biol. Chem., **50**, 227 (1986).
224. WORRELL, N.R., A.K. MALLETT, W.M. COOK, N.C.P. BALDWIN, and M.J. SHEPHERD: The Role of Gut Micro-Organisms in the Metabolism of Deoxynivalenol Administered to Rats. Xenobiotica, **19**, 25 (1989).
225. LAKE, B.G., J.C. PHILLIPS, D.G. WALTERS, D.L. BAYLEY, M.W. COOK, L.V. THOMAS, J. GILBERT, J.R. STARTIN, N.C.P. BALDWIN, B.W. BYCROFT, and P.M. DEWICK: Studies on the Metabolism of Deoxynivalenol in the Rat. Food Chem. Toxicol., **25**, 589 (1987).
226. YOSHIZAWA, T., K. OKAMOTO, T. SAKAMOTO, and K. KUWAMURA: *In vivo* Metabolism of T-2 Toxin, a Trichothecene Mycotoxin. On the Formation of Deepoxydation Products. Proc. Jap. Assoc. Mycotox., **21**, 9 (1985); Chem. Abstr., **103**, 210640 (1985).
227. YOSHIZAWA, T., T. SAKAMOTO, and K. KUWAMURA: Structures of Deepoxytrichothecene Metabolites from 3′-HydroxyHT-2 Toxin and T-2 Tetraol in Rats. Appl. Environ. Microbiol., **50**, 676 (1985).
228. CHATTERJEE, K., A. VISCONTI, and C.J. MIROCHA: Deepoxy T-2 Tetraol: A Metabolite of T-2 Toxin Found in Cow Urine. J. Agric. Food Chem., **34**, 695 (1986).
229. ONJI, Y., Y. DOHI, Y. AOKI, T. MORIYAMA, H. NAGAMI, M. UNO, T. TANAKA, and Y. YAMAZOE: Deepoxynivalenol: A New Metabolite of Nivalenol Found in the Excreta of Orally Administered Rats. J. Agric. Food Chem., **37**, 478 (1989).
230. KING, R.R., R.E. MCQUEEN, D. LEVESQUE, and R. GREENHALGH: Transformation of Deoxynivalenol (Vomitoxin) by Rumen Microorganisms. J. Agric. Food Chem., **32**, 1181 (1984).
231. COTE, L.-M., A.M. DAHLEM, T. YOSHIZAWA, S.P. SWANSON, and W.B. BUCK: Excretion of Deoxynivalenol and Its Metabolite, DOM-1, in Milk, Urine and Feces of Lactating Dairy Cattle. J. Dairy Sci., **69**, 2416 (1986).
232. COTE, L.-M., J. NICOLETTI, S.P. SWANSON, and W.B. BUCK: Production of Deepoxydeoxynivalenol (DOM-1), a Metabolite of Deoxynivalenol, by *in vitro* Rumen Incubation. J. Agric. Food Chem., **34**, 458 (1986).
233. HE, P., L.G. YOUNG, and C. FORSBERG: Microbial Transformation of Deoxynivalenol (Vomitoxin). Appl. Environ. Microbiol., **58**, 3857 (1992).
234. SWANSON, S.P., H.D. ROOD, J.C. BEHRENS, and P.E. SANDERS: Preparation and Characterization of the Deepoxy Trichothecenes: Deepoxy HT-2, Deepoxy T-2 Triol, Deepoxy T-2 Tetraol, Deepoxy 15-Monoacetoxyscirpenol, and Deepoxy Scirpentriol. Appl. Environ. Microbiol., **53**, 2821 (1987).
235. SWANSON, S.P., C. HELASZEK, W.B. BUCK, H.D. ROOD, and W.M. HASCHEK: The Role of Intestinal Microflora in the Metabolism of Trichothecene Mycotoxins. Food Chem. Toxicol., **26**, 823 (1988).
236. CORLEY, R.A., S.P. SWANSON, G.J. GULLO, L. JOHNSON, V.R. BEASLEY, and W.B. BUCK: Disposition of T-2 Toxin, a Trichothecene Mycotoxin, in Intravascularly Dosed Swine. J. Agric. Food Chem., **34**, 868 (1986).
237. ROUSH, W.R., M.A., MARLETTA, S. RUSSO-RODRIGUEZ, and J. RECCHIA: Trichothecene Metabolism Studies: Isolation and Structure Determination of 15-Acetyl-3α-(1′β-D-gluco-pyranosiduronyl)-scirpen-3,4β,15-triol. J. Am. Chem. Soc., **107**, 3354 (1985).

238. ROUSH, W.R., M.A., MARLETTA, S. RUSSO-RODRIGUEZ, and J. RECCHIA: Trichothecene Metabolism Studies, 2: Structure of 3α-(1'β-D-Glucopyranosiduronyl)-8α-isovaleryloxy-scirpen-3,4β,15-triol 15-Acetate Produced from T-2 Toxin *in vitro*. Tetrahedron Lett., **26**, 5231 (1985).

239. PACE, J.G., and M.R. WATTS: Hepatic Subcellular Distribution of [^{3}H] T-2 Toxin. Toxicon, **27**, 1307 (1989).

240. SEWALD, N., J.L. VON GLEISSENTHALL, M. SCHUSTER, G. MULLER, and R.T. APLIN: Structure Elucidation of a Plant Metabolite of 4-Desoxynivalenol. Tetrahedron Asymm., **3**, 953 (1992).

241. PRELUSKY, D.B., D.M. VEIRA, H.L. TRENHOLM, and K.E. HARTIN: Excretion Profiles of the Mycotoxin Deoxynivalenol, Following Oral and Intravenous Administration to Sheep. Fund. Appl. Toxicol., **6**, 356 (1986).

242. AVENT, A.G., J.F. GROVE, and J.R. HANSON: ^{13}C NMR Spectra of Some Trichothecene Mycotoxins and Derivatives. Magn. Reson. Chem., **26**, 475 (1988).

243. GREENHALGH, R., B.A. BLACKWELL, and M.E. SAVARD: The NMR Spectra of Trichothecenes and Related Fungal Metabolites. Tetrahedron, **45**, 2373 (1989).

244. GROVE, J.F.: Phytotoxic Compounds Produced by *Fusarium equiseti*, Part 8: Acid Catalyzed Rearrangement of 12,13-Epoxytrichothec-9-enes. J. Chem. Soc., Perkin Trans. 1, 647 (1986).

245. JEKER, N., and CH. TAMM: Apotrichothecene Rearrangement in Macrocyclic Trichothecene Derivatives. Tetrahedron Lett., **30**, 6001 (1989).

246. BLACKWELL, B.A., R. GREENHALGH, and A.D. BAIN: Carbon-13 and Proton Nuclear Magnetic Resonance Spectral Assignments of Deoxynivalenol and Other Mycotoxins from *Fusarium graminearum*. J. Agric. Food Chem., **32**, 1078 (1984).

247. TRIPATHI, D.N., L.R. CHAUHAN, and A. BHATTACHARYA: Separation and Identification of Mycotoxins by Thin-Layer Chromatography/Fast Atom Bombardment Mass Spectrometry. Anal. Sci., **7**, 423 (1991).

247a. YOUNG, J.C., and D.E. GAMES: Analysis of *Fusarium* Mycotoxins by Gas Chromatography–Fourier Transform Infrared Spectroscopy. J. Chromatogr., **A663**, 211 (1994).

248. BLACK, R.M., R.J. CLARKE, and R.W. READ: Detection of Trace Levels of Trichothecene Mycotoxins in Human Urine by Gas Chromatography-Mass Spectrometry. J. Chromatogr., **367**, 103 (1986).

249. BLACK, R.M., R.J. CLARKE, and R.W. READ: Detection of Trace Levels of Trichothecene Mycotoxins in Environmental Residues and Foodstuffs Using Gas Chromatography with Mass Spectrometric or Electron-Capture Detection. J. Chromatogr., **388**, 365 (1987).

250. KOSTIAINEN, R., and A. RIZZO: The Characterization of Trichothecenes as Their Heptafluorobutyrate Esters by Negative-Ion Chemical Ionization Tandem Mass Spectrometry. Anal. Chim. Acta, **204**, 233 (1988).

251. RAZA, S.K., S.A. HOWELL, and A.I. MALLET: Identification of Mycotoxins in Keratomycosis-Derived *Fusarium* Isolates by Gas Chromatography-Mass Spectrometry. J. Chromatogr., **620**, 243 (1993).

252. KANHERE, S.R., and P.M. SCOTT: Heptafluorobutyrylation of Trichothecenes Using a Solid-Phase Catalyst. J. Chromatogr., **511**, 384 (1990).

253. WREFORD, B.J., and K.J. SHAW: Analysis of Deoxynivalenol as Its Trifluoroacetyl Ester by Gas Chromatography-Electron Ionization Mass Spectrometry. Food Addit. Contam., **5**, 141 (1987).

254. SCHWADORF, K., and H.-M. MULLER: Determination of Trichothecenes in Cereals by Gas Chromatography with Ion Trap Detection. Chromatogr. **32**, 137 (1991).

255. KOSTIAINEN, R., and A. HESSO: Characterization of Trichothecenes by Ammonia Chemical Ionization and Tandem Mass Spectrometry. Biomed. Environ. Mass Spectrom., **15**, 79 (1988).

256. KOSTIAINEN, R.: Characterization of Trichothecenes by Tandem Mass Spectrometry Using Reactive Collisions with Ammonia. Biomed. Environ. Mass Spectrom., **16**, 197 (1988).

257. KOSTIAINEN, R.: Effect of Collision Gas Pressure and Collision Energy on Reactions Between Ammonia and Protonated Trichothecenes in the Collision Cell of a Triple-Quadrupole Mass Spectrometer. Biomed. Environ. Mass Spectrom., **18**, 116 (1989).

258. HEWETSON, D.W., and C.J. MIROCHA: Development of Mass Spectral Library of Trichothecenes Based on Positive Chemical Ionization Mass Spectra. J. Assoc. Off. Anal. Chem., **70**, 647 (1987).

259. KOSTIAINEN, R., and S. NOKELAINEN: Use of M-Series Retention Index Standards in the Identification of Trichothecenes by Electron Impact Mass Spectrometry. J. Chromatogr., **513**, 31 (1990).

260. PLATTNER, R.D., M.N. BEREMAND, and R.G. POWELL: Analysis of Trichothecene Mycotoxins by Mass Spectrometry and Tandem Mass Spectrometry. Tetrahedron, **45**, 2251 (1989).

261. MIROCHA, C.J., R.J. PAWLOSKY, and H.K. ABBAS: Analysis of T-2 Toxin in a Biological Matrix Using Multiple Reaction Monitoring. Arch. Environ. Contam. Toxicol., **18**, 349 (1989).

262. KOSTIAINEN, R., A. RIZZO, and A. HESSO: The Analysis of Trichothecenes in Wheat and Human Plasma Samples by Chemical Ionization Tandem Mass Spectrometry. Arch. Environ. Contam. Toxicol., **18**, 356 (1989).

262a. BURROWS, E.P.: Dimethyl Ether Chemical Ionization Mass Spectrometry of Trichothecene Biotoxins. Biol. Mass Spectrom., **23**, 492 (1994).

263. KOSTIAINEN, R., K. MATSUURA, and K. NOJIMA: Identification of Trichothecenes by Frit-Fast Atom Bombardment Liquid Chromatography-High-Resolution Mass Spectrometry. J. Chromatogr., **538**, 323 (1991).

264. RAJAKYLA, E., K. LAASASENAHO, and P.J.D. SAKKERS: Determination of Mycotoxins in Grain by High-Performance Liquid Chromatography and Thermospray Liquid Chromatography-Mass Spectrometry. J. Chromatogr., **384**, 391 (1987).

265. VOYKSNER, R.D., W.M. HAGLER, and S.P. SWANSON: Analysis of Some Metabolites of T-2 Toxin, Diacetoxyscirpenol and Deoxynivalenol by Thermospray High-Performance Liquid Chromatography-Mass Spectrometry. J. Chromatogr., **394**, 183 (1987).

266. KOSTIAINEN, R.: Identification of Trichothecenes by Thermospray, Plasmaspray and Dynamic Fast-Atom Bombardment Liquid Chromatography-Mass Spectrometry. J. Chromatogr., **562**, 555 (1991).

267. KOSTIAINEN, R., and P. KURONEN: Use of 1-[p-(2,3-Dihydroxypropoxy)phenyl]-1-alkanones as Retention Index Standards in the Identification of Trichothecenes by Liquid Chromatography-Thermospray and Dynamic Fast Atom Bombardment Mass Spectometry. J. Chromatogr., **543**, 39 (1991).

268. ROACH, J.A.G., J.A. SPHON, J.A. EASTERLING, and E.M. CALVEY: Capillary Supercritical Fluid Chromatography/Negative Ion Chemical Ionization Mass Spectrometry of Trichothecenes. Biomed. Environ. Mass Spectrom., **18**, 64 (1989).

269. YOUNG, J.C., and D.E. GAMES: Supercritical Fluid Chromatography of *Fusarium* Mycotoxins. J. Chromatogr., **627**, 247 (1992).

(*Received October 19, 1994, updated April 20, 1996*).

Cardiac Glycosides

D. DEEPAK, S. SRIVASTAVA, N. K. KHARE, and A. KHARE
Department of Chemistry, Lucknow University, Lucknow 226007, India

Contents

1. Introduction

Steroidal cardiac glycosides (*1*) constitute a fascinating series of plant natural products (*2*, *3*), some of whose representatives are important clinical agents (*4*). They comprise one of the most interesting medicinally efficacious groups of naturally occurring substances and are widely used to influence the vital blood pumping mechanism.

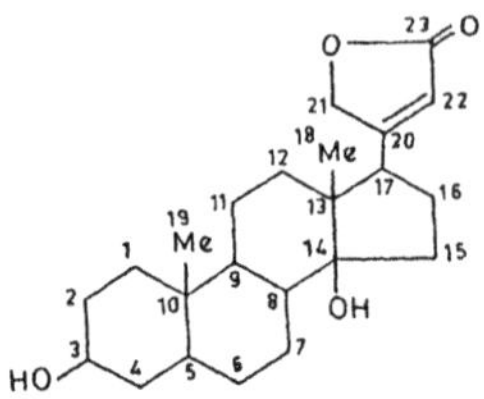

Cardenolide

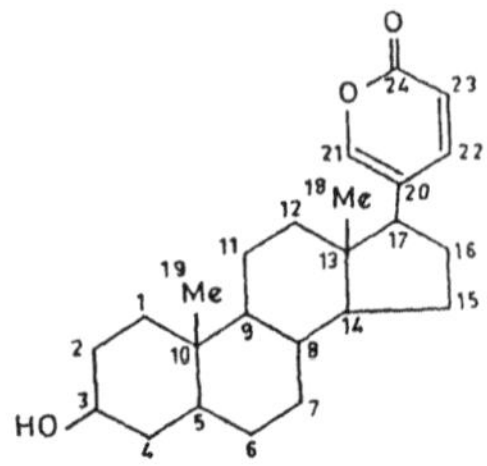

Bufadienolide

Fig. 1

Isolation of these glycosides has been reported from more than ten dozen species belonging to ten different plant families (*5*, *6*). They are found in almost every plant part but occur primarily in the seeds, the amount depending on maturity, time of collection and plant origin (*1*). They are steroidal glycosides with an α,β-unsaturated lactone ring attached to C-17 of the perhydrocyclopen-tanophenanthrene nucleus, the sugar being attached to the aglycone by an acetal linkage. The nature of the aglycone part leads to their classification into two different groups, the cardenolides and the bufadienolides (Fig. 1). Those aglycones containing 23 carbons with a five membered α,β-unsaturated γ-lactone are known as cardenolides (*7*), whereas those with a doubly unsaturated six membered lactone (α-pyrone) ring attached to C-17 of the steroidal moiety constitute the bufadienolides (*2*).

Some cardiac glycosides are used successfully as heart stimulating drugs (*4*, *8*). The literature discloses that the cardenolides are more useful therapeutic agents than the bufadienolides (*1*). Several review articles dealing with different aspects of the cardiac glycosides exist (*58*, *59*, *60*). A comprehensive review by Rastogi *et al.* (*1*) covered the literature till 1967. The present review covers the isolation, identification, spectroscopy and biological activity of naturally occurring cardiac glycosides from 1968 to 1993.

2. Isolation and Identification

In most instances, the cardiac glycosides which contain one to five sugar residues (*9*, *10*) have been isolated from the extracts of different plant parts. The seeds are specially rich in these compounds. Isolation of the glycosides in the pure state was traditionally very difficult due to their very similar behaviour on chromatography and their high sensitivity towards acids and bases. For preparative separation the most successfully employed method is column chromatography using different solid supports. However recent advances in isolation and purification techniques such as H.P.L.C. and gas chromatography have introduced a new dimension for detection, isolation and identification.

The conventional method of structure elucidation of a glycoside involved its acid hydrolysis (*11*) followed by identification of the aglycone and sugar residues, whereas the site of glycosidation was usually determined by comparing the UV spectrum of the glycoside with that of the aglycone in the presence or absence of various shift reagents (*12*). However these methods usually required large amounts of substances which are ordinarily not available. Recent years have seen a tremendous increase in publications utilizing more modern physicochemical techniques such as I.R., NMR (^1H, ^{13}C and 2D), mass spectroscopy (EI, FD and FAB), etc. as tools for structure elucidation of cardiac glycosides, which offer a non-destructive method for establishing their structure.

2.1. Structural Features of Cardiac Genins

About 84 cardiac genins have so far been isolated from plants. Digitoxigenin (*13*) and bufalin are representatives of the cardenolide and bufadienolide groups, respectively, with the other cardenolides and bufadienolides being considered as derivatives of these parent compounds. Some of the characteristic features found in cardenolide and bufadienolide genins reported so far may be summarized as follows:

1. The configuration of substituents at C-3, C-5 and C-17 may be α or β.
2. Double bonds may be present at C-4, C-5 or C-16.
3. Hydroxyl groups may be in the 1β, 2α, 3β, 5β, 11α, 11β, 12β(α), 15β, 16β and 19 positions. In a few cases the hydroxyl group is esterified.
4. Carbonyl functions are found at C-11, C-12, or C-19.
5. Epoxy groups occur at three sites, namely, 7-8β, 8-14β and 11-12β.

In some cardenolides the angular methyl group at C-10 is oxidized to an aldehyde (*14*), a hydroxymethyl (*15*) or a carboxylic acid group (*16*). The C-11 hydroxyl group when present is mostly α-oriented. Structural variations of the typical structural pattern are numerous but usually involve epimerisation at C-3, C-5 and C-17 and sometimes the presence of oxygen functions at other positions. An aglycone containing a 17β-butenolide ring undergoes an internal rearrangement (*17*) in the presence of alkali due to reaction of the 14β-hydroxyl with the lactone to form

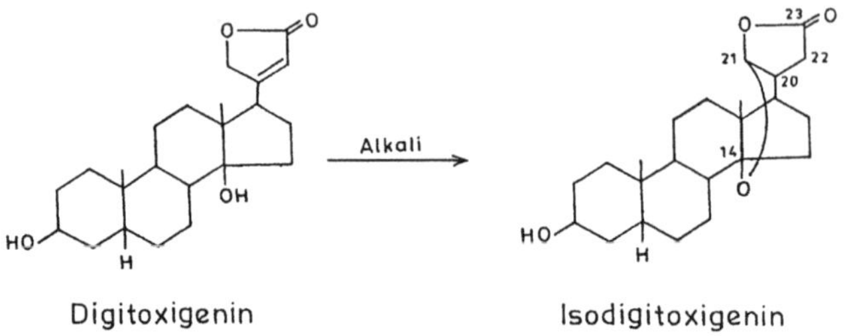

Fig. 2

a C-14-21 oxide called isogenin (Fig. 2). The tertiary 14β-hydroxyl group is so labile (*18*) that in the presence of acids it yields the corresponding anhydrogenin.

2.2. Sugars of Cardiac Glycosides

The sugars isolated by hydrolysis of these glycosides are mostly rare 6-deoxy and 2,6-dideoxy hexoses in addition to some common sugars or their di- or polysaccharides. Natural cardiac glycosides contain mono- or oligosaccharide-sugar chains (*9*, *10*). The sugars are attached to the genin by a glycosidic linkage preferably involving the hydroxyl at C-3. Investigations by earlier workers have also revealed the presence of oligosaccharides containing common sugars. These generally consist of a linear arrangement rather than a branched sugar chain. When the oligosaccharide moiety contains both a common and a deoxy sugar, it is the deoxy sugar which is directly linked (*2*) to the genin. A detailed study of the sugar linkages revealed that in a β-D-type sugar the hexopyranose ring is present in the $^4C_1$ conformation (*19*) and the aglycone is equatorially orientated on the sugar ring whereas in α-L-type sugars the hexopyranose exists in $^1C_4$ conformation (*19*) and the aglycone is preferentially held in the axial orientation (Fig. 3).

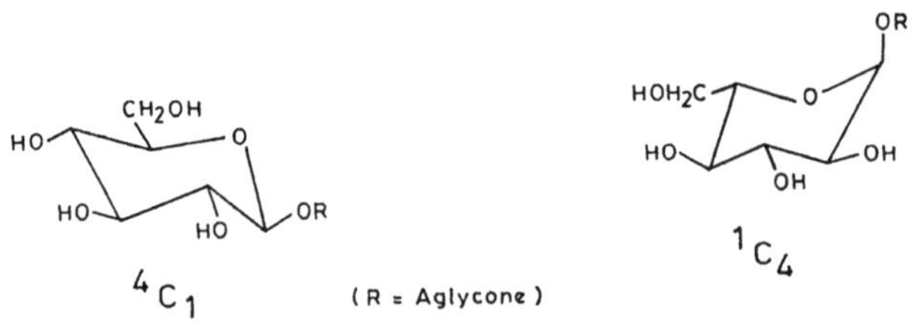

Fig. 3

2.3. Sequence, Number and Identification of Monosaccharides

Information about the number and sequence of monosaccharides and their nature had traditionally been obtained by acid hydrolysis of the glycosides under conditions which depend on the nature of the sugar. When the common sugar is glycosidically linked to aglycone, the glycoside is hydrolysed by Kiliani's method *(20)*. In glycosides containing 2-deoxy or 2,6-dideoxy sugars, milder conditions developed by RANGASWAMI and REICHSTEIN *(11)* and MANNICH and SIEWERT *(21)* are used. In order to establish the sequence of sugars in di-to polyglycosides very mild acid hydrolysis *(22)* (0.05 *M* H_2SO_4) which causes sugars to break off from the end. Sugars isolated after hydrolysis are characterized by co-chromatography (PC, TLC and HPLC) and by comparison of physical properties traditionally m.p. and $[\alpha]_D$ with those of authentic samples. In recent years, however, physicochemical and spectroscopic methods, such as IR, NMR (^1H, ^{13}C) and mass spectrometry have proved indispensable for this purpose as they generally require only small amounts of the substances.

2.4. I.R. Spectroscopy of Cardiac Glycosides

The use of I.R. spectroscopy in structure elucidation of cardiac glycosides is not very important but cannot be ignored. The absorptions used for detection of the lactone ring lie in the region 1625–1790 cm^{-1} which is characteristic of the double bond and carbonyl group of the lactone ring *(23)*. Other functionalities which can be identified are hydroxyls, double bonds as well as methyl groups *(24)*. The deformation bands for methyl group appear at 1360 cm^{-1}, while the stretching bands for hydroxy groups are visible at 3350 cm^{-1} (3546 cm^{-1} for C-14 hydroxy and 3472–3436 cm^{-1} for C-3 hydroxy group).

2.5. Mass Spectrometry of Cardiac Glycosides

Mass spectrometry has proved to be very helpful in assigning structures of cardiac glycosides *(25, 26)* but its application was not routine because of their low volatility. In recent times this problem has been overcome by better inlet techniques. Although within limits stereoisomeric sugars can be identified by means of mass spectral fragments *(25)*, a limitation of this approach is the inaccessibility of finer stereochemical details such as the configuration of glycosidic linkages.

Field desorption (FD) techniques (*28*, *29*) used in conjunction with the usual electron impact (EI) (*25*) method offer appreciable advantages; in particular high mass ion peaks are more evident in FD spectra (in the case of oligoglycosides), but the method does not produce fragment ions as extensively as do the EI mass spectra, which often provide more valuable structural information. Of late, Fast Atom Bombardment (FAB) (*56*, *57*) is increasingly gaining acceptance as one of the most important ionization methods in mass spectrometry. FAB mass spectra of cardiac glycosides invariably give a protonated or an alkali metal cationized molecular ion along with peaks arising from fragmentation of the molecule. The location of the glycosidic linkage between the sugar and the aglycone residue can also be ascertained by the mass fragments (*30*). In the case of oligoglycosides containing both deoxy and normal sugars, mass spectral fragmentation shows that it is the deoxy sugar which is always linked (*26*) to the genin moiety. Other mass spectroscopic fragments can lead to the identification of possible isomers of a cardiac aglycone (*25*).

2.6. NMR Spectrometry of Cardiac Glycosides

As in all other aspects of natural product chemistry NMR spectroscopy has played a most valuable role in structure elucidation and assignment of stereochemistry of cardiac glycosides.

^{1}H NMR spectroscopy, particularly at the high fields now commonly employed (400–600 MHz) provides information on stereochemical details such as the configuration of glycosidic linkages and the conformation of the steroidal portion. The characteristic splitting pattern of the anomeric proton signals of the sugar moieties indicates not only the configuration of the glycosidic linkages but also indicates the size and conformation of the monosaccharide rings. In the case of 2-deoxy sugars the anomeric protons appear as a double doublet in the region δ 4.6–5.2. Splitting patterns with coupling constants on the order of 8 and 2 Hz establish the presence of a β-glycosidic linkage in the $^4C_1$ conformation where H-1 is axial, whereas smaller couplings of 3 and 1 Hz indicate that the glycosidic linkage is α with the sugar in the $^1C_4$ conformation and H-1 is equatorial. In normal sugars the anomeric proton appears only as a doublet (*31*) whose coupling constant depends on the orientation of H-1 as well as of H-2. In the higher field region the characteristic signal of the secondary methyl group in 6-deoxy and 2,6-dideoxy sugars appears as a doublet ($J = 6$ Hz) in the region δ 1.5–1.0. The methylene protons of a 2-deoxy sugar appear as two sets of multiplets in the region δ 2.5–2.0

and 2.0–1.5 for the equatorial and axial protons, respectively. The signals of H-3, H-4 and H-5 in hexoses appear between δ3.5 and 4.5. Characteristic signals of the cardenolide portion are one resonance (*34*) in the region δ6.0–5.8 (H-22) and two double doublets (*33*, *34*) or two doublets (*35*) in the region δ5.1 to 4.7 for the H-21 of butenolide ring while the characteristic signals of the lactone ring of a bufadienolide are a doublet for H-21 and a double doublet for H-22 in the region δ7.2–8.0 (*150*). The H-23 signal appears as a doublet in the region δ6.0–6.5 (*150*).

The two angular methyl groups of a cardiac genin appear as singlets in the region δ1.5–1.0 while the methylene protons appear as multiplets in the region δ2.0–1.5 and the methine under a hydroxyl carbinol in the glycoside produces a signal in the region δ3–4. On acetylation the signal of the corresponding methine proton is shifted about 1 ppm downfield compared with the parent alcohol. The number of hydroxyl group may be established by reaction with the TAI reagent (*32*).

^{13}C NMR spectroscopy has played a significant role in structure elucidation and assignment of stereochemistry of the cardiac glycosides. The technique is particularly convenient when assignment of signals is facilitated by the availability of systematic data from a series of related compounds. Various authors have published articles giving details of chemical shifts of 5α- (*36*) and 5β- (*37*) cardiac genins and sugars. The main information obtained from the ^{13}C NMR spectrum is the number of sugars present in the molecule which is indicated by the number of anomeric carbon signals in the region (*31*, *34*, *38*) δ95–110. Furanoses and pyranoses can be readily differentiated on the basis of ^{13}C shielding data (*39*), as C-1, C-2 and C-4 of furanoses generally appear 4–14 ppm downfield from the chemical shift exhibited by the pyranose isomer (*39*). The hydroxylated carbons of the aglycone involved in glycosidation is generally found at lower field by 3–6 ppm (*40*). The upfield shift of the adjacent carbons seems to be less consistent but ranges between 0.5–4.0 ppm (*40*). The sugar sequence can also be ascertained by measurement of relaxation times. The difference in peak intensities of the signals of the carbons associated with the inner and terminal sugars observed in the partially relaxed Fourier transform (*41*) spectrum provides a method for sugar sequencing in glycosides. More recent methods for establishing the sequence and configuration of glycosides in the sugar moiety include the use of two-dimensional (*42*) correlation (COSY), two dimensional nuclear Overhauser effect (NOESY), two-dimensional *J* resolved (2D-J) and NOE difference spectroscopy.

The ^{13}C NMR data give important information regarding the nature of the cardiac genin present in the glycosides. The diagnostically most important signals of a cardenolide appear in the region δ170–176 for

C-23 and C-20 and in the region δ116–118 for C-22. The characteristic signals of a bufadienolide genin appear in the regions δ120–124 (C-20), 148–149 (C-21), 147–149 (C-22), 115–115.5 (C-23) and 162–164 (C-24).

The C-18 signal of the aglycone (cardenolide and bufadienolides) appears in the region δ15.5–20.0; however, the chemical shift of the other tertiary methyl group, *i.e.* C-19, depends on the nature of the substituent (H or OH) and its configuration at C-5. When H-5 is α C-19 appears at δ12.0–12.5, but when it is β, the signal of C-19 is observed in the range δ23.7–24.0. However when a β-OH group is present at C-5 the signal for C-19 is seen in the range of δ15.0–17.0. The carbons carrying OH group exhibit signals in the range of δ60.0–90.0.

3. Biological Activity

Cardiac glycosides are potent cardioactive agents that exert a positive inotropic effect. These glycosides comprise one of the most valuable group of therapeutic agents (*54, 55*) (*e.g.* digoxin) which are widely used to influence the vital blood pumping mechanism of the living organism. Na^+K^+-ATPase serves as receptor for cardiac glycosides (*43*) and forms a complex with it. The pharmacological qualities (*i.e.* good inotropic effect with little toxicity) of cardiac glycosides depend on the structure of the glycoside. It is generally accepted that an α, β-unsaturated lactone attached to C-17, *cis*, *trans* and *cis* fusions of rings A/B, B/C and C/D, respectively, a 14β-hydroxyl along with a sugar at 3β are essential for optimal inotropic activity (*44*). Introduction of extra hydroxyls onto the aglycone generally reduces its activity (*45*). Considerable loss of activity results from varying any of the above characteristics. The presence of a glycosidic moiety in a aglycone increases the inotropic effect compared with the corresponding aglycone. Change of configuration at C-5 to A/B *trans* yields compounds with decreased activity (*46*). The *cis*-fusion of ring C and D plays a very important role in the activity of heart glycosides. Compounds having *trans*-fused C/D rings, *i.e.* 14α-compounds (14-epidigitoxigenin) (*47*) are inactive in contrast to their 14β-analogues. Hydrogenation of the double bond at $\Delta^{20,22}$ of the lactone ring also led to the loss of activity (*46*).

Binding of the steroid aglycone to the enzyme is accepted as the initial step in reaction of the receptor with a cardiac glycoside. The glycosidic portion plays a secondary role in stabilizing the glycoside receptor complex (*45*). In the case of oligoglycosides, the first sugar attached to the genin has the greatest effect on binding and activity (*48*).

References, pp. 148–155

Recent studies on naturally occurring plant cardiac glycosides showed varied activities. 14-O-Acetyl acovenoside C isolated from *Acokanthera spectabilis* (*61*) increased the intra ventricular left pressure significantly and *dp*/*dt* about 15%; it also increased the myocardial contractile force by 20%. Glycosides of strophanthidin A isolated from *Erysimum linifolium* (*49*) showed cardiotonic activity. Gluconerigoside from *Nerium* (*76*) leaves exhibited cardiotonic and diuretic activity. Three cardiac glycosides of acovenosigenin isolated from *Euonymus alata* (*99*) showed antitumor activity against Mouse leukemia cells, human leukemia cells and human lung cancer cells. Some of the Asclepiadaceae glycosides (*50*, *51*), including calotropin, were reported to be cytotoxic. Ghalakinoside from *Pergularia tomentosa* (*98*) showed strong cytotoxic activity against KB cells. A crystalline compound isolated from the ether extract of *Thevetia ahonia* (*52*) exhibited strong cytotoxic activity against the human epidermoid carcinoma of nasopharynx (KB) test system, and was characterized as 3-O-methyl evomonoside (*81*). Strebloside and mansonin isolated from *Streblus asper* (*132*) displayed significant cytotoxic activity. Cymarin, strophanthidin and a strophanthidin glycoside from *Paraquetina nigrescens* (*53*) were found to be cytotoxic but inactive in the P-388 lymphocytic leukemia (3PS) system. Affinoside A, 4,5-dehydro-12-oxo-affinoside E and affinoside M isolated from *Anodendron affine* (*10*) exhibited potent growth inhibitory activity against the 2nd larval stage of *Bombyx mori*.

Acknowledgements

The authors are thankful to D.S.T., U.G.C. and C.S.I.R., New Delhi for financial assistance.

Table 1. *Cardenolide Glycosides Isolated from Plants*

Species (Family)	Cardenolide glycoside (Glycoside no.)	Genin	Sugar	Ref.
Acokanthera spectabilis (Apocynaceae)	Acobioside A (**1**)	Acovenosigenin	D-Glucose + L-Acovenose	(*61*)
	14-O-Acetyl-acovenoside C (**2**)	14-O-Acetyl-acovenosigenin	L-Acovenose + Gentiobiose	(*61*)
Adenium olesum (Apocynaceae)	Somalin (**3**)	Digitoxigenin	D-Cymarose	(*62*)
	Hongheloside A (**4**)	Oleandrigenin	D-Cymarose	(*62*)
	Δ^{16}-Somalin (**5**)	Δ^{16}-Digitoxigenin	D-Cymarose	(*62*)

Table 1 (*continued*)

Species (Family)	Cardenolide glycoside (Glycoside no.)	Genin	Sugar	Ref.
Adenium olesum (Apocynaceae)	Honghelin (**6**)	Digitoxigenin	D-Thevetose	(*62*)
	Obeside B (**7**)	Oleandrigenin	D-Thevetose	(*62*)
	Obeside C (**8**)	Gitoxigenin	D-Thevetose	(*62*)
	Obeside D (**9**)	Δ^{16}-Digitoxigenin	D-Thevetose	(*62*)
	16-O-Acetyl-strospeside (**10**)	Oleandrigenin	D-Digitalose	(*62*)
	Strospeside (**11**)	Gitoxigenin	D-Digitalose	(*62*)
	Oleandrigenin-β-D-glucoside (**12**)	Oleandrigenin	D-Glucose	(*62*)
	Hongheloside C (**13**)	Oleandrigenin	D-Glucose + D-Cymarose	(*62*)
	Obebioside A (**14**)	Digitoxigenin	D-Glucose + D-Thevetose	(*62*)
	Obebioside B (**15**)	Oleandrigenin	D-Glucose + D-Thevetose	(*62*)
	Obebioside C (**16**)	Gitoxigenin	D-Glucose + D-Thevetose	(*62*)
	Obebioside D (**17**)	Δ^{16}-Digitoxigenin	D-Glucose + D-Thevetose	(*62*)
	Odorobioside G (**18**)	Digitoxigenin	D-Glucose + D-Digitalose	(*62*)
	16-O-Acetyl-digitalinum verum (**19**)	Oleandrigenin	D-Glucose + D-Digitalose	(*62*)
	Digitalinum verum (**20**)	Gitoxigenin	D-Glucose + D-Digitalose	(*62*)
	Δ^{16}-Digitalinum verum (**21**)	Δ^{16}-Digitoxigenin	D-Glucose + D-Digitalose	(*62*)
	Echujin (**22**)	Digitoxigenin	Gentiobiose + D-Cymarose	(*62*)
	Honghelotrioside A (**23**)	Oleandrigenin	Gentiobiose + D-Cymarose	(*62*)
	Δ^{16}-Digitoxigenin-gentiobiosyl-(1 → 4)-D-cymaroside (**24**)	Δ^{16}-Digitoxigenin	Gentiobiose + D-Cymarose	(*62*)

Table 1 (*continued*)

Species (Family)	Cardenolide glycoside (Glycoside no.)	Genin	Sugar	Ref.
Adenium olesum (Apocynaceae)	Obetrioside A (**25**)	Digitoxigenin	Gentiobiose + D-Thevetose	(*62*)
	Obetrioside B (**26**)	Oleandrigenin	Gentiobiose + D-Thevetose	(*62*)
	Odoroside G (**27**)	Digitoxigenin	Gentiobiose + D-Digitalose	(*62*)
	16-O-Acetylneogitostin (**28**)	Oleandrigenin	Gentiobiose + D-Digitalose	(*62*)
Anodendron affine (Apocynaceae)	Affinoside A (**29**)	2α,3β,11α,14-Tetrahydroxy-12-oxo-14β-card-4,20(22)-dienolide-7β,8β-epoxide	4,6-Dideoxy-3-O-methyl-D-glycero-2-hexosulopyranose	(*63*)
	Affinoside B (**30**)	2α,3β,11β,14-Tetrahydroxy-12-oxo-5β,14β-card-9,20(22)-dienolide	4,6-Dideoxy-3-O-methyl-D-glycero-2-hexosulopyranose	(*63*)
	Affinoside J (**31**)	2α,3β,12β,14-Tetrahydroxy-11-oxo-14β-card-4,16,20(22)-trienolide	4,6-Dideoxy-3-O-methyl-D-glycero-2-hexosulopyranose	(*63*)
	Affinoside C (**32**)	2α,3β,14-Trihydroxy-11-oxo-5β,14β-card-20(22)-enolide	4,6-Dideoxy-3-O-methyl-D-glycero-2-hexosulopyranose	(*63*)
	Affinoside E (**33**)	2α,3β,12α,14-Tetrahydroxy-11-oxo-5β,14β-card-20(22)-dienolide	4,6-Dideoxy-3-O-methyl-D-glycero-2-hexosulopyranose	(*63*)
	Affinoside G (**34**)	2α,3β,5β,12α,14-Pentahydroxy-11-oxo-5β,14β-card-20(22)-enolide	4,6-Dideoxy-3-O-methyl-D-glycero-2-hexosulopyranose	(*63*)
	Affinoside F (**35**)	2α,3β,12β,14-Tetrahydroxy-11-oxo-14β-card-4,20(22)-dienolide	4,6-Dideoxy-3-O-methyl-D-glycero-2-hexosulopyranose	(*63*)

Table 1 (*continued*)

Species (Family)	Cardenolide glycoside (Glycoside no.)	Genin	Sugar	Ref.
Anodendron affine (Apocynaceae)	Affinoside H (**36**)	2α,3β,11α,14-Tetrahydroxy-12-oxo-14β-card-4,16-20(22)-trienolide-7β,8β-epoxide	4,6-Dideoxy-3-O-methyl-D-glycero-2-hexosulopyranose	(*63*)
	Affinoside D (**37**)	2α,3β,14-Trihydroxy-11-oxo-5β,14β-card-16,20(22)-dienolide	4,6-Dideoxy-3-O-methyl-D-glycero-2-hexosulopyranose	(*63*)
	Affinoside S-I (**38**)	Affinogenin C	D-Digitalose	(*64*)
	Affinoside S-II (**39**)	Affinogenin D-I	D-Digitalose	(*64*)
	Affinoside S-III (**40**)	Affinogenin D-I	6-Deoxy-D-glucose	(*64*)
	Affinoside S-IV (**41**)	Affinogenin D-J	D-Glucose	(*64*)
	Affinoside S-V (**42**)	Affinogenin D-II	D-Glucose	(*64*)
	Affinoside S-VI (**43**)	Affinogenin D-II	D-Glucose	(*64*)
	Affinoside S-VII (**44**)	Affinogenin D-I	4′-Acetyl-D-digitalose	(*64*)
	Affinoside M (**45**)	2α,3β,11α,14-Tetrahydroxy-12-oxo-14β-card-4,20(22)-dienolide	4,6-Dideoxy-3-(α)-O-methyl-hexosuloside	(*65*)
	Affinoside K (**46**)	2α,3β,11α,14-Tetrahydroxy-12-oxo-14β-card-4,20(22)-dienolide	4,6-Dideoxy-hexosuloside	(*65*)
	Affinoside A (**47**)	2α,3β,11α,14-12-Oxo-14β-card-4,20(22)-dienolide	4,6-Dideoxy-3-O-methylhexosuloside	(*65*)
	Affinoside L_a (**48**)	Affinogenin C (− 2α,3β,14-tri-hydroxy-11-oxo-5β,14β-card-20(22)-enolide)	6-Deoxy-3-O-methyl-D-gulosulose	(*66*)
	Affinoside L_b (**49**)	2α,3β,11β,14-Tetrahydroxy-12-oxo-5β,14β-Δ^9(11)-22(20)-dienolide	6-Deoxy-3-O-methyl-D-gulosulose	(*66*)

Table 1 (*continued*)

Species (Family)	Cardenolide glycoside (Glycoside no.)	Genin	Sugar	Ref.
Anodendron affine (Apocynaceae)	Affinoside L_c (**50**)	2α,3β,12α,14-Tetrahydroxy-11-oxo-5β,14β-card-20(22)-enolide	6-Deoxy-3-O-methyl-D-gulosulose	(*66*)
	Affinoside L_d (**51**)	2α,3β,12β,14-Tetrahydroxy-11-oxo-5β,14β-card-22(20)-enolide	6-Deoxy-3-O-methyl-D-gulosulose	(*66*)
	Affinoside L_e (**52**)	2α,3β,14-Trihydroxy-11-oxo-5β,14β-card-22(20)-enolide	6-Deoxy-D-gulosulose	(*66*)
	Affinoside O (**53**)	Affinogenin L_a (2α,3β,12β,14-tetrahydroxy-11-oxo-5β,14β-card-20(22)-enolide)	4,6-Dideoxy-3-O-methyl-D-hexosulose	(*67*)
	Affinoside N (**54**)	Affinogenin H (2α,3β,11β,14-tetrahydroxy-12-oxo-14β-card-4,16,20(22)-tri-enolide-7β,8β-epoxide)	4,6-Dideoxy-3-O-methyl-D-allose	(*67*)
	Affinoside I [16-Affinoside M ≡ 3′-epi-affinoside A] (**55**)	2α,3β,11β,14-Tetrahydroxy-12-oxo-14β-card-4,16,20(22)-tri-enolide-7β,8β-epoxide	4,6-Dideoxy-3-(α)-O-methylhexose	(*67*)
	Affinoside L_f (**56**)	2α,3β,11α,14-Tetrahydroxy-12-oxo-14β-cardenolide	6-Deoxy-3-O-methyl-D-gulosulose	(*68*)
	Affinoside L_g (**57**)	2α,3β,5β,11α,14-Pentahydroxy-12-oxo-14β-cardenolide	4,6-Dideoxy-3-O-methyl-D-gulosulose	(*68*)

Table 1 (*continued*)

Species (Family)	Cardenolide glycoside (Glycoside no.)	Genin	Sugar	Ref.
Anodendron affine (Apocynaceae)	Δ^{16}-Digitoxigenin-β-D-glucoside (**58**)	Δ^{16}-Digitoxigenin	D-Glucose	(*68*)
	3-epi-Δ^{16}-Digitoxigenin-β-D-glucoside (**59**)	3-epi-Δ^{16}-Digitoxigenin	D-Glucose	(*68*)
	Affinoside S-IX (**60**)	Sarmentogenin	6-Deoxy-D-gulose	(*69*)
	Affinoside S-X (**61**)	Sarmentogenin	4,6-Dideoxy-D-gulose	(*69*)
	Affinoside S-XI (**62**)	Affinogenin A	4,6-Dideoxy-D-gulose	(*69*)
	Affinoside P (**63**)	Affinogenin F	4,6-Dideoxy-3-O-methyl-D-hexosulose	(*69*)
	Affinoside Q (**64**)	Affinogenin F	4,6-Dideoxy-3-O-methyl-D-hexosulose	(*69*)
	Affinoside R (**65**)	Affinogenin R	4,6-Dideoxy-3-O-methyl-D-hexosulose	(*69*)
	Affinoside S (**66**)	Affinogenin R	4,6-Dideoxy-3-O-methyl-D-hexosulose	(*69*)
	Affinoside T (**67**)	Affinogenin R	4,6-Dideoxy-3-O-methyl-D-hexosulose	(*69*)
	4,5-Dehydro-12-oxo-affinoside E (**68**)	2α,3β,11α,14-Tetrahydroxy-12-oxo-14β-card-4,20(22)-dienolide	4,6-Dideoxy-3-O-methyl-D-hexosulose	(*70*)
	12-Oxo-affinoside E (**69**)	2α,3β,11α,14-Tetrahydroxy-12-oxo-14β-card-20(22)-enolide	4,6-Dideoxy-3-O-methyl-D-hexosulose	(*70*)
	16β-Hydroxy-affinoside A (**70**)	2α′,3β,11β,14,16β-Pentahydroxy-12-oxo-7β,8β-epoxy-14β-card-4,20(22)-dienolide	4,6-Dideoxy-3-O-methyl-D-hexosulose	(*70*)

Table 1 (*continued*)

Species (Family)	Cardenolide glycoside (Glycoside no.)	Genin	Sugar	Ref.
Beaumontia brevituba and *Beaumontia murtonii* (Apocynaceae)	Δ^{16}-Digitoxigenin-β-D-glucosyl-α-L-cymaroside (**71**)	Δ^{16}-Digitoxigenin	Glucose + L-Cymarose	(*33*)
	17α-Digitoxigenin-β-gentiobiosyl-β-D-cymaroside (**72**)	17α-Digitoxigenin	Gentiobiose + D-Cymarose	(*33*)
	17α-Digitoxigenin-β-gentiobiosyl-β-L-cymaroside (**73**)	17α-Digitoxigenin	Gentiobiose + L-Cymarose	(*33*)
	Digitoxigenin-β-D-glucosyl-α-L-cymaroside (**74**)	Digitoxigenin	D-Glucose + L-Cymarose	(*33*)
	17α-Digitoxigenin-β-D-glucosyl-α-L-cymaroside (**75**)	17α-Digitoxigenin	D-Glucose + L-Cymarose	(*33*)
	Digitoxigenin-α-L-cymaroside (**76**)	Digitoxigenin	L-Cymarose	(*33*)
	Δ^{16}-Digitoxigenin-β-gentiobiosyl-β-D-cymaroside (**77**)	Δ^{16}-Digitoxigenin	Gentiobiose + D-Cymarose	(*33*)
	Δ^{16}-Digitoxigenin-β-gentiobiosyl-α-L-cymaroside (**78**)	Δ^{16}-Digitoxigenin	Gentiobiose + L-Cymarose	(*33*)
	Digitoxigenin-β-gentiobiosyl-β-D-cymaroside (**79**)	Digitoxigenin	Gentiobiose + D-Cymarose	(*33*)
	Digitoxigenin-β-gentiobiosyl-α-L-cymaroside (**80**)	Digitoxigenin	Gentiobiose + L-Cymarose	(*33*)
	Oleandrigenin-β-gentiobiosyl-β-D-cymaroside (**81**)	Oleandrigenin	Gentiobiose + D-Cymarose	(*33*)
	Oleandrigenin-β-gentiobiosyl-α-L-cymaroside (**82**)	Oleandrigenin	Gentiobiose + L-Cymarose	(*33*)

Table 1 (*continued*)

Species (Family)	Cardenolide glycoside (Glycoside no.)	Genin	Sugar	Ref.
Carissa spinarum (Apocynaceae)	Odoroside H (**83**)	Digitoxigenin	D-Digitalose	(*71*)
	Evomonoside (**84**)	Digitoxigenin	L-Rhamnose	(*71*)
	Odoroside G (**85**)	Digitoxigenin	D-Glucose + D-Digitalose	(*71*)
Cerbera manghus (Apocynaceae)	β-Gentiobiosyl-(1″ → 4′)-deacetyl-tanghinin (**86**)	Tanghinigenin	Gentiobiose + L-Thevetose	(*72*)
	Tanghinigenin-β-D-glucosyl-(1″ → 4′)-β-L-thevetoside (**87**)	Tanghinigenin	D-Glucose + L-Thevetose	(*72*)
	17βH-Neriifolin (**88**)	17α-Digitoxigenin	L-Thevetose	(*72*)
	17βH-Tanghinigenin β-D-glucosyl-(1″ → 4′)-β-L-thevetoside (**89**)	17α-Tanghinigenin	D-Glucose + L-Thevetose	(*72*)
	17βH-Tanghinigenin β-L-thevetoside (**90**)	17α-Tanghinigenin	L-Thevetose	(*72*)
	Neriifolin (**91**)	Digitoxigenin	L-Thevetose	(*72*)
	Cerberin (**92**)	Digitoxigenin	2′-Acetyl-L-thevetose	(*72*)
	Thevetin B (**93**)	Digitoxigenin	L-Thevetose + Gentiobiose	(*72*)
	2′-O-Acetyl-thevetin B (**94**)	Digitoxigenin	2′-O-Acetyl-L-thevetose + Gentiobiose	(*72*)
Cerbera manghus and *Cerbera odollum* (Apocynaceae)	17α-Digitoxigenin-L-acofrioside (17α-Solanoside) (**95**)	17α-Digitoxigenin	L-Acofriose	(*73*)
	Tanghinigenin-L-acofrioside (**96**)	Tanghinigenin	L-Acofriose	(*73*)
	17α-Neriifolin (**97**)	Digitoxigenin	L-Thevetose	(*73*)
	17β-Neriifolin (**98**)	17α-Digitoxigenin	L-Thevetose	(*73*)
	17α-Deacetyl-tanghinin (**99**)	Tanghinigenin	L-Thevetose	(*73*)

References, pp. 148–155

Table 1 (*continued*)

Species (Family)	Cardenolide glycoside (Glycoside no.)	Genin	Sugar	Ref.
Cerbera manghus and *Cerbera odollum* (Apocynaceae)	17β-Deacetyl-tanghinin (**100**)	17α-Tanghinigenin	L-Thevetose	(*73*)
	Oleagenin-thevetoside (Cerleaside A) (**101**)	Oleagenin	L-Thevetose	(*73*)
	8β-Hydroxy-17β-digitoxigenin-L-thevetoside (Cerdollaside) (**102**)	8β-Hydroxy-17β-digitoxigenin	L-Thevetose	(*73*)
	8β-Hydroxy-17α-digitoxigenin-L-thevetoside (17α-Cerdollaside) (**103**)	8β-Hydroxy-17α-digitoxigenin	L-Thevetose	(*73*)
	17β-Digitoxigenin-acofrioside (Solanoside) (**104**)	17β-Digitoxigenin	L-Acofriose	(*73*)
	17α-Tanghinigenin-β-D-glucose-3-ulosyl-(1 → 4)-α-L-thevetoside (**105**)	17α-Tanghinigenin	D-Glucose-3-ulose + L-Thevetose	(*74*)
	Oleagenin-β-D-glucose-(1 → 4)-α-L-thevetoside (Cerleaside B) (**106**)	Oleagenin	D-Glucose + L-Thevetose	(*74*)
	Digitoxigenin-β-D-gentiotriosyl-(1 → 4)-α-L-thevetoside (**107**)	Digitoxigenin	D-gentiotriose + L-Thevetose	(*74*)
	17β-Digitoxigenin-β-D-glucosyl-(1 → 4)-α-L-thevetoside (**108**)	17β-Digitoxigenin	D-Glucose + L-Thevetose	(*74*)
	17α-Digitoxigenin-β-D-glucosyl-(1 → 4)-α-L-thevetoside (**109**)	17α-Digitoxigenin	D-Glucose + L-Thevetose	(*74*)
	17β-Tanghinigenin-β-D-glucosyl-(1 → 4)-α-L-thevetoside (**110**)	17β-Tanghinigenin	D-Glucose + L-Thevetose	(*74*)

Table 1 (*continued*)

Species (Family)	Cardenolide glycoside (Glycoside no.)	Genin	Sugar	Ref.
Cerbera manghus and *Cerbera odollum* (Apocynaceae)	17α-Tanghinigenin-β-D-glucosyl-(1 → 4)-α-L-thevetoside (**111**)	17α-Tanghinigenin	D-Glucose + L-Thevetose	(*74*)
	Digitoxigenin-gentiobiosyl-(1 → 4)-α-L-thevetoside (**112**)	Digitoxigenin	Gentiobiose + L-Thevetose	(*74*)
	Tanghinigenin-gentiobiosyl-(1 → 4)-α-L-thevetoside (**113**)	Tanghinigenin	Gentiobiose + L-Thevetose	(*74*)
	17α-Digitoxigenin-β-D-Glucose-3-ulosyl-(1 → 4)-α-L-thevetoside (**114**)	17α-Digitoxigenin	D-Glucose-3-ulose + L-Thevetose	(*74*)
Mandevilla pentlandiana (Apocynaceae)	Digitoxigenin-3-O-β-D-cymaroside (**115**)	Digitoxigenin	D-Cymarose	(*9*)
	Oleandrigenin-3-O-β-D-diginoside (**116**)	Oleandrigenin	D-Diginose	(*9*)
	Digitoxigenin-3-O-β-D-diginoside (**117**)	Digitoxigenin	D-Diginose	(*9*)
	Oleandrigenin-3-O-β-D-digitaloside (**118**)	Oleandrigenin	D-Digitalose	(*9*)
	Digitoxigenin-3-O-β-D-digitaloside (**119**)	Digitoxigenin	D-Digitalose	(*9*)
	Oleandrigenin-3-O-β-D-glucopyranosyl-(1 → 6)-β-D-glucopyranosyl-(1 → 4)-β-D-diginopyranoside (**120**)	Oleandrigenin	D-Glucose + D-Diginose	(*9*)

Table 1 (*continued*)

Species (Family)	Cardenolide glycoside (Glycoside no.)	Genin	Sugar	Ref.
Mandevilla pentlandiana (Apocynaceae)	Oleandrigenin-3-O-α-L-rhamnopyranosyl-(1 → 6)-β-D-glucopyranosyl-(1 → 4)-β-D-cymaropyranoside (**121**)	Oleandrigenin	L-Rhamnose + D-Glucose + D-Cymarose	(*9*)
	Digitoxigenin-3-O-β-D-glucopyranosyl-(1 → 6)-β-D-glucopyranosyl-(1 → 4)-β-D-cymaropyranoside (**122**)	Digitoxigenin	D-Glucose + D-Cymarose	(*9*)
	Digitoxigenin-3-O-α-L-rhamnopyranosyl-(1 → 6)-β-D-glucopyranosyl-(1 → 4)-β-D-cymaropyranoside (**123**)	Digitoxigenin	L-Rhamnose + D-Glucose + D-Cymarose	(*9*)
	Digitoxigenin-3-O-β-D-glucopyranosyl-(1 → 6)-β-D-glucopyranosyl-(1 → 4)-β-D-cymaropyranosyl-(1 → 4)-β-D-cymaropyranoside (**124**)	Digitoxigenin	D-Glucose + D-Cymarose	(*9*)
	Digitoxigenin-3-O-β-D-glucopyranosyl-(1 → 6)-β-D-glucopyranosyl-(1 → 4)-β-D-diginopyranosyl-(1 → 4)-β-D-cymaropyranosyl-(1 → 4)-β-D-cymaropyranoside (**125**)	Digitoxigenin	D-Glucose + D-Diginose + D-Cymarose	(*9*)
Melodinus monogynus (Apocynaceae)	Glucoevonogenin (**126**)	Evonogenin	D-Glucose	(*75*)
	Glucoevonoloside (**127**)	Cannogenol	L-Rhamnose + D-Glucose	(*75*)

Table 1 (*continued*)

Species (Family)	Cardenolide glycoside (Glycoside no.)	Genin	Sugar	Ref.
Nerium (Apocynaceae)	Gluconerigoside (**128**)	Oleandrigenin	D-Glucose + D-Diginose	(*76*)
Nerium odorum (Apocynaceae)	Glucosyl nerigoside (**129**)	Oleandrigenin	D-Glucose + D-Diginose	(*77*)
	Gentiobiosyl oleandrin (**130**)	Oleandrigenin	L-Oleandrose + gentiobiose	(*77*)
	Oleandrigenin-β-D-glucoside (**131**)	Oleandrigenin	D-Glucose	(*77*)
	Gentiobiosyl odoroside A (**132**)	Digitoxigenin	Gentiobiose + D-Diginose	(*77*)
	16-O-Acetyldigit-olinium verum (**133**)	Oleandrigenin	D-Digitalose + D-Glucose	(*77*)
	Odoroside G (**134**)	Digitoxigenin	Gentiobiose + D-Digitalose	(*77*)
Nerium oleander (Apocynaceae)	Keneroside (**135**)	2α,3β, Dihydroxy-8,14-epoxy-5β-card-16:17,20(22)-dienolide	D-Diginose	(*78*)
Nerium odorum (Apocynaceae)	Digitoxigenin-β-gentiotriosyl-(1 → 4)-β-D-digitaloside (**136**)	Digitoxigenin	Gentiotriose + Digitalose	(*79*)
	Uzarigenin-β-gentiobiosyl-(1 → 4)-β-D-diginoside (**137**)	Uzarigenin	Gentiobiose + D-Diginose	(*79*)
	5α-Oleandrigenin-β-D-digitaloside (**138**)	5α-Oleandrigenin	D-Digitalose	(*79*)
	5α-Oleandrigenin-β-D-glucosyl-(1 → 4)-β-D-diginoside (**139**)	5α-Oleandrigenin	D-Glucose + D-Diginose	(*79*)
	5α-Oleandrigenin-β-D-glucosyl-(1 → 4)-β-D-digitaloside (**140**)	5α-Oleandrigenin	D-Glucose + D-Digitalose	(*79*)

References, pp. 148–155

Table 1 (*continued*)

Species (Family)	Cardenolide glycoside (Glycoside no.)	Genin	Sugar	Ref.
Strophanthus divaricatus (Apocynaceae)	Musaroside (**141**)	Sarmutogenin	D-Digitalose	(*80*)
	Lokundjoside (**142**)	Bipindigenin	L-Rhamnose	(*80*)
	Sarmentoloside (**143**)	Sarmentologenin	6-Deoxy-L-talose	(*80*)
	Sarhamnoloside (**144**)	Sarmentologenin	L-Rhamnose	(*80*)
	Sarmutogenin-glucosyl oleandroside (**145**)	Sarmentogenin	D-Glucose + L-Oleandrose	(*80*)
	Sarmutogenin-glucosyl diginoside (**146**)	Sarmentogenin	D-Glucose + L-Diginose	(*80*)
	Sarmentogenin-glucosyl oleandroside (**147**)	Sarmentogenin	D-Glucose + L-Oleandrose	(*80*)
	Sarmentogenin-glucosyl diginoside (**148**)	Sarmentogenin	D-Glucose + L-Diginose	(*80*)
	Divaricoside (**149**)	Sarmentogenin	L-Oleandrose	(*80*)
	Divaricoside (**150**)	Sarmentogenin	L-Diginose	(*80*)
	Decoside (**151**)	Decogenin	L-Oleandrose	(*80*)
	Sarnovide (**152**)	Sarmentogenin	D-Digitalose	(*80*)
Thevetia ahouia (Apocynaceae)	3′-O-Methyl-evomonoside (**153**)	Digitoxigenin	3-O-Methyl-L-rhamnose	(*81*)
Thevetia neriifolia (Apocynaceae)	Uzarigenin-β-D-glucosyl-α-L-thevetoside (**154**)	Uzarigenin	D-Glucose + L-Thevetose	(*82*)
	Uzarigenin-β-gentiobiosyl-(1 → 4)-α-L-thevetoside (**155**)	Uzarigenin	Gentiobiose + L-Thevetose	(*82*)
	Uzarigenin-β-gentiobiosyl-(1 → 4)-α-L-acofrioside (**156**)	Uzarigenin	Gentiobiose + L-Acofriose	(*82*)
	Thevetioside H (**157**)	Thevetiogenin	Gentiobiose + L-Acofriose	(*82*)

Table 1 (*continued*)

Species (Family)	Cardenolide glycoside (Glycoside no.)	Genin	Sugar	Ref.
Thevetia neriifolia (Apocynaceae)	Thevetioside I (**158**)	Thevetiogenin	L-Rhamnose	(*82*)
	(20*R*)-18,20-Epoxy-digitoxigenin-α-L-thevetoside (**159**)	(20*R*)-18,20-Epoxy-digitoxigenin	L-Thevetose	(*82*)
	(20*S*)-18,20-Epoxy-digitoxigenin-α-L-thevetoside (**160**)	(20*S*)-18,20-Epoxy-digitoxigenin	L-Thevetose	(*82*)
	(20*R*)-18,20-Epoxy-digitoxigenin-β-D-glucosyl-(1 → 4)-α-L-thevetoside (**161**)	(20*R*)-18,20-Epoxy-digitoxigenin	D-Glucose + L-Thevetose	(*82*)
	Neriifolin (**162**)	Digitoxigenin	L-Thevetose	(*10*)
	Solanoside (**163**)	Digitoxigenin	L-Acofriose	(*10*)
	Digitoxigenin-3-O-β-gentiobiosyl-(1 → 4)-α-L-acofriose (**164**)	Digitoxigenin	Gentiobiose + L-Acofriose	(*10*)
	Thevetioside A (**165**)	Thevetiogenin	L-Thevetose	(*10*)
	Thevetioside B (**166**)	Thevetiogenin	L-Acofriose	(*10*)
	Thevetioside C (**167**)	Thevetiogenin	D-Glucose + L-Thevetose	(*10*)
	Thevetioside D (**168**)	Thevetiogenin	D-Glucose + L-Acofriose	(*10*)
	Thevetioside E (**169**)	Thevetiogenin	D-Glucose + 2-O-Acetyl-L-thevetose	(*10*)
	Thevetioside F (**170**)	Thevetiogenin	Gentiobiose + L-Thevetose	(*10*)
	Thevetioside G (**171**)	Thevetiogenin	Gentiobiose + 2-O-Acetyl-L-thevetose	(*10*)
	Thevetin B (**172**)	Digitoxigenin	Gentiobiose + L-Thevetose	(*10*)

References, pp. 148–155

Table 1 (*continued*)

Species (Family)	Cardenolide glycoside (Glycoside no.)	Genin	Sugar	Ref.
Thevetia neriifolia (Apocynaceae)	Thevetiogenin-β-gentiobiosyl-(1 → 4)-α-L-acofrioside (**173**)	Thevetiogenin	Gentiobiose + L-Acofriose	(*10*)
Asclepias curassavica (Asclepiadaceae)	Corotoxigenin-3-O-β-cellobiosyl-O-β-D-allomethyloside (**174**)	Corotoxigenin	Cellobiose + D-Allomethylose	(*83*)
	Coroglaucigenin-3-β-Cellobiosyl-O-β-D-allomethyl-oside (**175**)	Coroglaucigenin	Cellobiose + D-Allomethylose	(*83*)
	12β-Hydroxy-coroglaucigenin-3-O-β-cellobiosyl-β-D-allomethyloside (**176**)	12β-Hydroxy-coroglaucigenin	Cellobiose + D-Allomethylose	(*83*)
	16α-Hydroxy-calotropin (**177**)	16α-Hydroxy-calotropagenin	4,6-Dideoxy-hexosone	(*83*)
	3′-*epi*′-19-Norafroside (**178**)	2α,3β,14β,15β-Tetrahydroxy-19-nor-5α-card-20(22)-enolide	4,6-Dideoxy-hexosulose	(*84*)
Asclepias fruticosa (Asclepiadaceae)	Afroside (**179**)	2α,3β,14β,15β-Tetrahydroxy-5α-card-20(22)-enolide	4,6-Dideoxy-hexosulose	(*85*)
	19-Deoxyuscharin (**180**)	Gomphogenin	3-Thiazoline-4,6-dideoxy-2-hexosulose	(*86*)
	4-β-Hydroxy-gomphoside (**181**)	Gomphogenin	6-Deoxy-2-hexosulose	(*86*)
	3′-Didehydro-gomphoside (**182**)	Gomphogenin	3-Oxy-4,6-dideoxy-2-hexosulose	(*86*)
	3′-Epigomphoside (**183**)	Gomphogenin	3-*epi*-4,6-Dideoxy-2-hexosulose	(*86*)
	3′-Epiafroside (**184**)	Afrogenin	3-*epi*-4,6-Dideoxy-2-hexosulose	(*86*)

Table 1 (*continued*)

Species (Family)	Cardenolide glycoside (Glycoside no.)	Genin	Sugar	Ref.
Asclepias fruticosa (Asclepiadaceae)	3′-Epigomphoside-3′-acetate (**185**)	Gomphogenin	3-*epi*-(O-Acetyl)-4,6-dideoxy-2-hexosulose	(*86*)
	3′-Didehydro-afroside (**186**)	Afrogenin	3-Oxy-4,6-dideoxy-2-hexosulose	(*86*)
	3′-Epiafroside-3′-acetate (**187**)	Afrogenin	3-*epi*-(O-Acetyl)-4,6-dideoxy-2-hexosulose	(*86*)
Asclepias humistrata (Asclepiadaceae)	Humistratin (**188**)	2α,3β,14-Tri-hydroxy-19-oxo-5α,14β-card-7,20(22)-dienolide	4,6-Dideoxy-β-D-glycero-D-glycero-2-hexosulo-pyranose	(*87*)
Asclepias speciosa (Asclepiadaceae)	Aspecioside (**189**)	12β-Hydroxy-5α-tanghinigenin	6-Deoxy-D-allose	(*88*)
Asclepias subulata (Asclepiadaceae)	Calactin (**190**)	Calotropagenin	4,6-Dideoxy-hexosone	(*89*)
	Calotropin (**191**)	Calotropagenin	4,6-Dideoxy-hexosone	(*89*)
	Uscharidin (**192**)	Calotropagenin	6-Deoxyhexosdione	(*89*)
	Uzarigenin-3-O-D-glucoside (**193**)	Uzarigenin	D-Glucose	(89)
	Corotoxigenin-3-β-D-glucoside (**194**)	Corotoxigenin	D-Glucose	(*89*)
	Frugoside-4′-β-D-glucoside (**195**)	Coroglaucigenin	D-Allomethylose + D-Glucose	(89)
	16α-Hydroxy-calactin (**196**)	16α-Hydroxycalo-tropagenin	4,6-Dideoxy-hexosone	(*89*)
	Coroglaucigenin-3β-glucoside (**197**)	Coraglaucigenin	D-Glucose	(*89*)
Asclepias vestita (Asclepiadaceae)	16α-Hydroxy-asclepin (**198**)	16α-Hydroxy-calotropagenin	3′-Acetyl-4,6-dideoxyhexosone	(*90*)
	16α-Acetoxy-asclepin (**199**)	16α-Acetyl-calotropagenin	3′-Acetyl-4,6-dide-oxyhexosone	(*90*)
	Asclepin (**200**)	Calotropagenin	3′-Acetyl-4,6-dide-oxyhexosone	(*90*)

References, pp. 148–155

Table 1 (*continued*)

Species (Family)	Cardenolide glycoside (Glycoside no.)	Genin	Sugar	Ref.
Asclepias vestita (Asclepiadaceae)	16α-Acetoxy-calotropin (**201**)	16α-Acetyl-calotropagenin	4,6-Dideoxy-hexosone	(*90*)
	5,6-Dehydrocalactin (**202**)	5,6-Dehydro-calotropagenin	4,6-Dideoxy-hexosone	(*91*)
	5,6-Dehydrocalo-tropin (**203**)	5,6-Dehydro-calotropagenin	4,6-Dideoxy-hexosone	(*91*)
	5,6-Dehydroascle-pin (**204**)	5,6-Dehydro-calotropagenin	3′-Acetyl-4,6-dideoxyhexosone	(*91*)
	5,6-Dehydrouscha-ridin (**205**)	5,6-Dehydro-calotropagenin	6-Deoxyhexo-dione	(*91*)
	5,6-Dehydrocalo-toxin (**206**)	5,6-Dehydro-calotropagenin	6-Deoxyhexosone	(*91*)
	5,6-Dehydrocalo-toxin-3′,4′-diacetate (**207**)	5,6-Dehydro-calotropagenin	3′,4′-Di-acetyl-6-deoxyhexosone	(*91*)
	16α-Hydroxy-5,6-dehydrocalotropin (**208**)	16α-Hydroxy-5,6-dehydrocalotro-pagenin	4,6-Dideoxy-hexosone	(*91*)
	16α-Acetoxy-5,6-dehydrocalotropin (**209**)	16α-Acetoxy-5,6-dehydrocalotro-pagenin	4,6-Dideoxy-hexosone	(*91*)
	16α-Hydroxy-5,6-dehydroasclepin (**210**)	16α-Hydroxy-5,6-dehydrocalotro-pagenin	4,6-Dideoxy-hexosone	(*91*)
	16α-Acetoxy-5,6-dehydroasclepin (**211**)	16α-Acetoxy-5,6-dehydrocalotro-pagenin	4,6-Dideoxy-hexosone	(*91*)
Cryptolepis buchanani (Asclepiadaceae)	Buchanin (**212**)	Sarverogenin	L-Oleandrose	(*92*)
	Cryptosin (**213**)	Sarverogenin	2-Deoxyglucose	(*93*)
	Cryptanoside A (**214**)	Sarverogenin	L-Oleandrose	(*94*)
	Cryptanoside C (**215**)	Sarverogenin	D-Glucose + L-Oleandrose	(*94*)

Table 1 (*continued*)

Species (Family)	Cardenolide glycoside (Glycoside no.)	Genin	Sugar	Ref.
Cryptostegia madagascariensis (Asclepiadaceae)	14,16-Dianhydro-gitoxigenin-3-rhamnoside (**216**)	14,16-Dianhydro-gitoxigenin	Rhamnose	(*95*)
	16-Anhydrogi-toxigenin-3-rhamnoside (**217**)	16-Anhydrogitoxi-genin	Rhamnose	(*95*)
	16-Propionylgi-toxigenin-3-rhamnoside (**218**)	16-Propionylgi-toxigenin	Rhamnose	(*95*)
Digitalis subalpina (Asclepiadaceae)	Subalpinoside (**219**)	Oleandrigenin	D-Digitoxose + D-Glucose	(*96*)
Glossostelma carsoni (Asclepiadaceae)	Xysmalogenin-β-D-glucoside (**220**)	Xysmalogenin	D-Glucose	(*96*)
	Periplogenin-D-glucoside (**221**)	Periplogenin	D-Glucose	(*96*)
	Strophanthidin-D-glucoside (**222**)	Strophanthidin	D-Glucose	(*96*)
Oxystelma esculentum (Asclepiadaceae)	Oxyline (**223**)	3-*epi*-Uzarigenin	Cymarose + Liliacinabiose + Digitoxose	(*22*)
	Oxystelmoside (**372**)	Uzarigenin	D-Xylose + D-Digitalose	(*160*)
	Oxystelmine (**373**)	Periplogenin	D-Digitalose + 6-Deoxy-D-glucose	(*160*)
Pergularia tomentosa (Asclepiadaceae)	Calactin (**224**)	Calotropagenin	4,6-Dideoxy-hexosone	(*98*)
	Ghalakinoside (**374**)	2α,3β,12β,9,14-Pentahydroxy-5α,14β-card-20(22)-enolide	4-Deoxyhexosulose	(*98*)
Enonymus alata (Celastraceae)	Acovenosigenin-3-O-β-D-6-deoxy-glucoside (**225**)	Acovenosigenin	6-Deoxy-D-glucose	(*99*)

References, pp. 148–155

Table 1 (*continued*)

Species (Family)	Cardenolide glycoside (Glycoside no.)	Genin	Sugar	Ref.
Enonymus alata (Celastraceae)	Acovenosigenin-3-O-β-D-glucosido-(1 → 4) - D-6-deoxyglucoside (**226**)	Acovenosigenin	D-Glucose + 6-Deoxy-D-glucose	(*99*)
	Acovenosigenin-3-O-β-gentiobiosido-(1 → 4)-β-D-6-deoxyglucoside (**227**)	Acovenosigenin	Gentiobiose + 6-Deoxy-D-glucose	(*99*)
Elaeodendron glaucum (Celastraceae)	Elaeodendroside A (**228**)	2α,3β,11α,14β-Tetrahydroxy-12-oxocard-4,20(22)-dienolide	2,3-Methylidene-pentopyranose	(*100*)
	Elaeodendroside D (**229**)	2α,3β,14β-Trihydroxy-card-4,20(22)-dienolide	2,3-Methylidene-pentopyranose	(*100*)
	Elaeodendroside E (**230**)	2α,3β,14β-Trihydroxy-16β-acetoxycard-4,20(22)-dienolide	2,3-Methylidene-pentopyranose	(*100*)
	Elaeodendroside H (**231**)	2α,3β,14β-Trihydroxy-card-4,16,20(22)-trienolide	2,3-Methylidene-pentopyranose	(*100*)
	Elaeodendroside I (**232**)	2α,3β,11α,14β-Tetrahydroxy-card-4,20(22)-dienolide	2,3-Methylidene-pentopyranose	(*100*)
	Elaeodendroside J (**233**)	2α,3β,12α,14β-Tetrahydroxy-11-oxocard-4,20(22)-dienolide	2,3-Methylidene-pentopyranose	(*100*)
Lophopetalum toxicum (Celastraceae)	* (**234**)	Antiarigenin	D-Gulose + 6-Deoxy-D-talose	(*101*)
	Antilloside (**235**)	Antiarigenin	6-Deoxy-D-allose	(*101*)
	Strophalloside (**236**)	Strophanthidin	6-Deoxy-D-allose	(*101*)

Table 1 (*continued*)

Species (Family)	Cardenolide glycoside (Glycoside no.)	Genin	Sugar	Ref.
	Strophanthidin chinovoside (**237**)	Strophanthidin	6-Deoxy-D-glucose	(*101*)
	β-Antiarin (**238**)	Antiarigenin	L-Rhamnose	(*101*)
	* (**239**)	Strophanthidin	D-Allose + 6-Deoxy-D-allose	(*101*)
	* (**240**)	Strophanthidin	D-Glucose + 6-Deoxy-D-talose	(*101*)
Cheiranthus allionii (Crucifereae)	Cheirantoside (**241**)	Cannogenin	6-Deoxy-D-glucose	(*102*)
	Strophanthidin-3β-O-β-L-rhamnopyranoside (**242**)	Strophanthidin	L-Rhamnose	(*103*)
	Digitoxigenin-3-β-O-α-L-rhamnopyranoside (**243**)	Digitoxigenin	L-Rhamnose	(*103*)
	Digitoxigenin-3-β-O-α-L-rhamnopyranosyl-4-O-β-D-glucopyranoside (**244**)	Digitoxigenin	L-Rhamnose + D-Glucose	(*103*)
	Cannogenol-3-β-O-β-D-glucopyranoside (**245**)	Cannogenol	D-Glucose	(*103*)
	Uzarigenin-3-β-O-β-D-glucopyranosyl-1-4′-O-β-D-glucopyranoside (**246**)	Uzarigenin	D-Glucose	(*104*)
	Uzarigenin-3-β-O-β-D-glucopyranoside (**247**)	Uzarigenin	D-Glucose	(*104*)
	Cheirotoxin (**248**)	Strophanthidin	D-Gulomethylose + D-Glucose	(*105*)
	Sarmentogulomethyloside (**249**)	Sarmentogenin	D-Gulomethylose	(*105*)

References, pp. 148–155

Table 1 (*continued*)

Species (Family)	Cardenolide glycoside (Glycoside no.)	Genin	Sugar	Ref.
Cheiranthus allionii (Crucifereae)	Gulosarmentoglucoside (**250**)	Sarmentogenin	D-Gulomethylose + D-Glucose	(*105*)
	Bipindogulomethyloside (**251**)	Bipindogenin	D-Gulomethylose	(*106*)
	Neouzarin (**252**)	Uzarigenin	D-Glucose	(*106*)
	Glucobipindogulomethyloside (**253**)	Bipindogenin	D-Glucose + D-Gulomethylose	(*106*)
	Digitoxigenin gulomethyloside (**254**)	Digitoxigenin	D-Gulomethylose	(*107*)
	Glucodigigulomethyloside (**255**)	Digitoxigenin	D-Glucose + D-Gulomethylose	(*107*)
	Glucoerycordin (**256**)	Cannogenol	D-Glucose + 6-Deoxy-D-glucose	(*108*)
Cheiranthus scoparius (Crucifereae)	Arguayoside (**257**)	16-β-Acetoxystrophanthidin	D-Digitoxose	(*109, 110*)
	Tancidoside (**258**)	16-β-Acetoxystrophanthidin	D-Boivinose	(*109, 110*)
Erysimum altacium (Crucifereae)	Erysimoside (**259**)	Strophanthidin	D-Digitoxose + D-Glucose	(*111*)
Erysimum contractium (Crucifereae)	Nigrescigenin digitoxoside (**260**)	Nigrescigenin	Digitoxose	(*112*)
Erysimum crepidifolium (Crucifereae)	Glucostrophalloside (**261**)	Strophanthidin	D-Allomethylose + D-Glucose	(*113*)
Erysimum cuspidatum (Crucifereae)	Glucolokundjoside (**262**)	Bipindogenin	D-Glucose + L-Rhamnose	(*114*)
Erysimum repandum (Crucifereae)	Glucostrophalloside (**263**)	Strophanthidin	D-Glucose + D-Allomethylose	(*115*)

Table 1 (*continued*)

Species (Family)	Cardenolide glycoside (Glycoside no.)	Genin	Sugar	Ref.
Erysimum marschallianum (Cruciferae)	Sinapoylglucoerysimoside (**264**)	Strophanthidin	*	(*116*)
Mallotus japonicus (Euphorbiaceae)	Corotoxigenin-3-O-β-D-glucopyranosyl-(1 → 4)-α-L-rhamnopyranoside (**265**)	Corotoxigenin	D-Glucose + L-Rhamnose	(*117*)
	Mallogenin-3-O-β-D-glucopyranosyl-(1 → 4)-α-L-rhamnopyranoside (**266**)	Mallogenin	D-Glucose + L-Rhamnose	(*117*)
	Coroglaucigenin 3-O-β-D-glucopyranosyl-(1 → 4)-α-L-rhamnopyranoside (**267**)	Coroglaucigenin	D-Glucose + L-Rhamnose	(*117*)
	Panogenin-3-O-β-D-glucopyranosyl-(1 → 4)-α-L-rhamnopyranoside (**268**)	Panogenin	D-Glucose + L-Rhamnose	(*117*)
	Corotoxigenin rhamnoside (**269**)	Corotoxigenin	L-Rhamnose	(*117*)
	Mallogenin rhamnoside (**270**)	Mallogenin	L-Rhamnose	(*117*)
	Coroglaucigenin rhamnoside (**271**)	Coroglaucigenin	L-Rhamnose	(*117*)
	Panogenin rhamnoside (**272**)	Panogenin	L-Rhamnose	(*117*)
	Corotoxigenin trioside (**273**)	Corotoxigenin	D-Glucose + L-Rhamnose	(*118*)
Coronilla glauca (Leguminoseae)	Glucocoroglaucigenin (**274**)	3β,14β,19-Trihydroxy-5α-card-20(22)-enolide (Coroglaucigenin)	D-Glucose	(*119*)

References, pp. 148–155

Table 1 (*continued*)

Species (Family)	Cardenolide glycoside (Glycoside no.)	Genin	Sugar	Ref.
Coronilla scorpioides (Leguminoseae)	Coronillobiosidal (**275**)	Coroglaucigenin	D-Glucose	(*120*)
Convallaria (Liliaceae)	Neoconvalloside (**276**)	Strophanthidin	L-Rhamnose + D-Glucose	(*121*)
Convallaria majalis (Liliaceae)	Cannogenol-3-O-α-L-rhamnoside (**277**)	Cannogenol	L-Rhamnose	(*122*)
	Cannogenol-3-O-β-D-allomethyloside (**278**)	Cannogenol	D-Allomethylose	(*122*)
	Sarhamnoloside (**279**)	Sarmentogenin	L-Rhamnose	(*123*)
	Canesceol (**280**)	Sarmentogenin	D-Glucomethylose	(*123*)
	Glucoperigulo-methyloside (**281**)	Periplogenin	D-Glucose + D-Glucomethylose	(*124*)
	Neoconvallatoxo-loside (**282**)	Strophanthidin	D-Glucose + L-Rhamnose	(*125*)
	Glycoside F (Strophanthidin-3-O-6′-deoxy-β-D-allosido-α-L-rhamnoside) (**283**)	Strophanthidin	6-Deoxy-D-allose + L-Rhamnose	(*126*)
	Glycoside U (Strophanthidin-3-O-6′-deoxy-β-D-allosido-α-L-arabinoside) (**284**)	Strophanthidin	6-Deoxy-D-allose + L-Arabinose	(*126*)
	Glycoside H_1 (Strophanthidin-3-O-α-L-rhamnosido-2′-β-D-glucoside) (**285**)	Strophanthidin	L-Rhamnose + D-Glucose	(*126*)

Table 1 (*continued*)

Species (Family)	Cardenolide glycoside (Glycoside no.)	Genin	Sugar	Ref.
Convallaria majalis (Liliaceae)	Glycoside S (Cannogenol-3-O-α-L-rhamnosido-β-D-glucoside) (**286**)	Cannogenol	L-Rhamnose + D-Glucose	(*126*)
	Glycoside Z (Cannogenol-3-O-6′-deoxy-β-D-allosido-α-L-rhamnoside) (**287**)	Cannogenol	6′ Deoxy-D-allose + L-Rhamnose	(*126*)
	Glycoside G (19-Hydroxysarmentogenin-3-O-α-L-rhamnoside) (**288**)	19-Hydroxy-sarmentogenin	L-Rhamnose	(*126*)
	Glycoside Y (Sarmentogenin-3-O-6′-deoxy-β-D-allosido-α-L-rhamnoside) (**289**)	Sarmentogenin	6′-Deoxy-D-allose + L-Rhamnose	(*126*)
	Glycoside A_2 (Sarmentogenin-3-O-6′-deoxy-β-D-glucoside) (**290**)	Sarmentogenin	6′-Deoxy-D-gulose	(*126*)
Rhodea japonica (Liliaceae)	Rhodexin A (**291**)	3β,11α,14β-Trihydroxy-5β-card-20(22)-enolide (Sarmentogenin)	L-Rhamnose	(*127*)
	Rhodexoside (**292**)	3β,11α,14β-Trihydroxy-5β-card-20(22)-enolide	D-Glucose + L-Rhamnose	(*128*)
Convallaria majalis (Liliaceae)	Canarigenin-3-O-α-L-rhamnopyranosyl-(1 → 5)-O-β-D-xylofuranoside (**293**)	Canarigenin	L-Rhamnose + D-Xylose	(*129*)
Streblus asper (Moraceae)	Vijaloside (**294**)	Periplogenin	D-Glucose + D-Xylose	(*130*)

Table 1 (*continued*)

Species (Family)	Cardenolide glycoside (Glycoside no.)	Genin	Sugar	Ref.
Streblus asper (Moraceae)	Strebloside (**295**)	Strophanthidin	2,3-Di-O-methyl-fucose	(*131*)
	Mansonin (**296**)	Strophanthidin	2,3-Di-O-methyl-6-deoxy-D-glucose	(*131*)
	Asperoside (**297**)	Digitoxigenin	2,3-Di-O-methyl-D-glucose	(*132*)
Adonis aestivalis (Ranunculaceae)	Strophanthidin-3-O-β-D-digitoxosido-α-L-cymarosido-β-D-glucoside (**298**)	Strophanthidin	D-Digitoxose + L-Cymarose + D-Glucose	(*133*)
	Strophanthidin-3-O-β-D-digitoxosido-β-D-diginosido-β-D-glucoside (**299**)	Strophanthidin	D-Digitoxose + D-Diginose + D-Glucose	(*133*)
Adonis aleppica (Ranunculaceae)	Periplorhamnoside (**300**)	Periplogenin	L-Rhamnose	(*134*)
	Strophanthidin-diginoside (**301**)	Strophanthidin	Diginose	(*134*)
	Alepposide A (**302**)	Strophanthidin	D-Glucose + D-Diginose + D-Digitoxose + D-Oleandrose	(*135*)
	Alepposide B (**303**)	Strophanthidin	D-Glucose + D-Oleandrose D-Digitoxose	(*135*)
Adonis distorta (Ranunculaceae)	Canarigenin-3-β-D-4′-acetyl-cymaroside (**304**)	Canarigenin	4′-Acetyl-D-cymarose	(*136*)
	Cymarin (**305**)	Strophanthidin	Cymarose	
Adonis vernalis (Ranunculaceae)	Adonitoxigenin-3-O-α-L-rhamnosido-β-D-xyloside (**306**)	Adonitoxigenin	L-Rhamnose + D-Xylose	(*137*)
	Adonitoxigenin-3-[O-α-L-(2′O-acetyl)-rhamnosido]-β-D-xyloside (**307**)	Adonitoxigenin	2′-O-Acetyl-L-rhamnose + D-Xylose	(*137*)

Table 1 (*continued*)

Species (Family)	Cardenolide glycoside (Glycoside no.)	Genin	Sugar	Ref.
Adonis vernalis (Ranunculaceae)	Adonitoxigenin-3-[O-α-L-(3′O-acetyl)-rhamnosido]-β-D-xyloside (**308**)	Adonitoxigenin	3′-O-Acetyl-L-rhamnose + D-Xylose	(*137*)
	Strophanthidin digitaloside (**309**)	Strophanthidin	D-Digitalose	(*138*)
	Strophanthidin-gulomethyloside (Deglucocheiro-toxin) (**310**)	Strophanthidin	D-Gulomethylose	(*138*)
	Strophanthidin fucoside (**311**)	Strophanthidin	Fucose	(*139*)
Digitalis cariensis (Scrophulariaceae)	3′-O-Acetyl gluco-evatromonoside (**312**)	Digitoxigenin	D-Glucose + 3′-O-Acetyl-D-digitoxose	(*140*)
Digitalis lanata (Scrophulariaceae)	Digoxigenin-3-O-β-D-digitoxosido-β-D-digitoxosido-β-D-2,6-dideoxy-glucoside (**313**)	Digoxigenin	D-Digitoxose + 2,6-Dideoxy-D-glucose	(*141*)
	Digoxigenin mono-digitaloside (**314**)	Digoxigenin	D-Digitalose	(*142*)
	Digitoxigenin-3-O-β-D-digitoxosido-β-D-xyloside (**315**)	Digitoxigenin	D-Digitoxose + D-Xylose	(*143*)
	Digitoxigenin-3-O-β-digitoxosido-β-D-digitoxosido-β-D-xyloside (**316**)	Digitoxigenin	D-Digitoxose + D-Xylose	(*143*)
	Digoxigenin-3-O-β-D-digitoxosido-β-D-2,6-dideoxygluco-side (**317**)	Digoxigenin	Digitoxose + 2,6-Dideoxyglucose	(*144*)
	Digoxigenin-3-O-β-D-digitaloside (**318**)	Digoxigenin	Digitalose	(*144*)

References, pp. 148–155

Table 1 (*continued*)

Species (Family)	Cardenolide glycoside (Glycoside no.)	Genin	Sugar	Ref.
Digitalis lanata (Scrophulariaceae)	Digoxigenin-3-O-β-D-digitoxosido-β-D-digitoxosido-β-D-gluco-methyloside (**319**)	Digoxigenin	Digitoxose + Glucomethylose	(*145*)
	Digoxigenin-3-O-β-D-digitoxosido-β-D-gluco-methyloside (**320**)	Digoxigenin	Digitoxose + Glucomethylose	(*145*)
	Diginatigenin-3-O-β-D-digitaloside (**321**)	Diginatigenin	Digitalose	(*145*)
Digitalis thapsi (Scrophulariaceae)	Digithapsin A (**322**)	Gitoxigenin	Digitoxose + 3′-O-Acetyl-D-digitoxose	(*146*)
	Digithapsin B (**323**)	Gitoxigenin	Digitoxose + 3′-O-Acetyl-D-digitoxose	(*146*)
Isoplexis chalcantha (Scrophulariaceae)	Uzarigenin-3-O-β-digitoxoside (**324**)	Uzarigenin	Digitoxose	(*34*)
	Uzarigenin-3-O-β-canaroside (**325**)	Uzarigenin	D-Canarose	(*34*)
	Digitoxigenin-3-O-β-digitoxoside (**326**)	Digitoxigenin	D-Digitoxose	(*34*)
Corchorus acutangulus (Tiliaceae)	Erysimoside (**327**)	Strophanthidin	D-Digitoxose + D-Glucose	(*147*)
	Olitoriside (**328**)	Strophanthidin	D-Boivinose + D-Glucose	(*147*)
Corchorus capsularis (Tiliaceae)	Gluco-(1 → 6)-olitoriside (**329**)	Strophanthidin	D-Glucose + 2,6-Dideoxy-D-gulose	(*148*)
Corchorus olitorius (Tiliaceae)	Olitoriusin (**330**)	Strophanthidin	D-Glucose + D-Digitoxose	(*149*)
	Erysimoside (**331**)	Strophanthidin	D-Glucose + D-Digitoxose	(*149*)

Table 2. *Bufadienolide Glycosides Isolated from Plants*

Species (Family)	Bufadienolide glycoside (Glycoside no.)	Genin	Sugar	Ref.
Cotyledon orbiculata and *Tylecodon grandiflorus* (Crassulaceae)	Tyledoside C (**332**)	2α,3β,11α,14β-Tetrahydroxy-12-oxo-7β,8β-epoxy-bufadienolide	3′-O-Acetyl-4,6-dideoxy-hexosulose	(*150*)
	Orbicuside A (**333**)	3β,11α,14β-Trihydroxy-2,12-dioxo-7β,8β-epoxy-bufadienolide	4,6-Dideoxy-hexose	(*150*)
	Orbicuside B (**334**)	3β,11α,12β,14β-Tetrahydroxy-2,12-dioxo-7β,8β-epoxybufa-dienolide	4,6-Dideoxy-hexose	(*150*)
	Orbicuside C (**335**)	3β,4β,11α,14β-Tetrahydroxy-2,12-dioxo-7β,8β-epoxybufa-dienolide	4,6-Dideoxy-hexose	(*150*)
	Tyledoside D (**336**)	2α,3β,11α,14β-Tetrahydroxy-12-oxo-7β,8β-epoxy-bufadienolide	4,6-Dideoxy-3-hexosulose	(*150*)
Kalanchoe lanceolata (Crassulaceae)	Lancetoxin A (**337**)	5-O-Acetyl-hellebrigenin	2,3,4,5-Tetra-hydroxyhexanoyl	(*151*)
	Lancetoxin B (**338**)	5-O-Acetyl-hellebrigenin	Rhamnose	(*151*)
Urginea aphylla (Liliaceae)	Scilliphaeosidin-3-O-β-D-glucoside (**339**)	Scilliphaeosidin	D-Glucose	(*152*)
	Scillarenin-3-O-β-D-glucoside (**340**)	Scillarenin	D-Glucose	(*152*)
	16-β-O-Acetylgama-bufotalin-3-O-α-L-rhamnoside (**341**)	16-β-O-Acetyl-gamabufotalin	L-Rhamnose	(*152*)
	5α,4,5-Dihydroscilli-rosidin-3-O-α-L-thevetosido-β-D-glucosido-β-D-glucoside (**342**)	5α,4,5-Dihydro-scillirosidin	L-Thevetose + D-Glucose	(*152*)

Table 2 (*continued*)

Species (Family)	Bufadienolide glycoside (Glycoside no.)	Genin	Sugar	Ref.
Urginea hesperia (Liliaceae)	Proscillaridine A (**343**)	Scillarenin	L-Rhamnose	(*153*)
	Scilliphaeoside (**344**)	Scilliphaeosidin	L-Rhamnose	(*153*)
	Gamabufotalin-3-O-α-L-rhamnoside (**345**)	Gamabufotalin	L-Rhamnose	(*153*)
	11α-Hydroxyscilliglucosidin-3-O-α-L-rhamnoside (**346**)	11α-Hydroxy-scilliglaucosidin	L-Rhamnose	(*153*)
	Scillarenin-3-O-α-L-2′3′-diacetylrhamnosido-4′-β-D-glucoside (**347**)	Scillarenin	2′3′-Diacetyl-L-rhamnose + D-Glucose	(*153*)
	Scillarenin-3-O-α-L-rhamnosido-4′-β-D-glucosido-3′-β-D-glucoside (**348**)	Scillarenin	L-Rhamnose + D-Glucose	(*153*)
	Scillarenin-3-O-α-L-rhamnosido-4′-β-D-glucosido-4″-β-D-glucoside (**349**)	Scillarenin	L-Rhamnose + D-Glucose	(*153*)
	Scillarenin-3-O-α-L-2′3′-diacetylrhamnosido-4′-β-D-glucosido-3″-β-D-glucoside (**350**)	Scillarenin	2′3′-Diacetyl-L-rhamnose + D-Glucose	(*153*)
	Scillarenin-3-O-α-L-2′3′-diacetylrhamnosido-4′-β-D-glucosido-4″-β-D-glucoside (**351**)	Scillarenin	2′3′-Diacetyl-L-rhamnose + D-Glucose	(*153*)
	Scilliphaeosidin-3-O-α-L-rhamnosido-4′-β-D-glucosido-3″-β-D-glucoside (**352**)	Scilliphaeosidin	L-Rhamnose + D-Glucose	(*153*)
	Scilliphaeosidin-3-O-α-L-rhamnosido-4′-β-D-glucoside-4″-β-D-glucoside (**353**)	Scilliphaeosidin	L-Rhamnose + D-Glucose	(*153*)

Table 2 (*continued*)

Species (Family)	Bufadienolide glycoside (Glycoside no.)	Genin	Sugar	Ref.
Urginea indica (Liliaceae)	Scillarene-A (**354**)	Scillarenin	L-Rhamnose + D-Glucose	(*154*)
	Scilliglaucosidin-3-O-α-L-rhamnoside (**355**)	Scilliglaucosidin	L-Rhamnose	(*154*)
	Scilliglaucosidin-3-O-β-D-glucoside (**356**)	Scilliglaucosidin	D-Glucose	(*154*)
Urginea maritima (Liliaceae)	Glucoscillarene A (**357**)	Scillarenin A	L-Rhamnose + D-Glucose	(*155*)
	Glucoscilliphaeoside (**358**)	Scilliphaeosidin	L-Rhamnose + D-Glucose	(*155*)
	12-epi-Scilliphaeoside (**359**)	12-epi-Scilliphaeosidin	L-Rhamnose	(*155*)
	16-β-O-Acetyl-gamabufotalin-3-O-α-L-rhamnoside (**360**)	16-β-O-Acetyl-gamabufotalin	L-Rhamnose	(*155*)
	Scilliglaucoside (**361**)	Scilliglaucigenin	D-Glucose	(*155*)
	5α,4,5-Dihydroproscillaridin A (**362**)	5α,4,5-Dihydroscillirosidin	L-Rhamnose	(*155*)
	5α,4,5-Dihydroglaucoscillaren A (**363**)	5α,4,5-Dihydroscillarenin	D-Glucose + L-Rhamnose	(*155*)
	Gamabufotalin-3-O-α-L-rhamnosido-β-D-glucoside (**364**)	Gamabufotalin	L-Rhamnose + D-Glucose	(*155*)
	19-Oxo-5α,4,5-dihydroproscillaridin A (**365**)	19-Oxo-5α,4,5-dihydroscillarenin	L-Rhamnose	(*155*)
Urginea pancration (Liliaceae)	(**366**)	3β,8β,12β,14-Tetrahydroxy-6-O-acetylbufadienolide	L-Thevetose + D-Glucose	(*156*)

References, pp. 148–155

Table 2 (*continued*)

Species (Family)	Bufadienolide glycoside (Glycoside no.)	Genin	Sugar	Ref.
Urginea physodes (Liliaceae)	Physodine A (**367**)	Hellebrigenin	D-Digitalose	(*157*)
Helleborus odorus (Ranunculaceae)	Hellebrin (**368**)	Hellebrigenin	L-Rhamnose + D-Glucose	(*158*)
	Desglucohellebrin (**369**)	Hellebrigenin	L-Rhamnose	(*158*)
	11α-Hydroxydes-glucohellebrin (**370**)	11α-Hydroxy-hellebrigenin	L-Rhamnose	(*158*)
Thesium lineatum (Santalaceae)	Thesiuside (**371**)	5-O-Acetyl-hellebrigenin	D-Glucose	(*159*)

Table 3. *Formulas of Cardiac Glycosides*

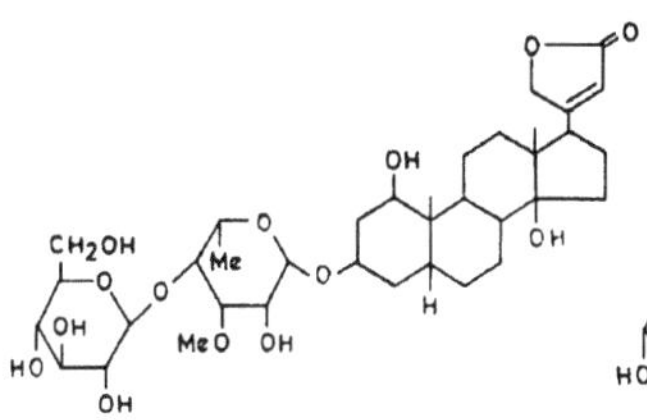

(1) $C_{36}H_{56}O_{14}$ (*61*)

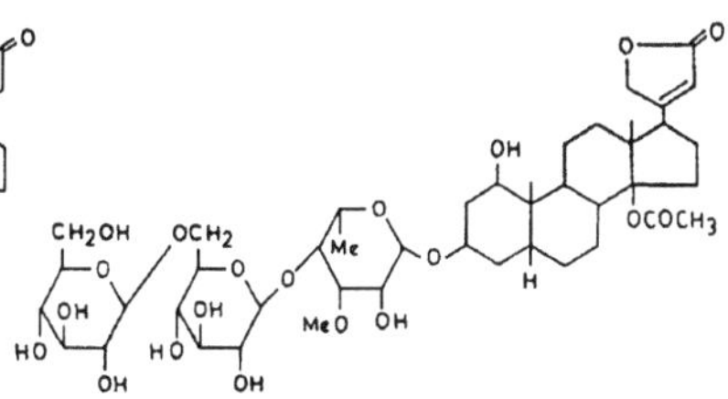

(2) $C_{44}H_{68}O_{20}$ (*61*)

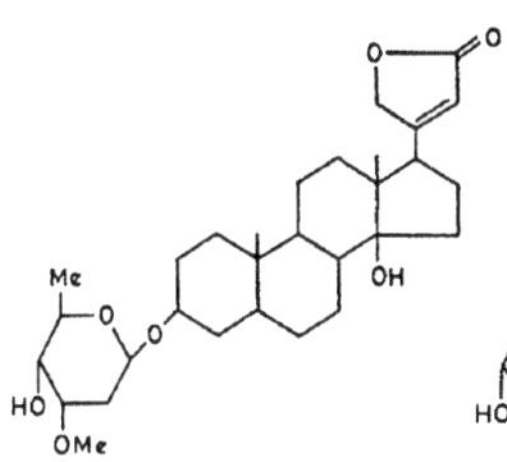

(3) $C_{30}H_{46}O_7$ (*62*)

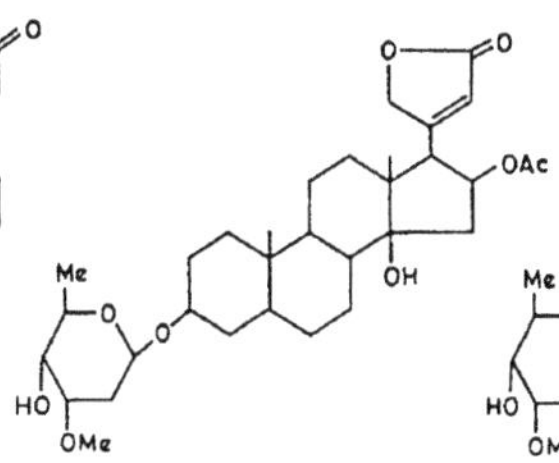

(4) $C_{32}H_{48}O_9$ (*62*)

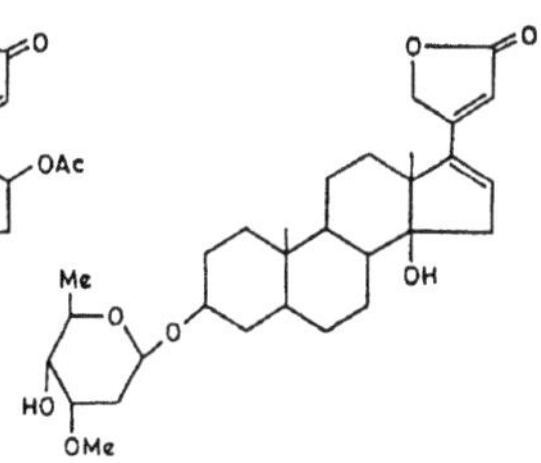

(5) $C_{30}H_{44}O_7$ (*62*)

Table 3 (*continued*)

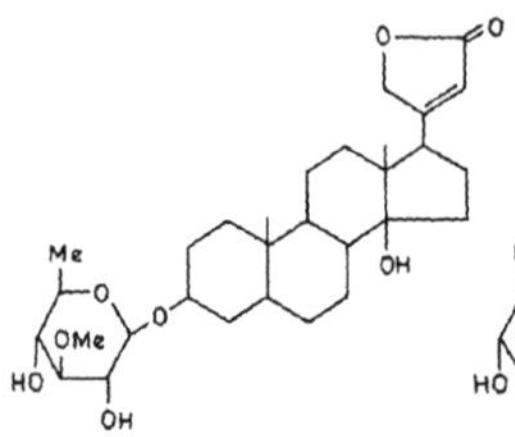

(6) $C_{30}H_{46}O_8$ *(62)*

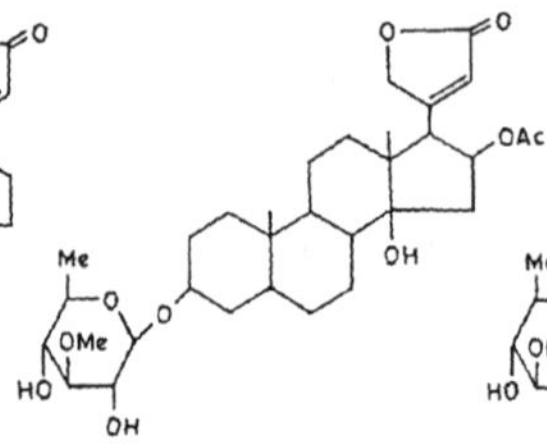

(7) $C_{32}H_{48}O_{10}$ *(62)*

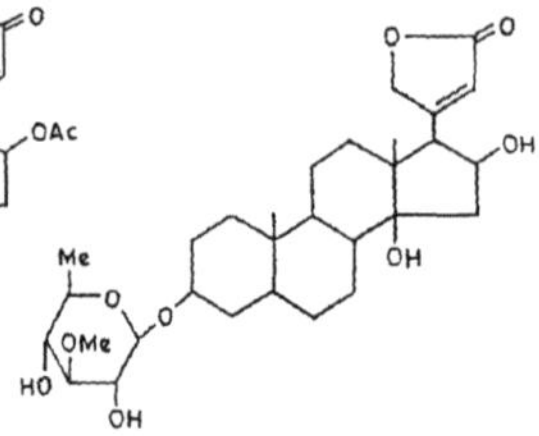

(8) $C_{30}H_{46}O_9$ *(62)*

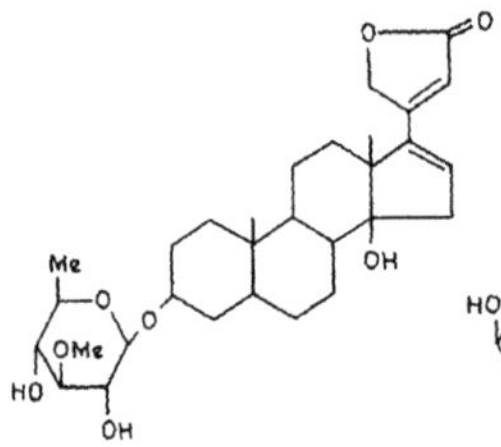

(9) $C_{30}H_{44}O_8$ *(62)*

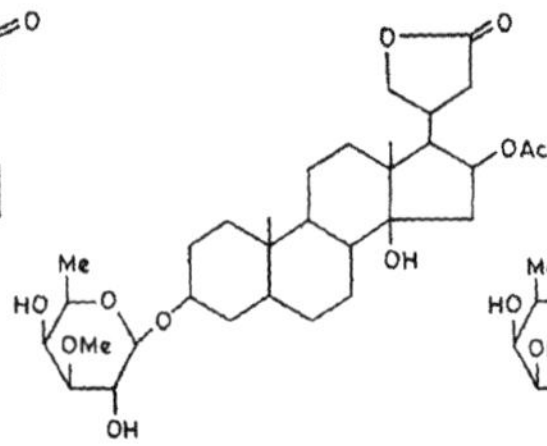

(10) $C_{32}H_{48}O_{10}$ *(62)*

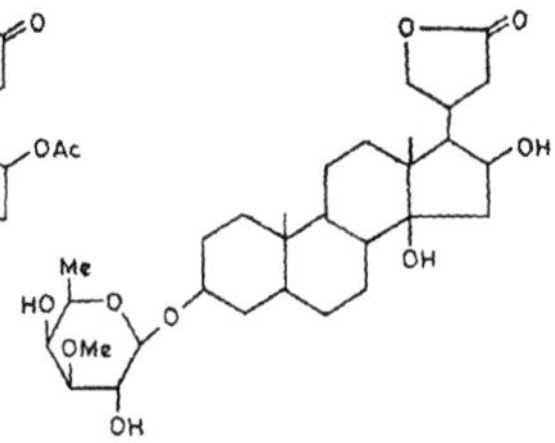

(11) $C_{30}H_{46}O_9$ *(62)*

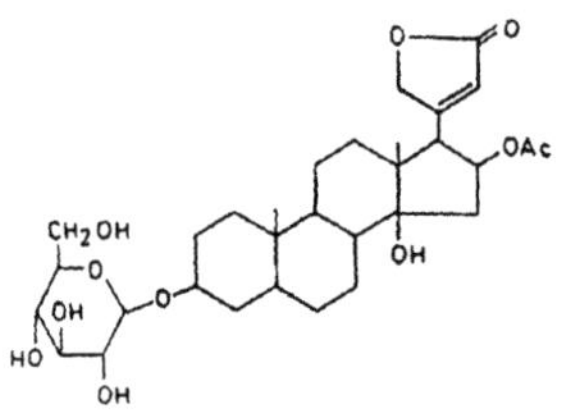

(12) $C_{31}H_{46}O_{11}$ *(62)*

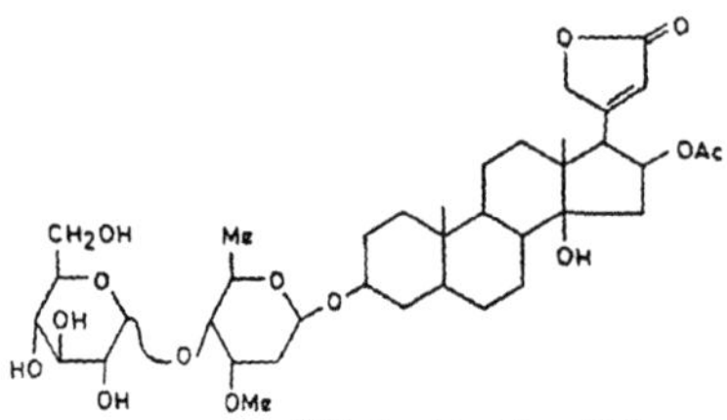

(13) $C_{38}H_{58}O_{14}$ *(62)*

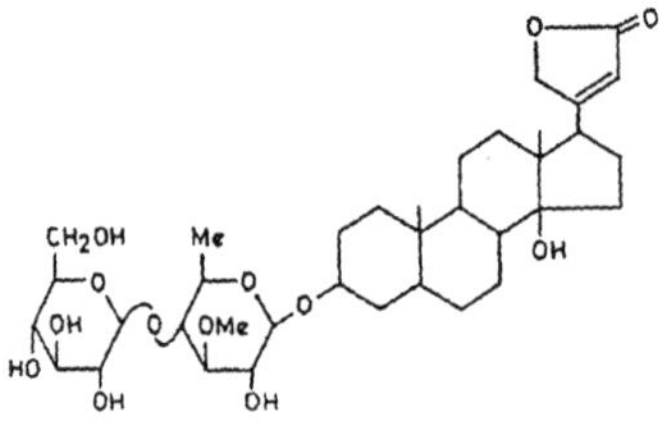

(14) $C_{36}H_{56}O_{13}$ *(62)*

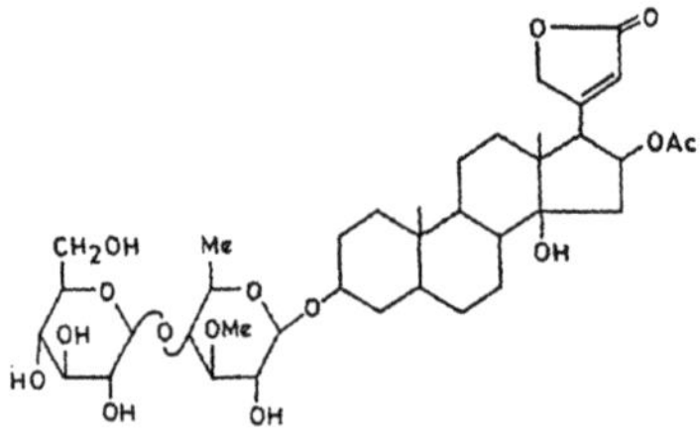

(15) $C_{38}H_{58}O_{15}$ *(62)*

References, pp. 148–155

Table 3 (*continued*)

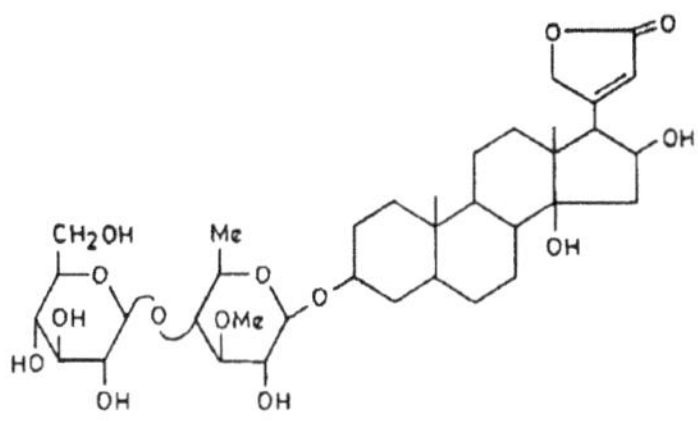

(16) $C_{36}H_{56}O_{14}$ (*62*)

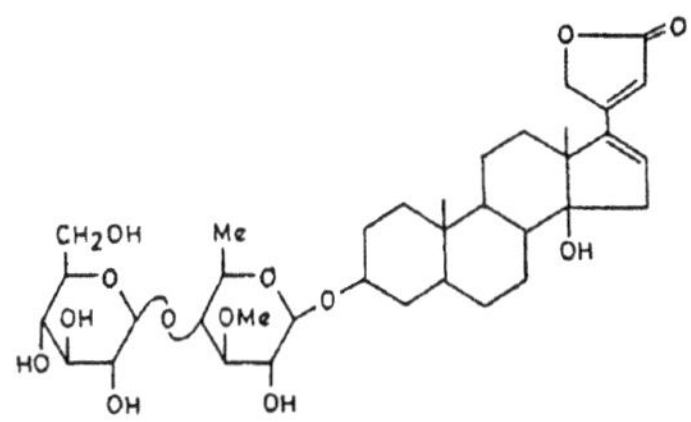

(17) $C_{36}H_{54}O_{13}$ (*62*)

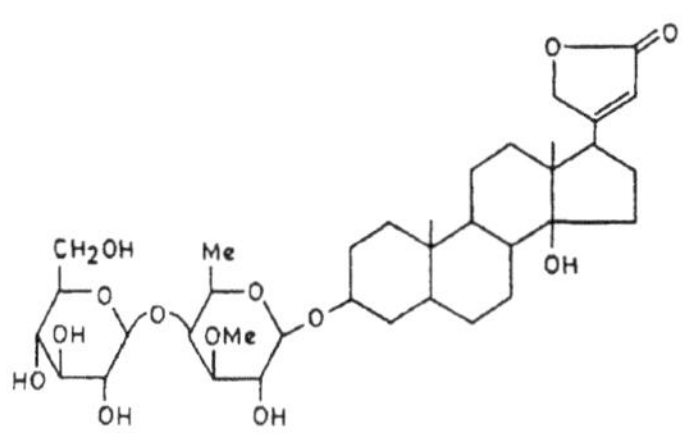

(18) $C_{36}H_{56}O_{13}$ (*62*)

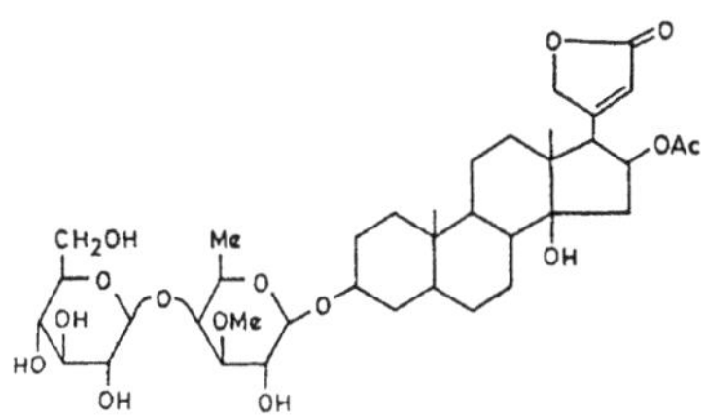

(19) $C_{38}H_{58}O_{15}$ (*62*)

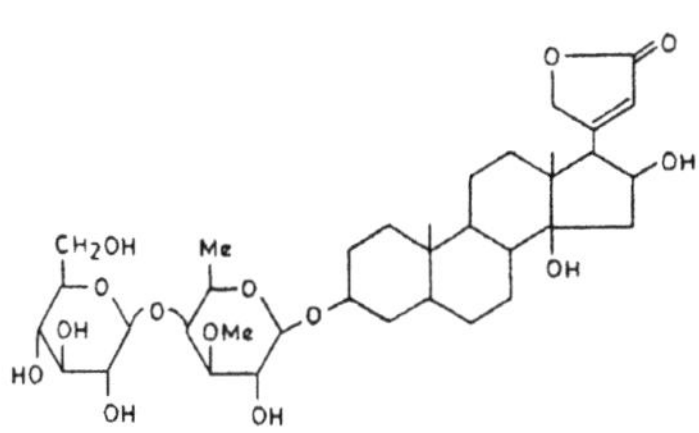

(20) $C_{36}H_{54}O_{14}$ (*62*)

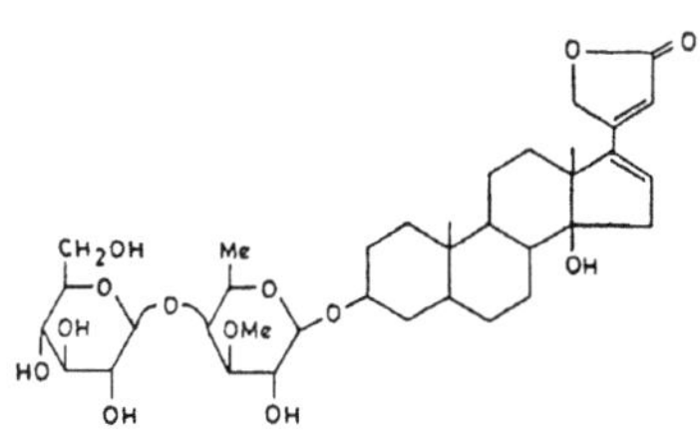

(21) $C_{36}H_{54}O_{13}$ (*62*)

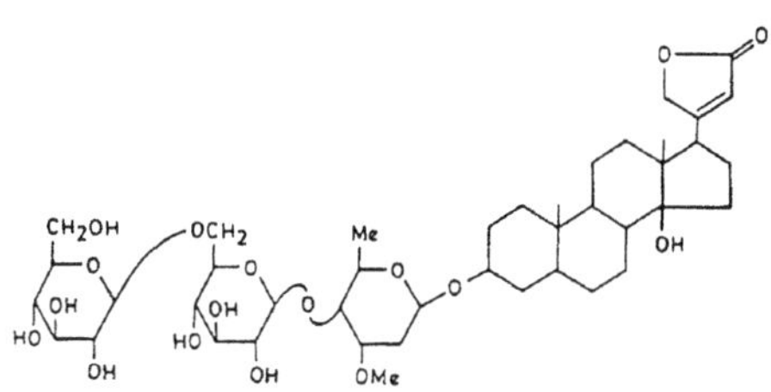

(22) $C_{42}H_{66}O_{17}$ (*62*)

Table 3 (*continued*)

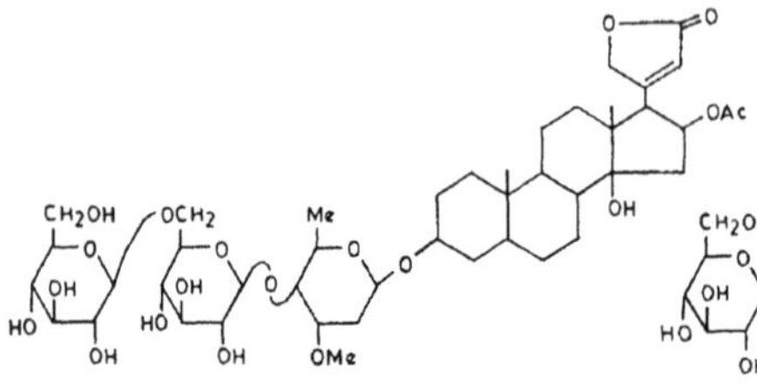

(23) $C_{44}H_{68}O_{19}$ (*62*)

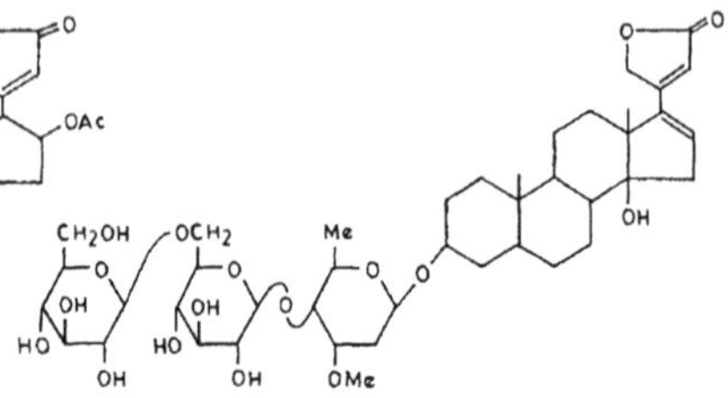

(24) $C_{42}H_{64}O_{17}$ (*62*)

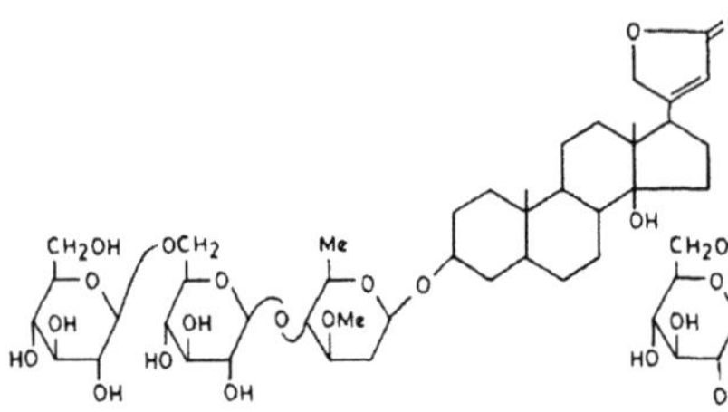

(25) $C_{42}H_{66}O_{17}$ (*62*)

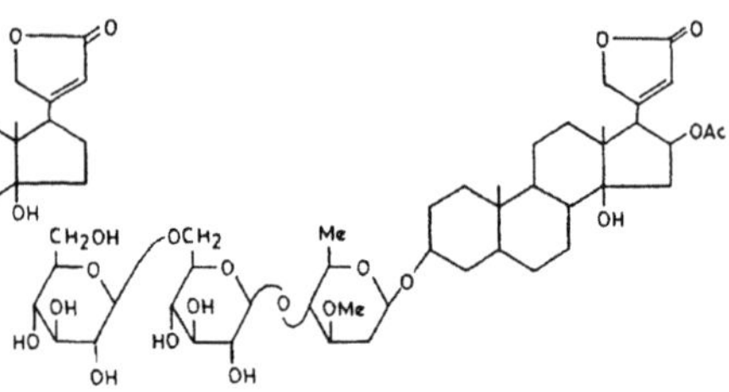

(26) $C_{44}H_{68}O_{19}$ (*62*)

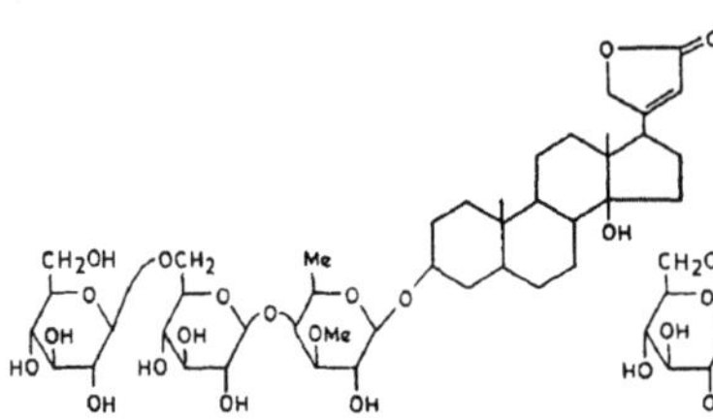

(27) $C_{42}H_{66}O_{18}$ (*62*)

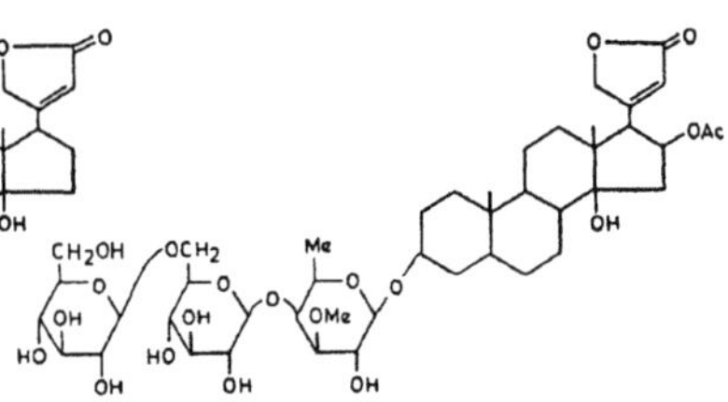

(28) $C_{44}H_{68}O_{20}$ (*62*)

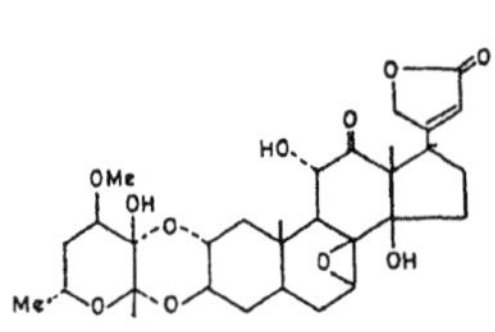

(29) $C_{30}H_{38}O_{11}$ (*63*)

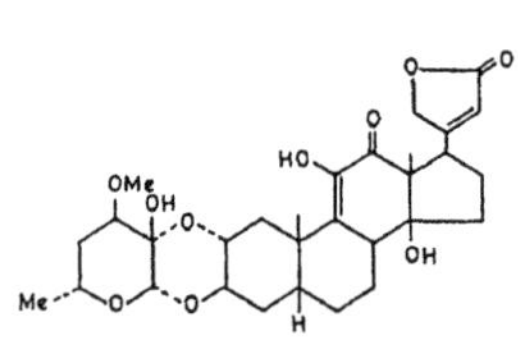

(30) $C_{30}H_{38}O_{10}$ (*63*)

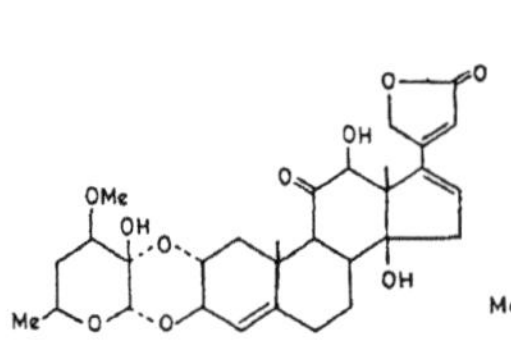

(31) $C_{30}H_{38}O_{10}$ (*63*)

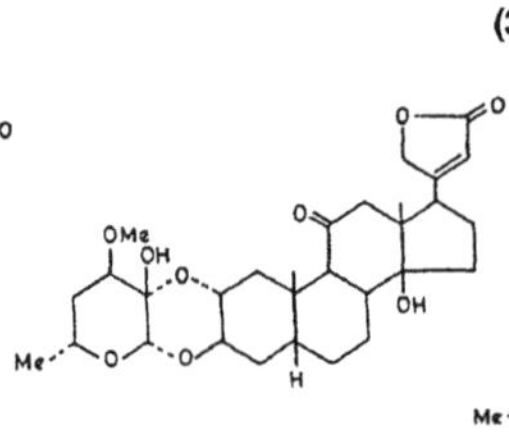

(32) $C_{30}H_{42}O_{9}$ (*63*)

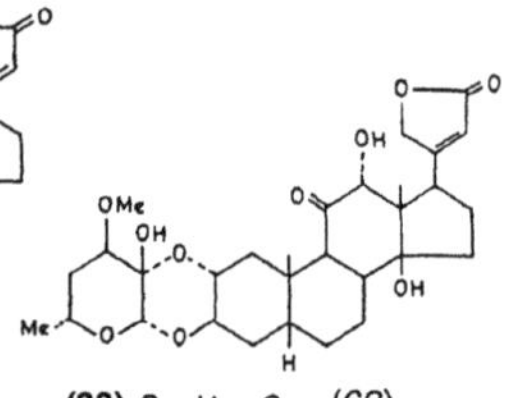

(33) $C_{30}H_{42}O_{10}$ (*63*)

Table 3 (*continued*)

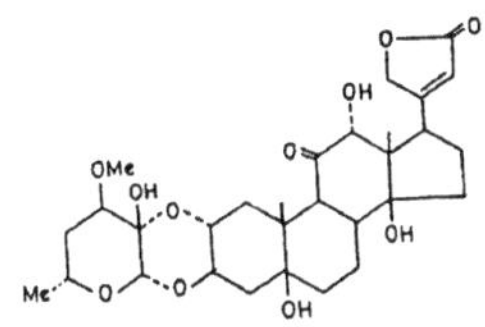

(34) $C_{30}H_{42}O_{11}$ *(63)*

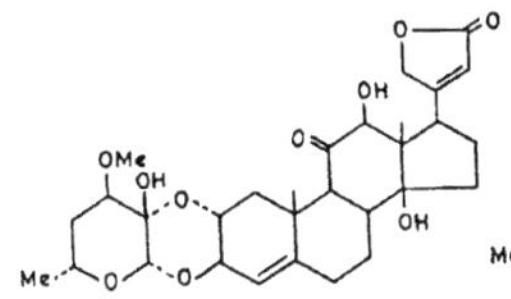

(35) $C_{30}H_{40}O_{10}$ *(63)*

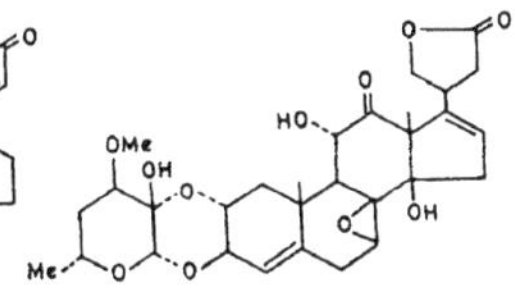

(36) $C_{30}H_{36}O_{11}$ *(63)*

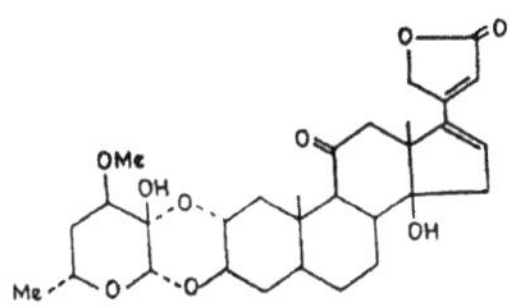

(37) $C_{30}H_{40}O_{9}$ *(63)*

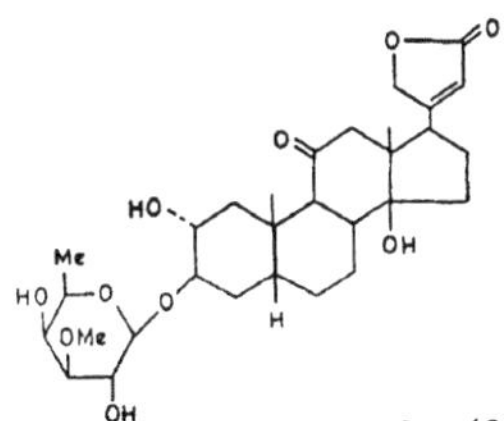

(38) $C_{30}H_{44}O_{10}$ *(64)*

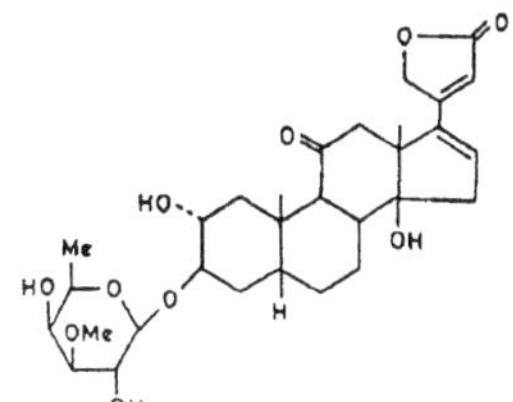

(39) $C_{30}H_{42}O_{10}$ *(64)*

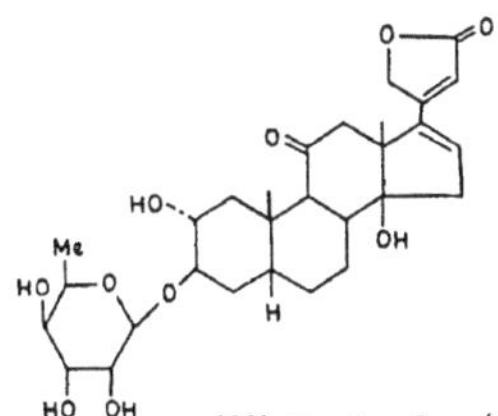

(40) $C_{29}H_{40}O_{10}$ *(64)*

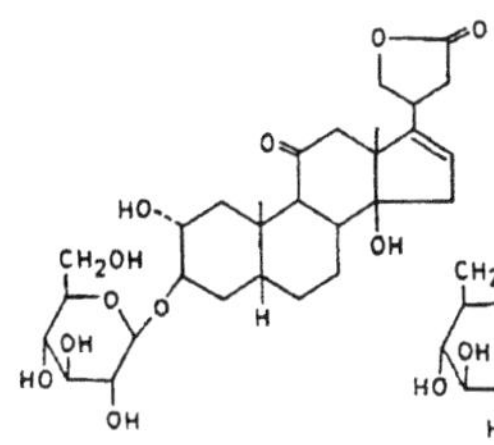

(41) $C_{29}H_{40}O_{11}$ *(64)*

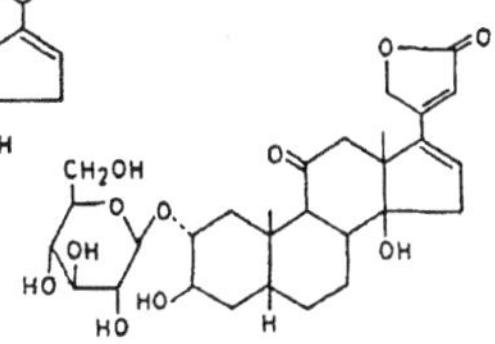

(42) $C_{29}H_{40}O_{11}$ *(64)*

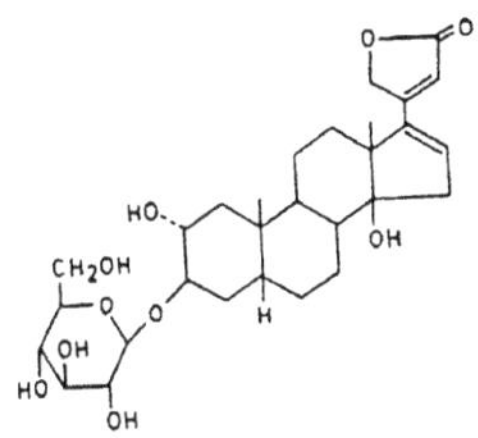

(43) $C_{29}H_{40}O_{10}$ *(64)*

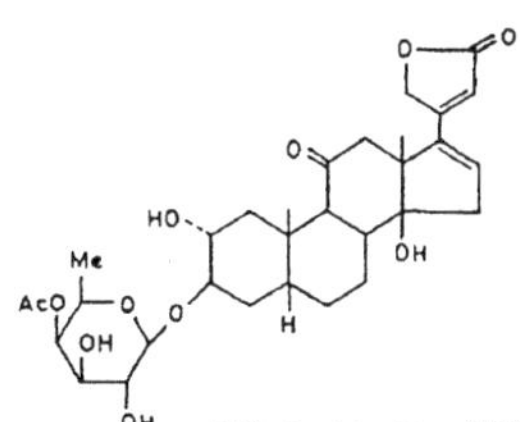

(44) $C_{31}H_{42}O_{11}$ *(64)*

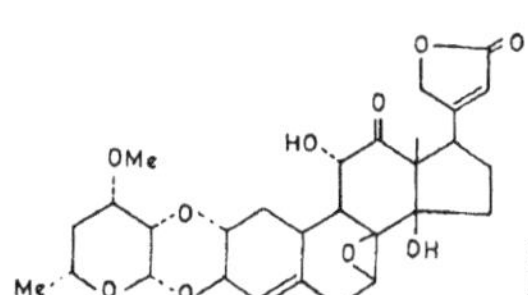

(45) $C_{30}H_{38}O_{10}$ *(65)*

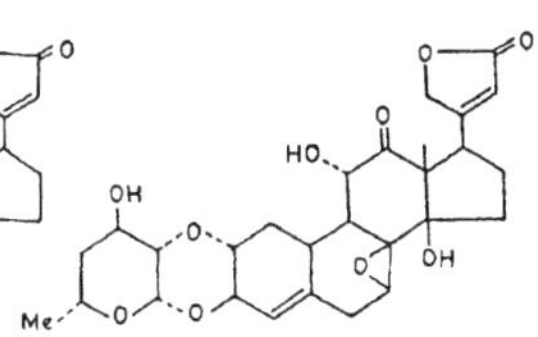

(46) $C_{29}H_{36}O_{10}$ *(65)*

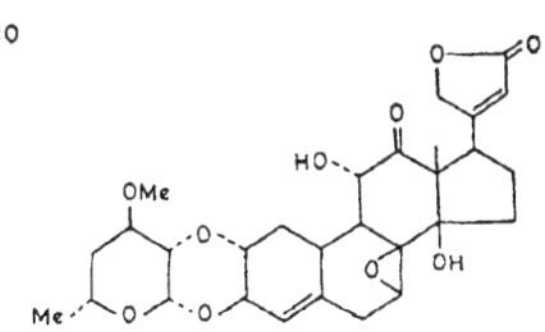

(47) $C_{30}H_{38}O_{10}$ *(65)*

Table 3 (*continued*)

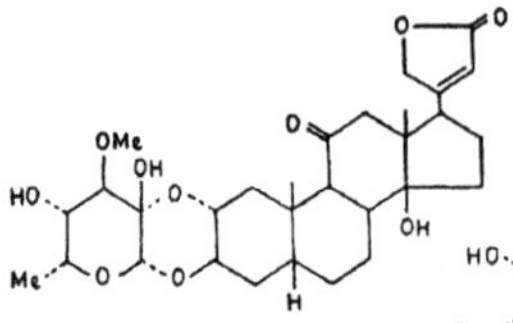

(48) $C_{30}H_{42}O_{10}$ *(66)*

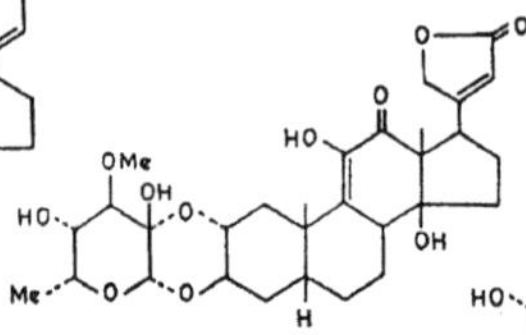

(49) $C_{30}H_{40}O_{11}$ *(66)*

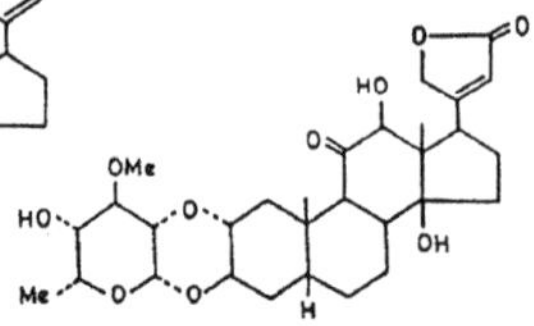

(50) $C_{30}H_{42}O_{10}$ *(66)*

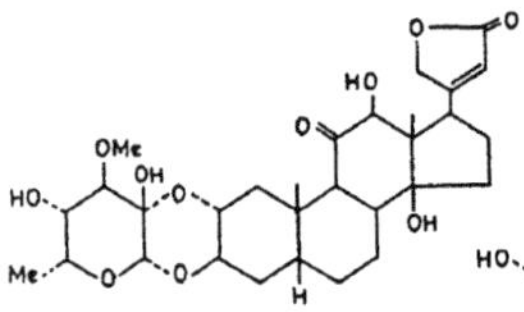

(51) $C_{30}H_{42}O_{11}$ *(66)*

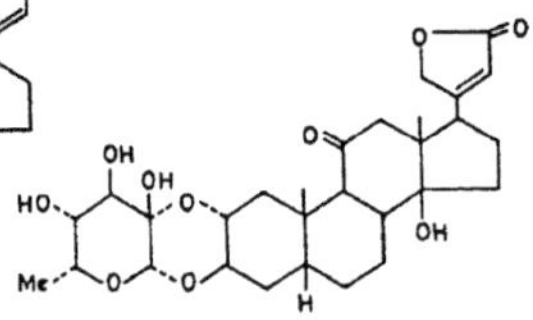

(52) $C_{29}H_{40}O_{10}$ *(66)*

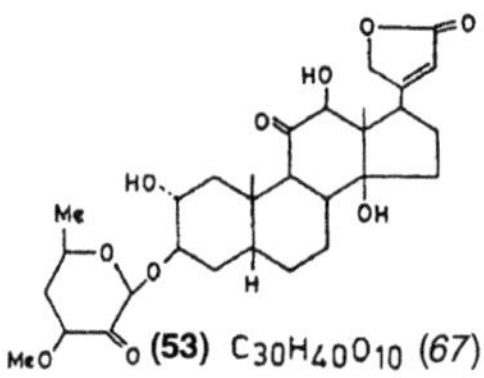

(53) $C_{30}H_{40}O_{10}$ *(67)*

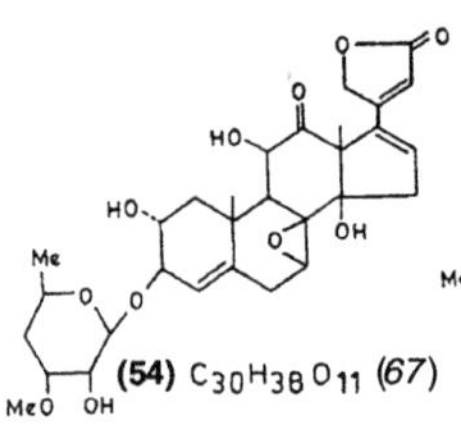

(54) $C_{30}H_{38}O_{11}$ *(67)*

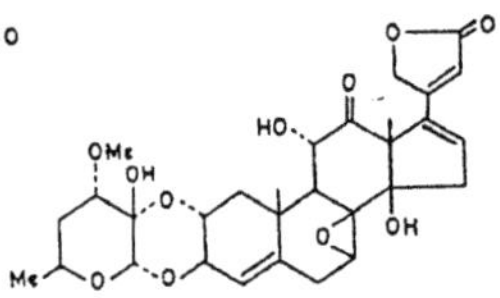

(55) $C_{30}H_{36}O_{11}$ *(67)*

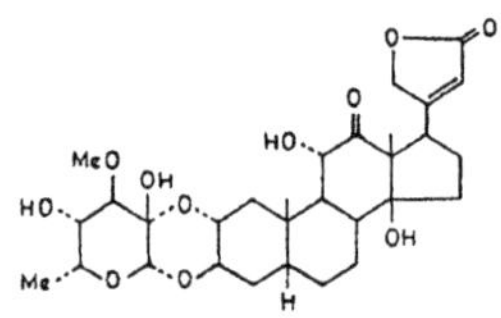

(56) $C_{30}H_{29}O_{11}$ *(68)*

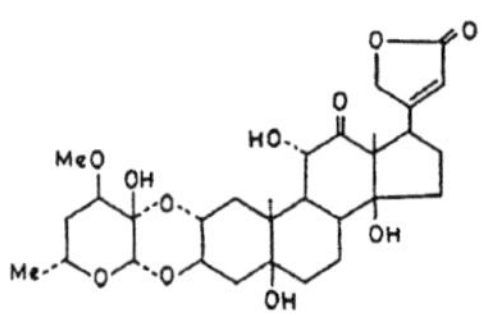

(57) $C_{30}H_{38}O_{11}$ *(68)*

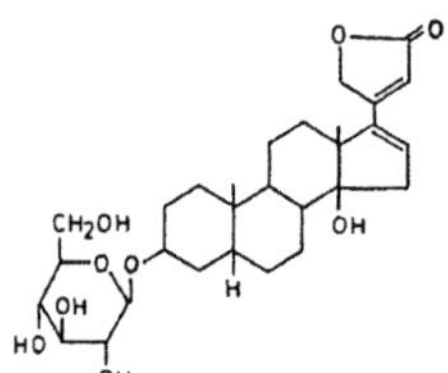

(58) $C_{29}H_{35}O_{9}$ *(68)*

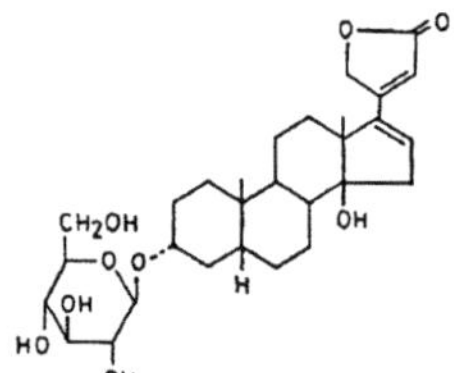

(59) $C_{29}H_{44}O_{9}$ *(68)*

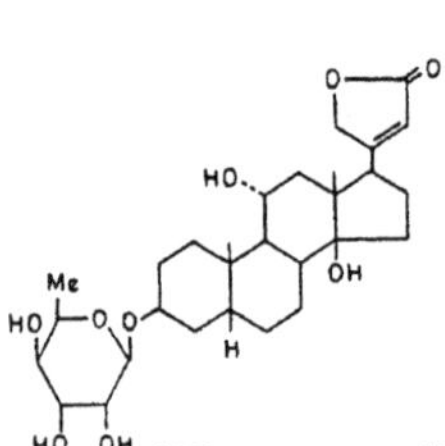

(60) $C_{29}H_{44}O_{9}$ *(69)*

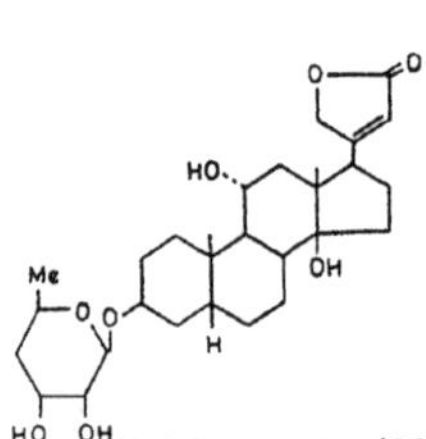

(61) $C_{29}H_{44}O_{8}$ *(69)*

References, pp. 148–155

Table 3 (*continued*)

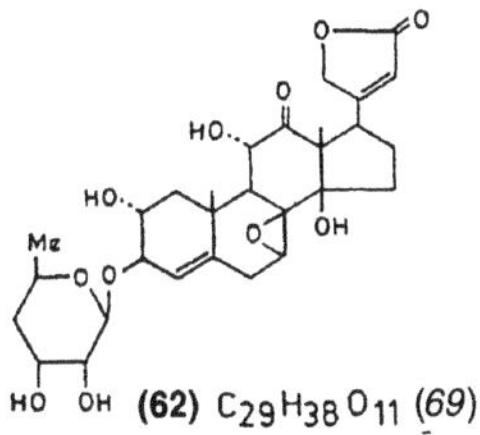

(62) $C_{29}H_{38}O_{11}$ *(69)*

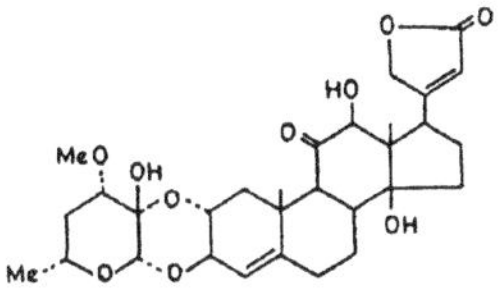

(63) $C_{30}H_{40}O_{10}$ *(69)*

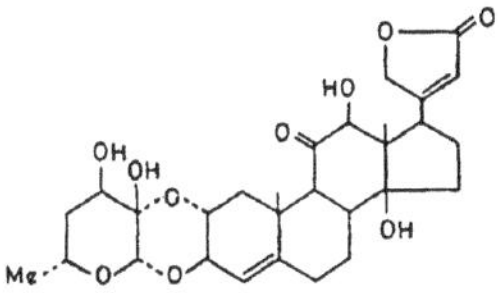

(64) $C_{29}H_{38}O_{10}$ *(69)*

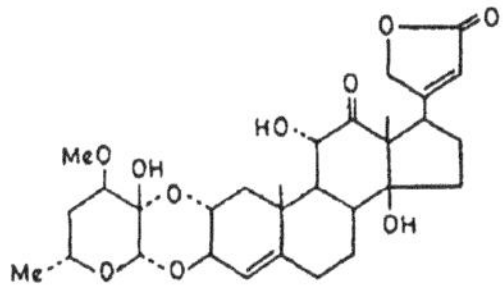

(65) $C_{30}H_{40}O_{10}$ *(69)*

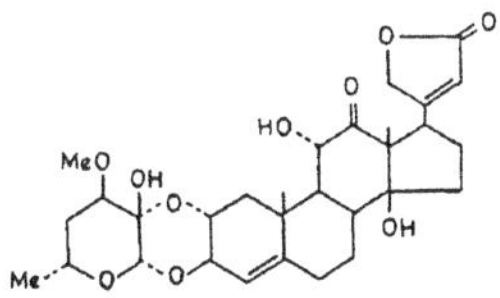

(66) $C_{30}H_{40}O_{10}$ *(69)*

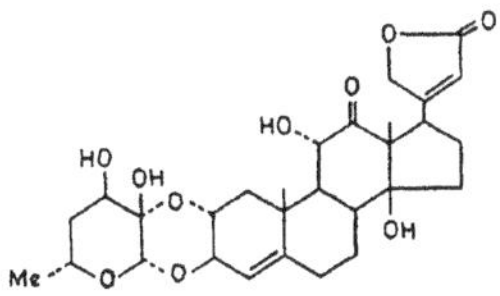

(67) $C_{29}H_{38}O_{10}$ *(69)*

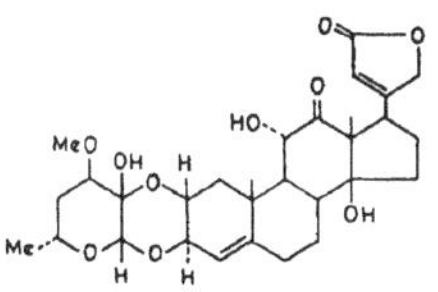

(68) $C_{30}H_{40}O_{10}$ *(70)*

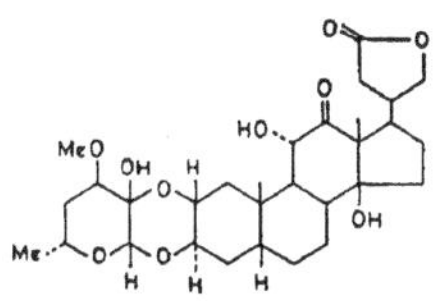

(69) $C_{30}H_{42}O_{10}$ *(70)*

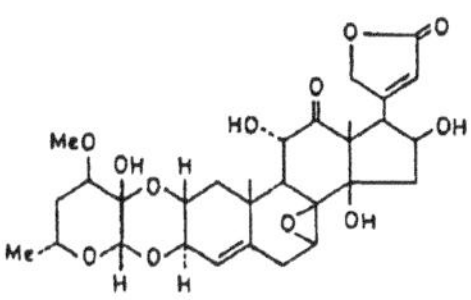

(70) $C_{30}H_{38}O_{12}$ *(70)*

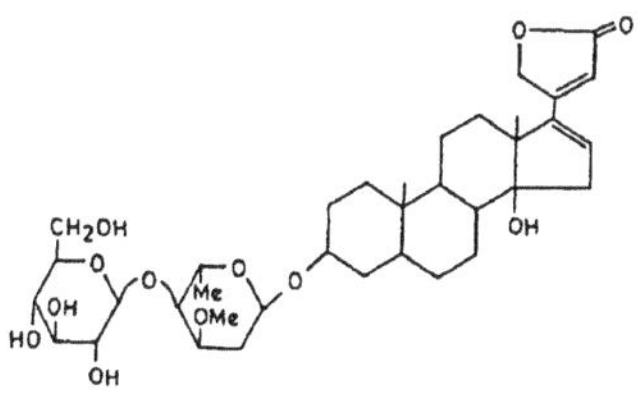

(71) $C_{36}H_{54}O_{12}$ *(33)*

Table 3 (*continued*)

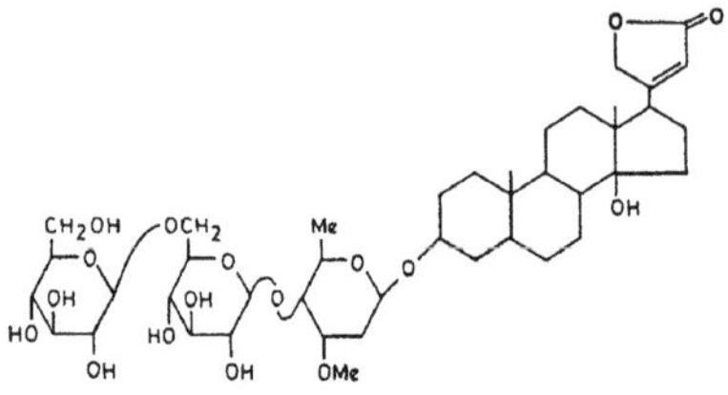

(72) $C_{42}H_{66}O_{17}$ *(33)*

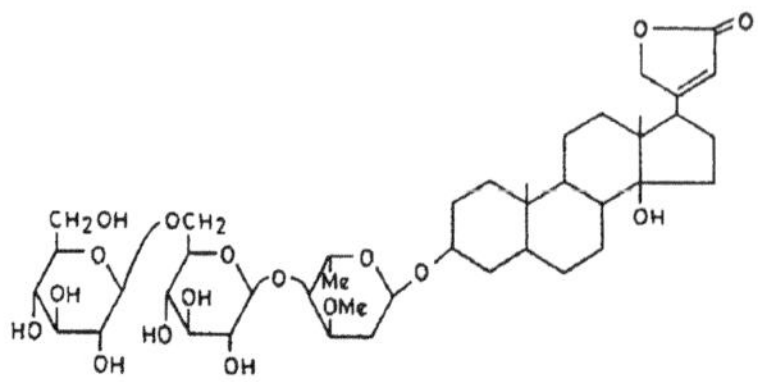

(73) $C_{42}H_{66}O_{17}$ *(33)*

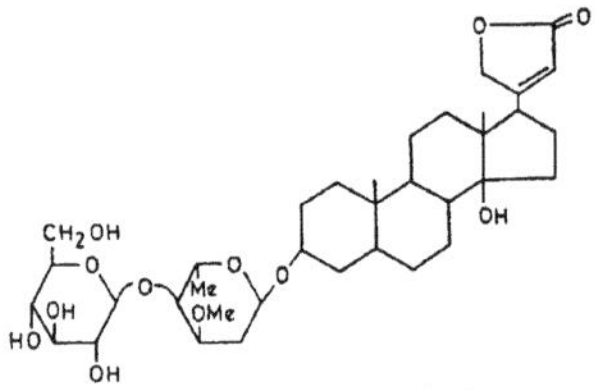

(74) $C_{36}H_{56}O_{12}$ *(33)*

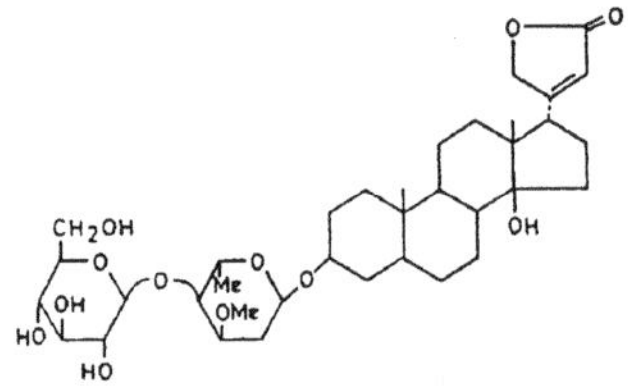

(75) $C_{36}H_{56}O_{12}$ *(33)*

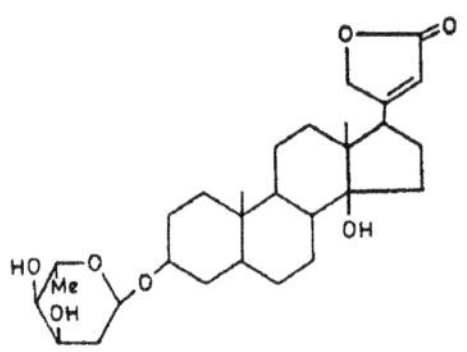

(76) $C_{29}H_{44}O_{7}$ *(33)*

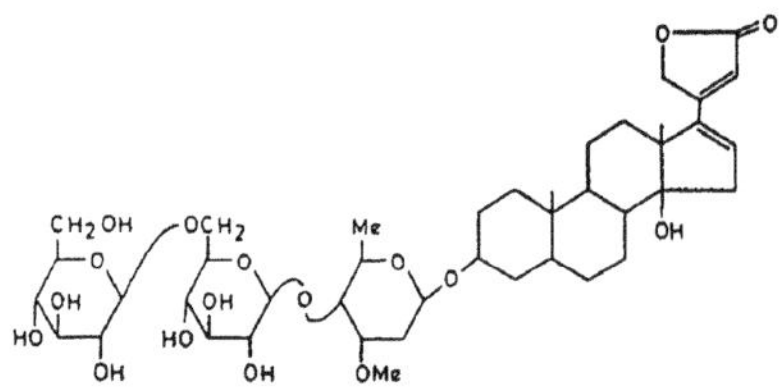

(77) $C_{42}H_{64}O_{17}$ *(33)*

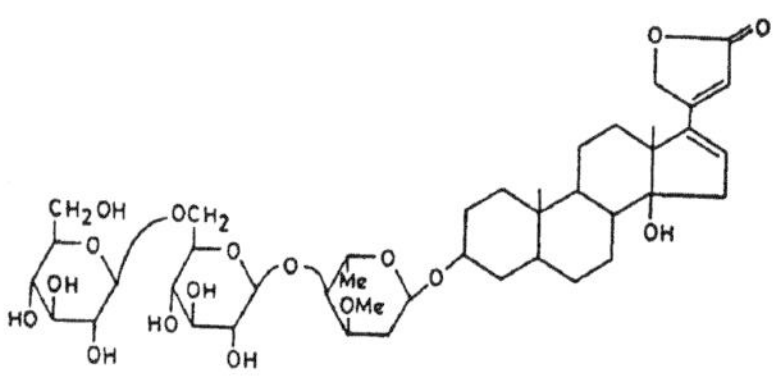

(78) $C_{42}H_{64}O_{17}$ *(33)*

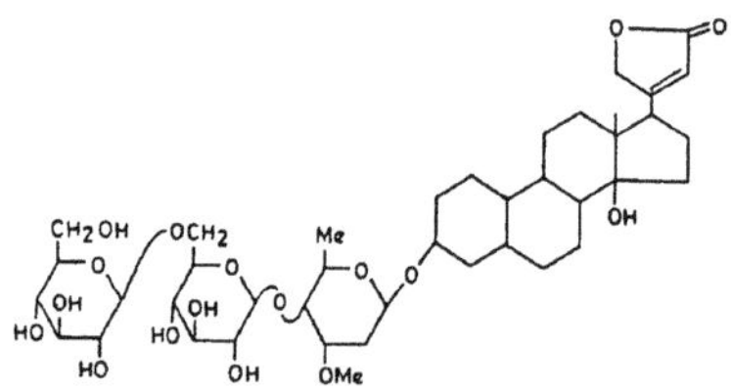

(79) $C_{42}H_{66}O_{17}$ *(33)*

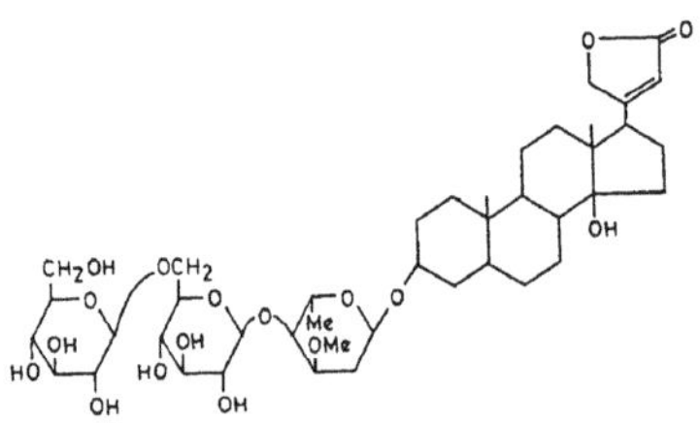

(80) $C_{42}H_{66}O_{17}$ *(33)*

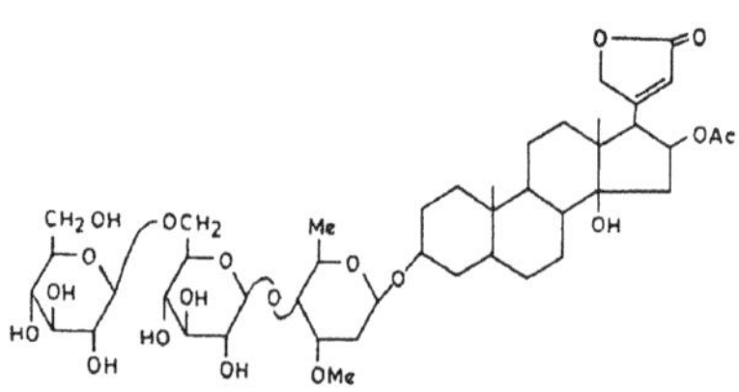

(81) $C_{44}H_{68}O_{19}$ *(33)*

Table 3 (*continued*)

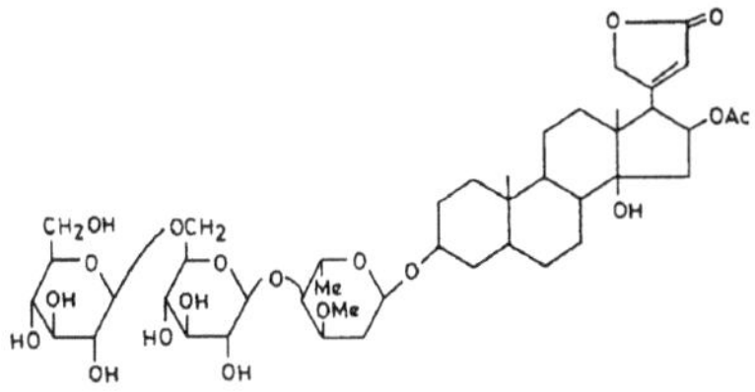

(82) $C_{44}H_{68}O_{19}$ (*33*)

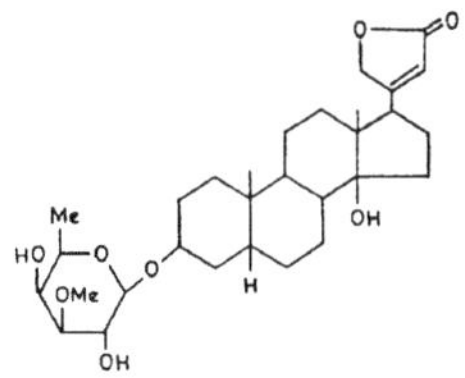

(83) $C_{30}H_{46}O_{8}$ (*71*)

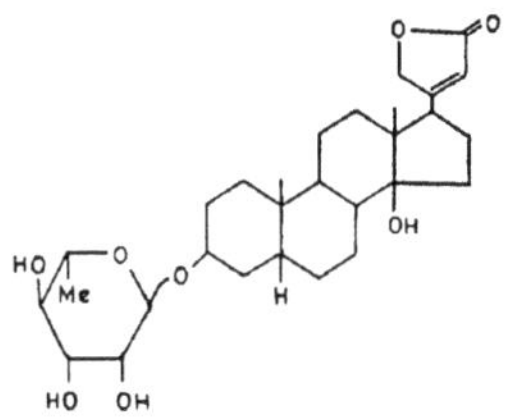

(84) $C_{29}H_{44}O_{8}$ (*71*)

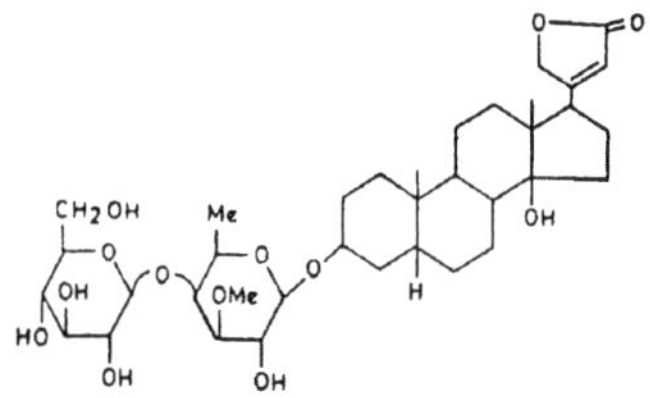

(85) $C_{36}H_{56}O_{13}$ (*71*)

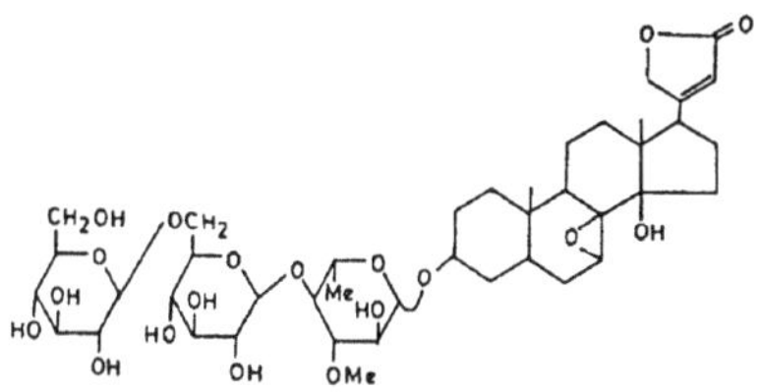

(86) $C_{42}H_{62}O_{19}$ (*72*)

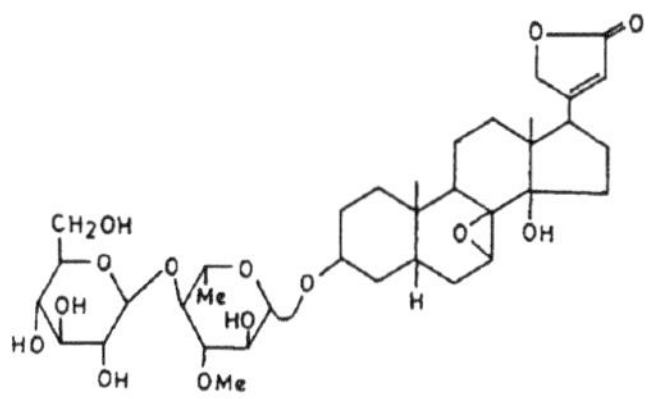

(87) $C_{36}H_{52}O_{14}$ (*72*)

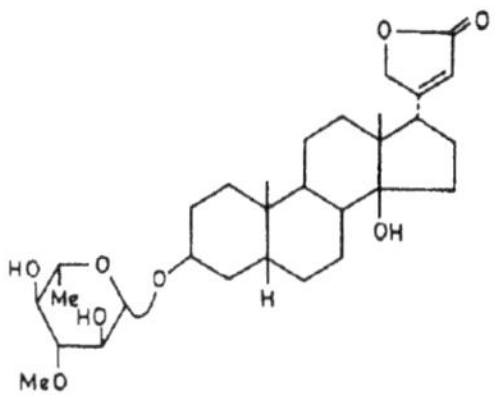

(88) $C_{30}H_{46}O_{8}$ (*72*)

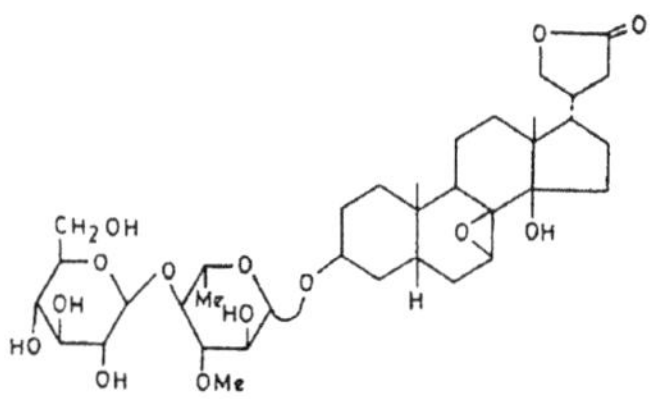

(89) $C_{36}H_{54}O_{14}$ (*72*)

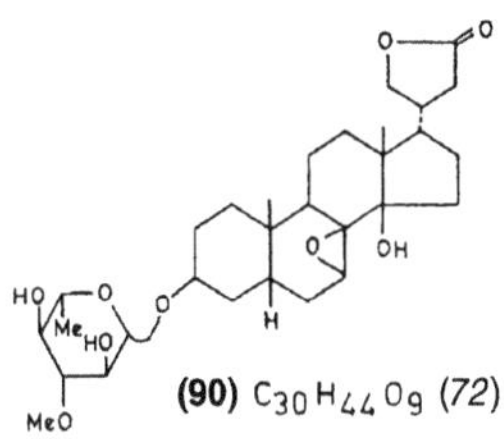

(90) $C_{30}H_{44}O_{9}$ (*72*)

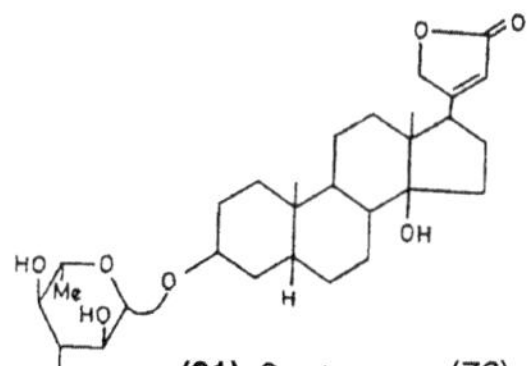

(91) $C_{30}H_{46}O_{8}$ (*72*)

Table 3 (*continued*)

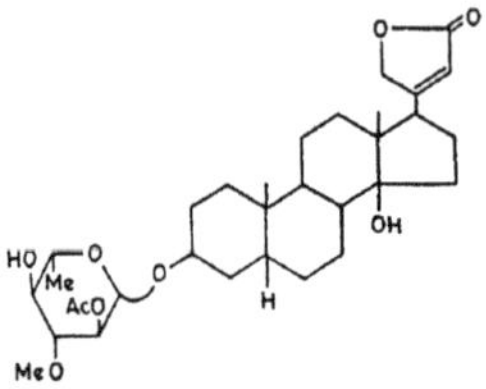

(92) $C_{32}H_{48}O_9$ (*72*)

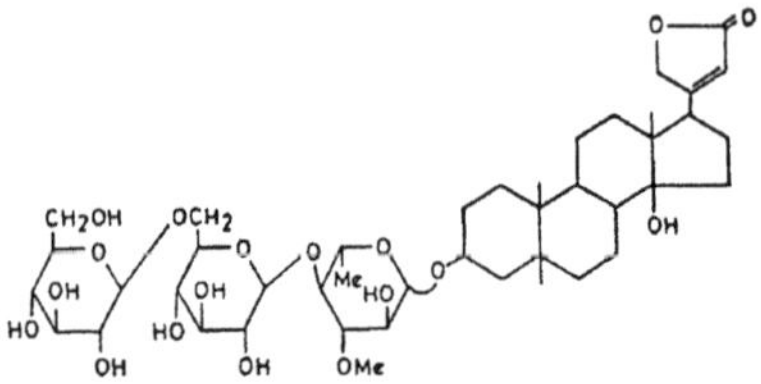

(93) $C_{42}H_{66}O_{18}$ (*72*)

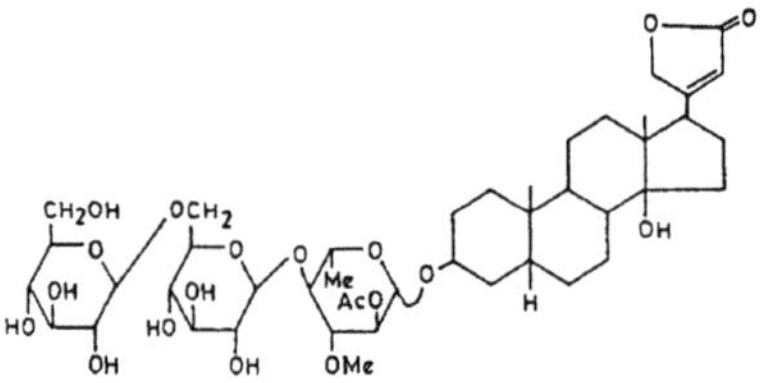

(94) $C_{44}H_{66}O_{19}$ (*72*)

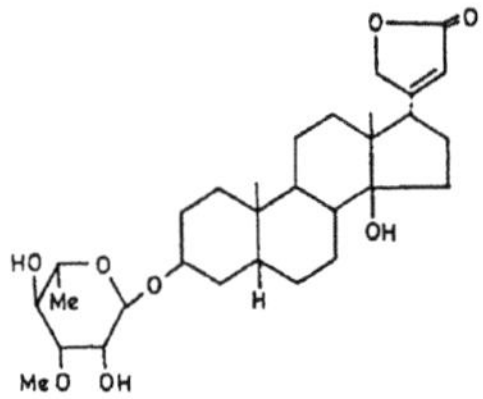

(95) $C_{30}H_{46}O_8$ (*73*)

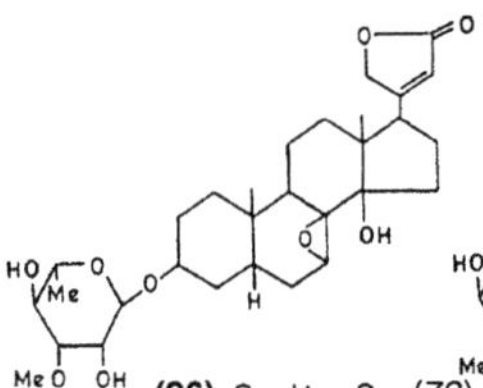

(96) $C_{30}H_{44}O_9$ (*73*)

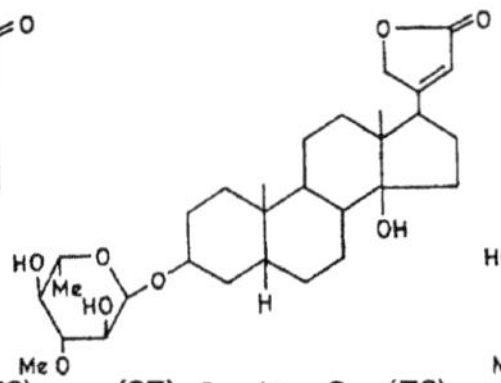

(97) $C_{30}H_{46}O_8$ (*73*)

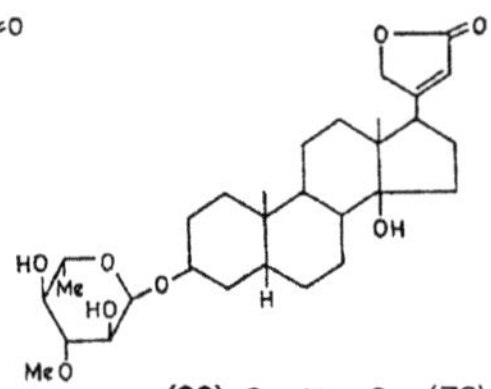

(98) $C_{30}H_{46}O_8$ (*73*)

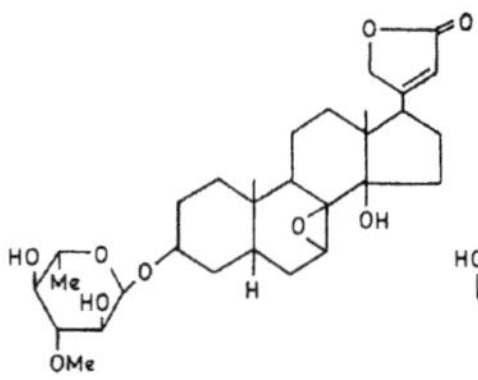

(99) $C_{30}H_{44}O_9$ (*73*)

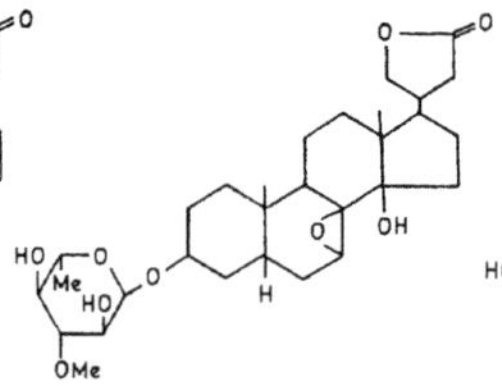

(100) $C_{30}H_{44}O_9$ (*73*)

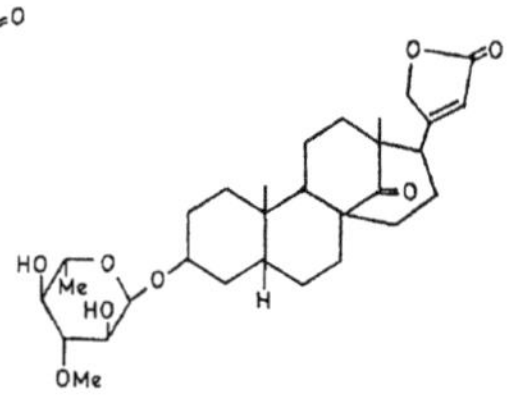

(101) $C_{30}H_{46}O_8$ (*73*)

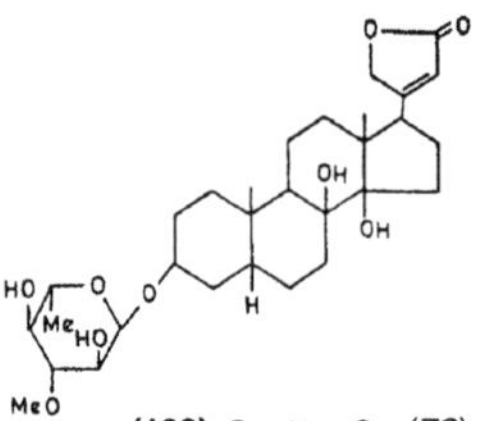

(102) $C_{30}H_{46}O_9$ (*73*)

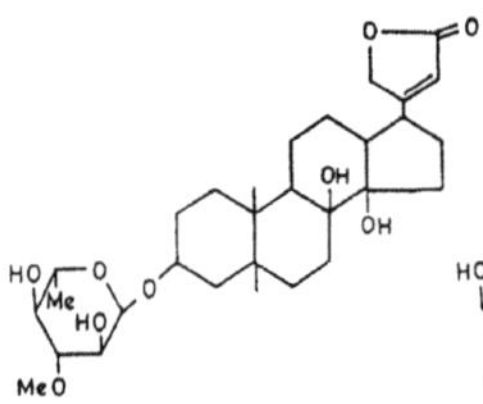

(103) $C_{30}H_{46}O_9$ (*73*)

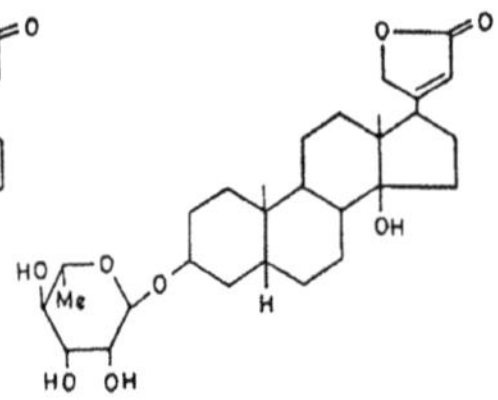

(104) $C_{29}H_{44}O_8$ (*73*)

References, pp. 148–155

Table 3 (*continued*)

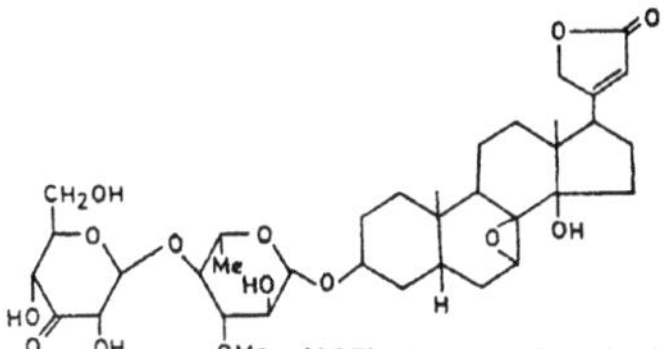

(105) $C_{36}H_{52}O_{14}$ (*74*)

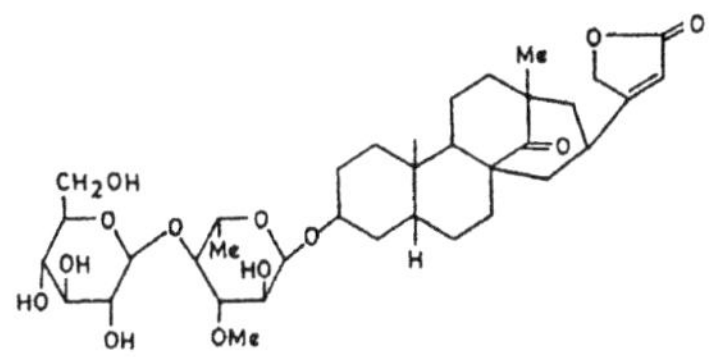

(106) $C_{36}H_{54}O_{13}$ (*74*)

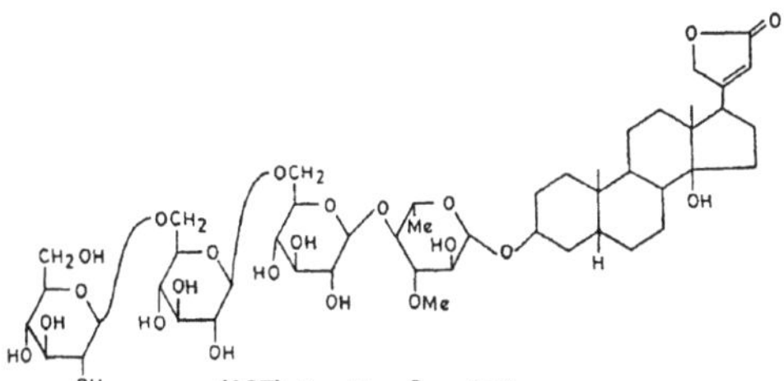

(107) $C_{48}H_{74}O_{23}$ (*74*)

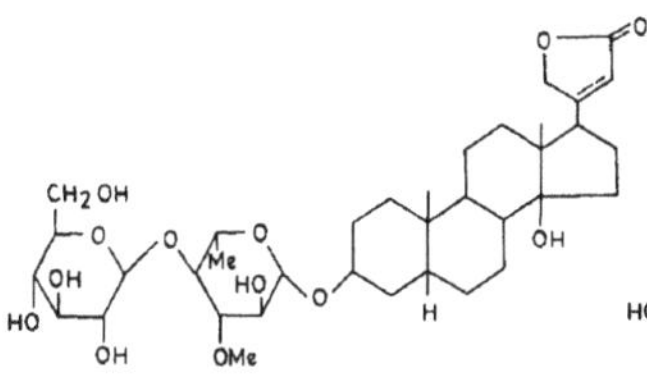

(108) $C_{36}H_{56}O_{13}$ (*74*)

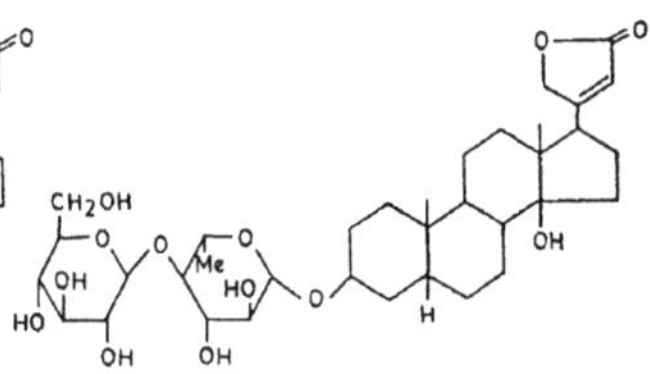

(109) $C_{35}H_{54}O_{13}$ (*74*)

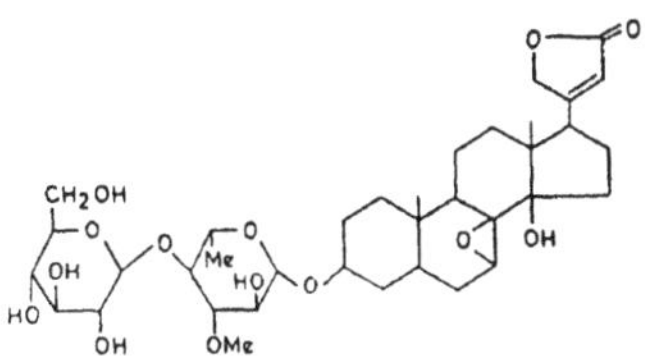

(110) $C_{36}H_{54}O_{14}$ (*74*)

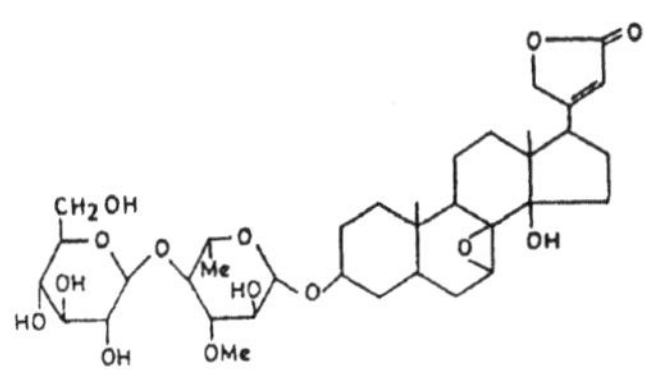

(111) $C_{36}H_{54}O_{14}$ (*74*)

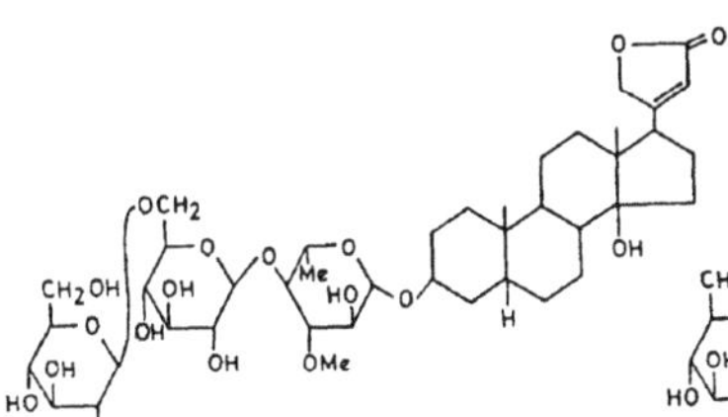

(112) $C_{42}H_{66}O_{18}$ (*74*)

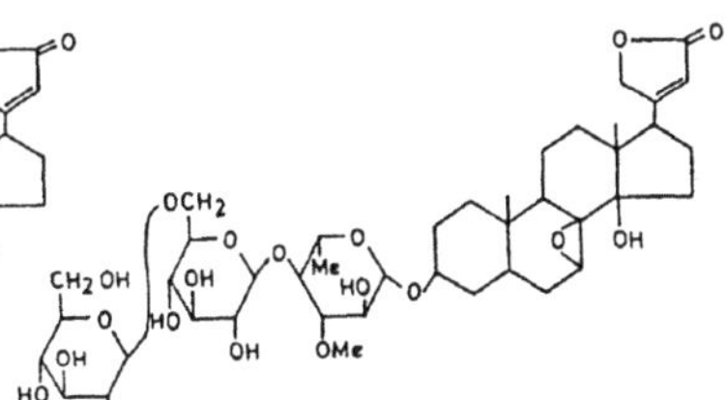

(113) $C_{42}H_{64}O_{19}$ (*74*)

Table 3 (*continued*)

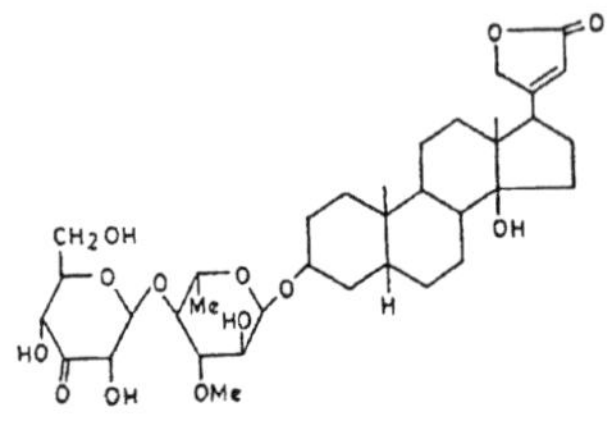

(114) $C_{36}H_{54}O_{13}$ (*74*)

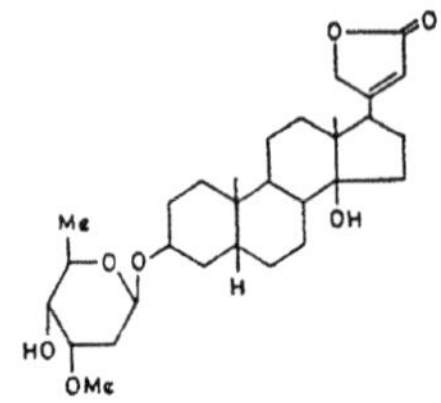

(115) $C_{30}H_{46}O_{7}$ (*9*)

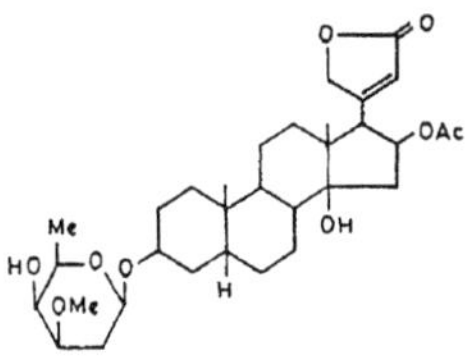

(116) $C_{32}H_{48}O_{9}$ (*9*)

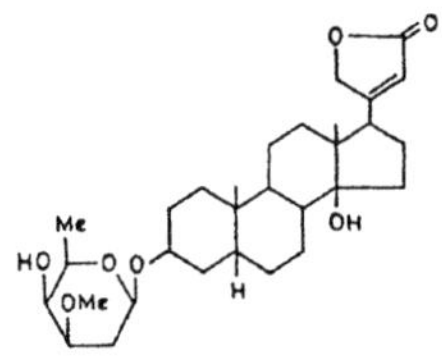

(117) $C_{30}H_{46}O_{7}$ (*9*)

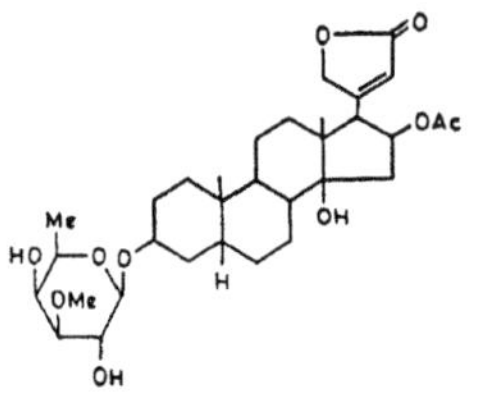

(118) $C_{32}H_{48}O_{10}$ (*9*)

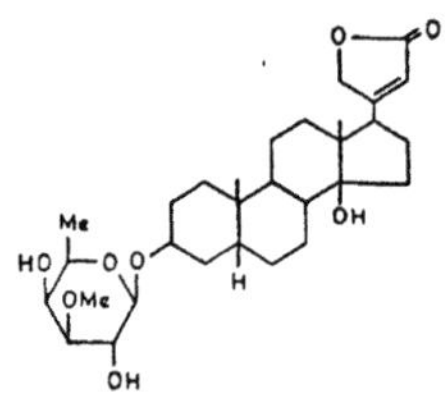

(119) $C_{30}H_{46}O_{8}$ (*9*)

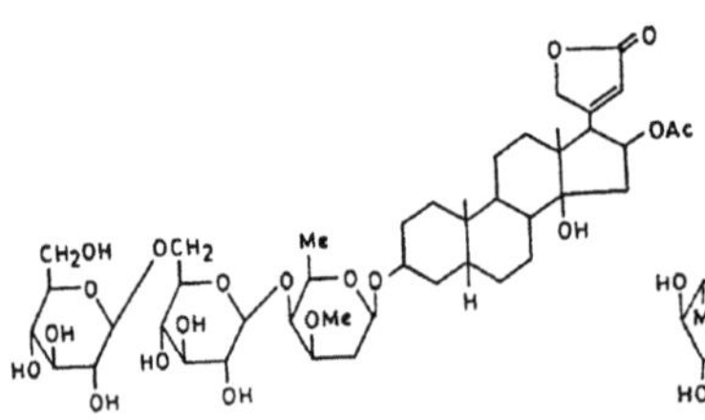

(120) $C_{44}H_{68}O_{19}$ (*9*)

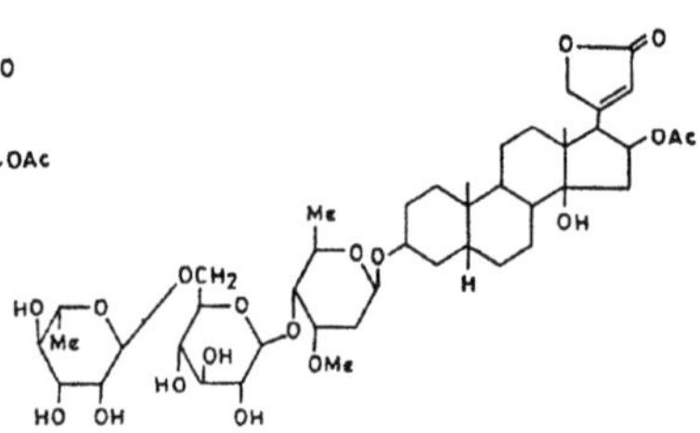

(121) $C_{44}H_{68}O_{18}$ (*9*)

Table 3 (*continued*)

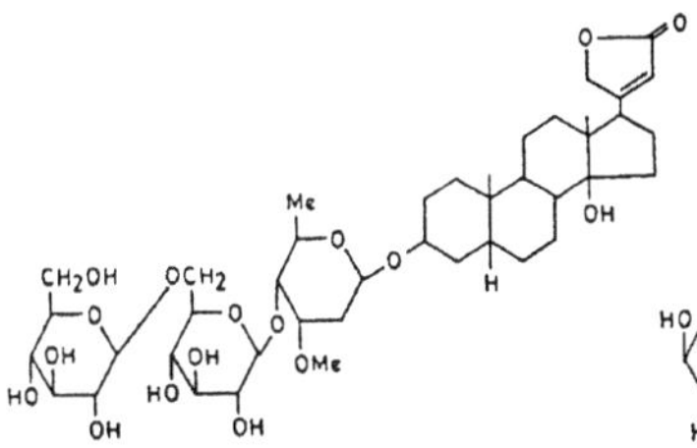

(122) $C_{42}H_{66}O_{17}$ (*9*)

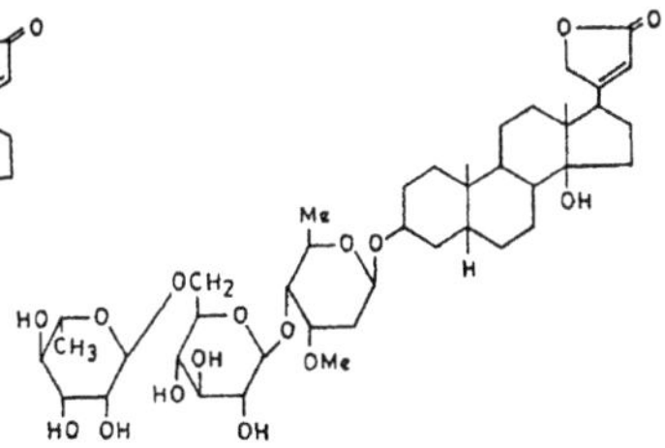

(123) $C_{42}H_{66}O_{16}$ (*9*)

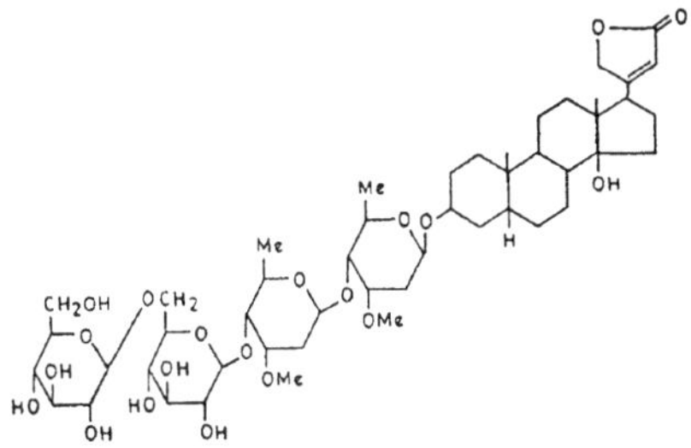

(124) $C_{49}H_{78}O_{20}$ (*9*)

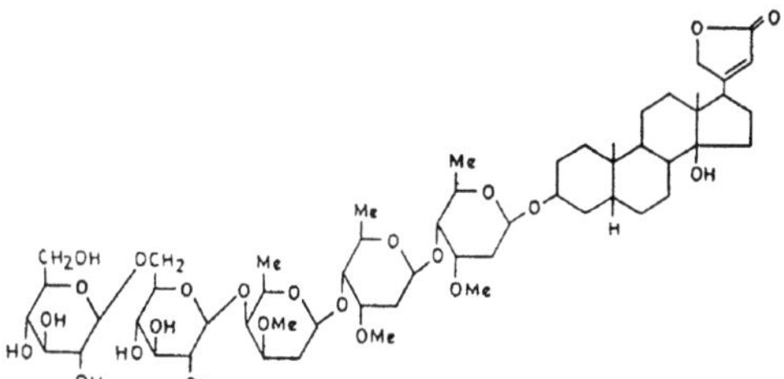

(125) $C_{56}H_{90}O_{23}$ (*9*)

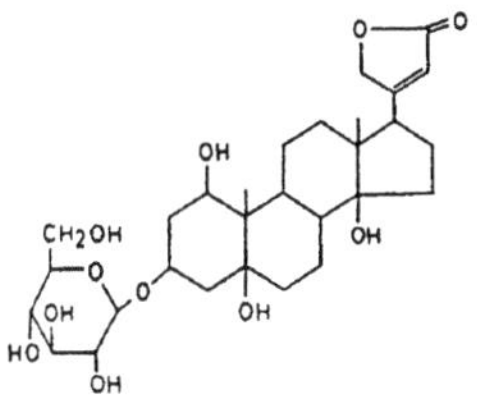

(126) $C_{29}H_{44}O_{11}$ (*75*)

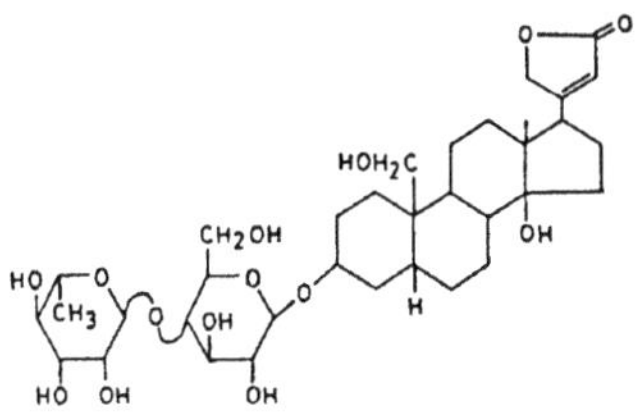

(127) $C_{35}H_{54}O_{14}$ (*75*)

Table 3 (*continued*)

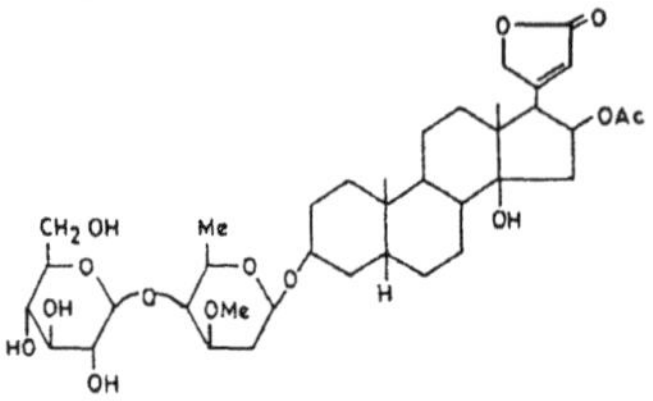

(128) $C_{38}H_{58}O_{14}$ (*76*)

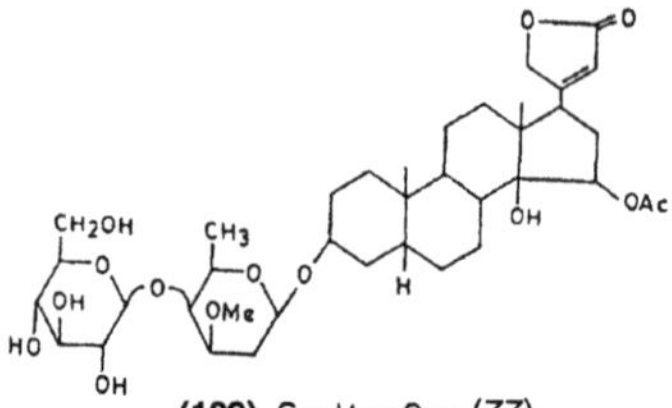

(129) $C_{38}H_{56}O_{14}$ (*77*)

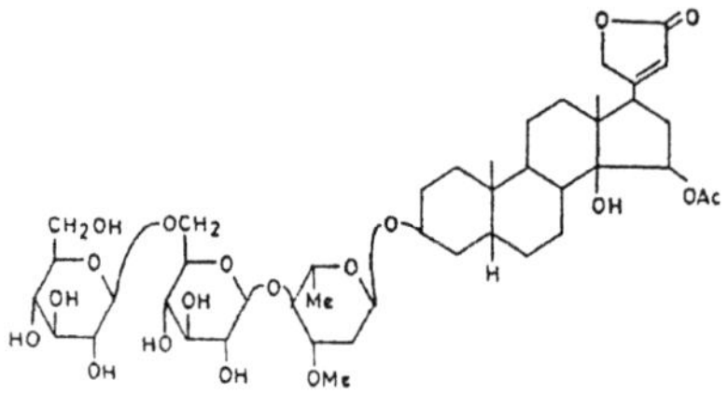

(130) $C_{44}H_{68}O_{19}$ (*77*)

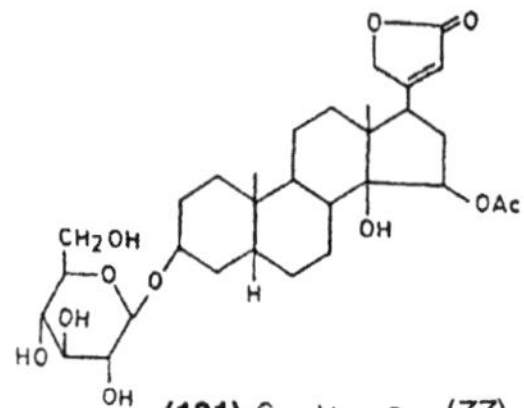

(131) $C_{31}H_{46}O_{11}$ (*77*)

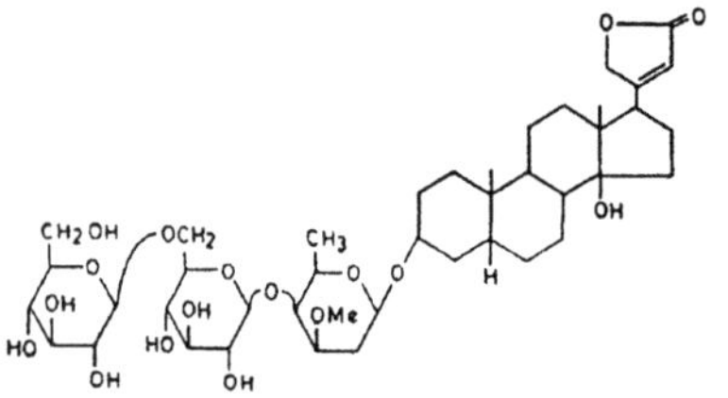

(132) $C_{42}H_{64}O_{17}$ (*77*)

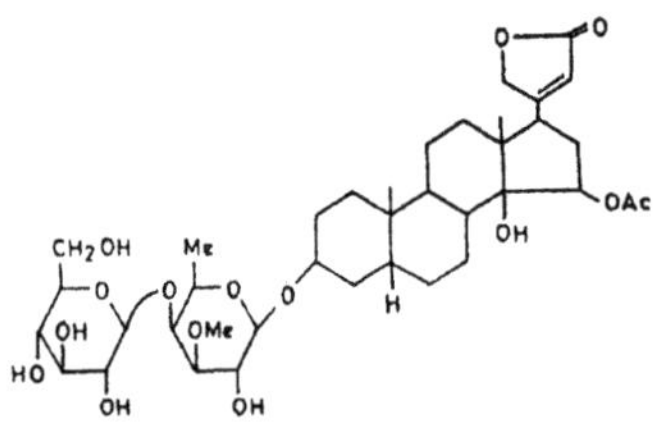

(133) $C_{38}H_{58}O_{15}$ (*77*)

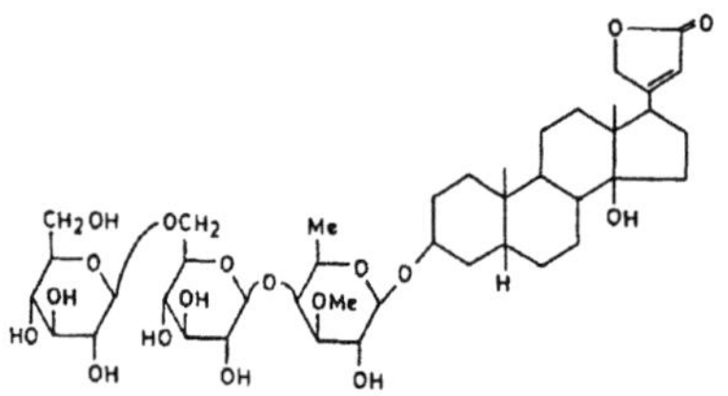

(134) $C_{42}H_{66}O_{18}$ (*77*)

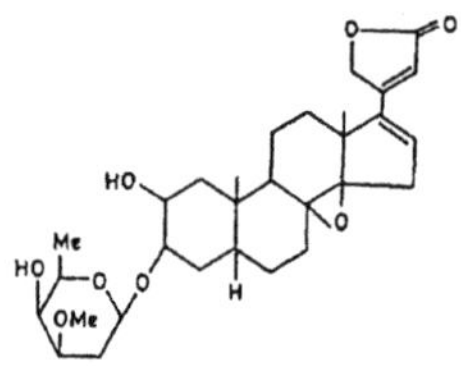

(135) $C_{30}H_{42}O_{8}$ (*78*)

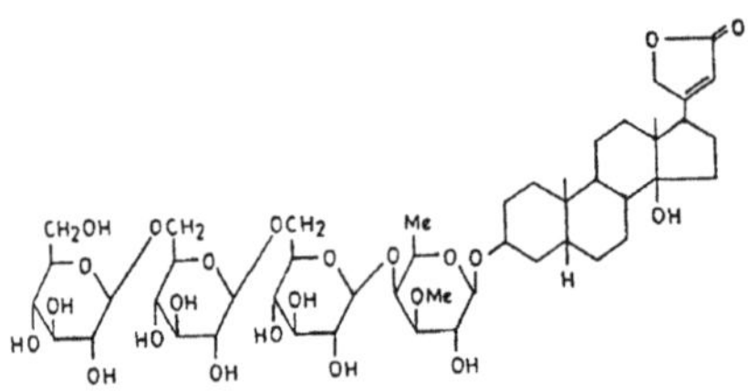

(136) $C_{48}H_{76}O_{23}$ (*79*)

Table 3 (*continued*)

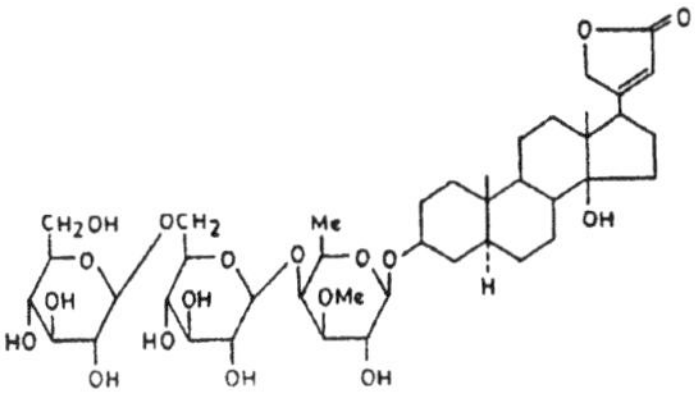

(137) $C_{42}H_{66}O_{18}$ *(79)*

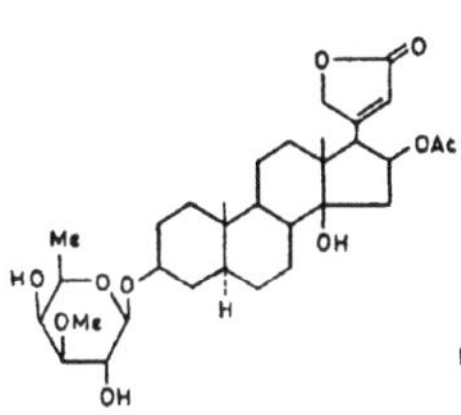

(138) $C_{32}H_{48}O_{10}$ *(79)*

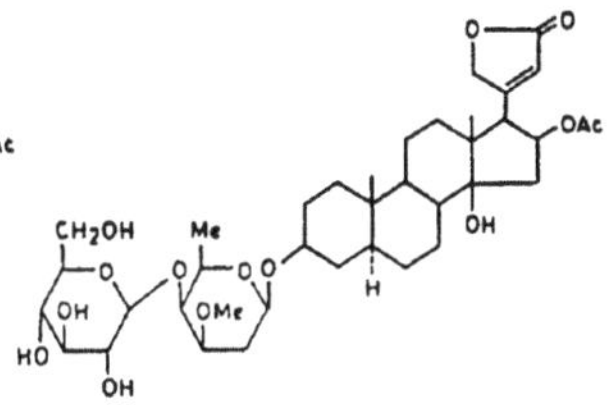

(139) $C_{38}H_{58}O_{14}$ *(79)*

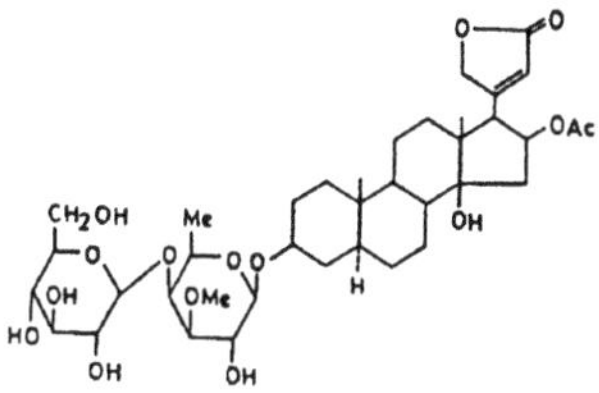

(140) $C_{38}H_{58}O_{15}$ *(79)*

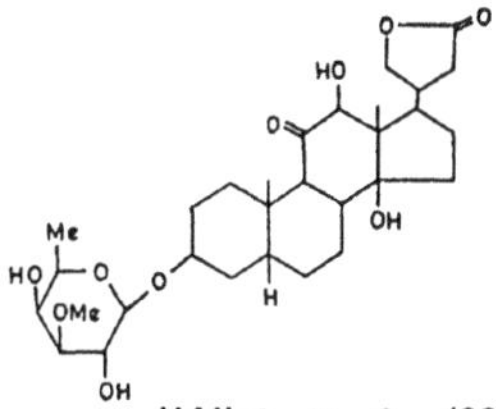

(141) $C_{30}H_{44}O_{10}$ *(80)*

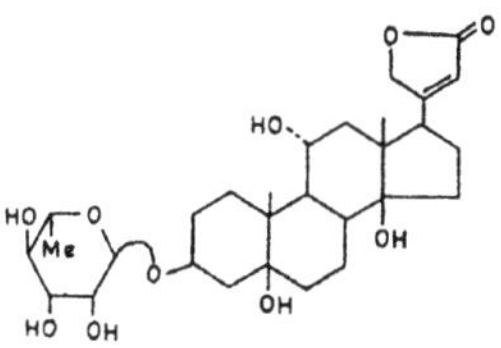

(142) $C_{29}H_{44}O_{10}$ *(80)*

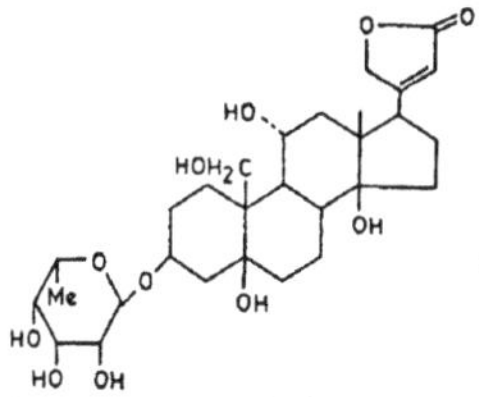

(143) $C_{29}H_{44}O_{11}$ *(80)*

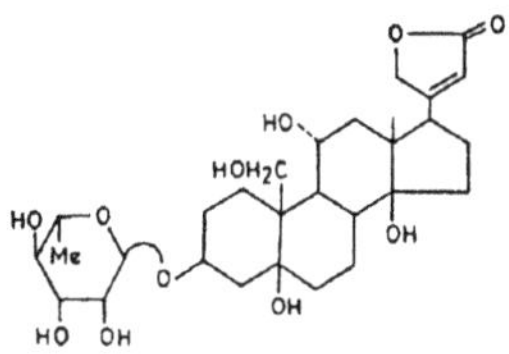

(144) $C_{29}H_{44}O_{11}$ *(80)*

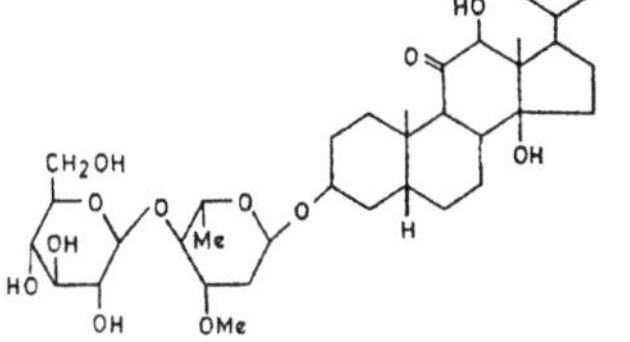

(145) $C_{36}H_{54}O_{14}$ *(80)*

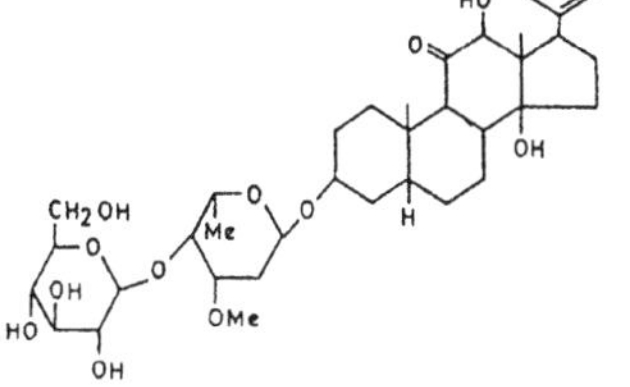

(146) $C_{36}H_{54}O_{14}$ *(80)*

Table 3 (*continued*)

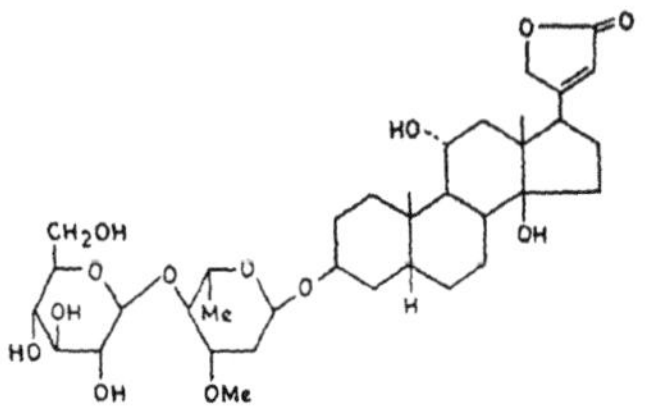

(147) $C_{36}H_{56}O_{13}$ *(80)*

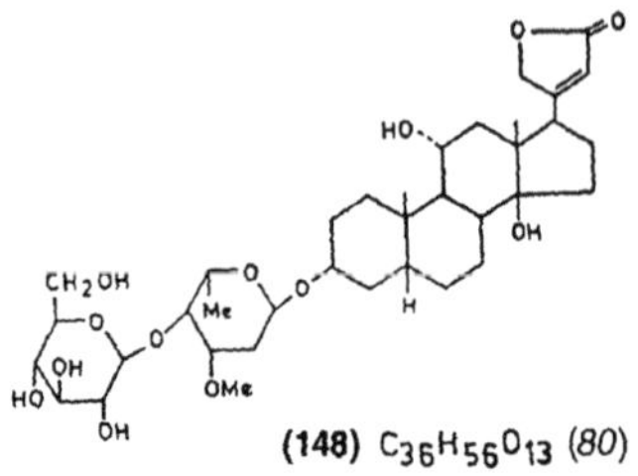

(148) $C_{36}H_{56}O_{13}$ *(80)*

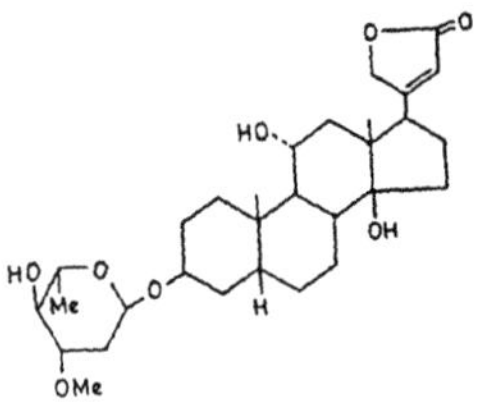

(149) $C_{30}H_{46}O_{8}$ *(80)*

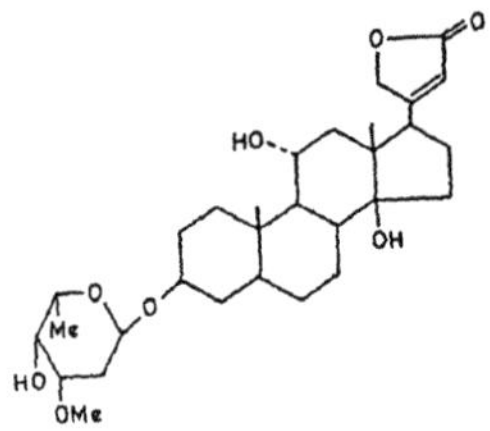

(150) $C_{30}H_{46}O_{8}$ *(80)*

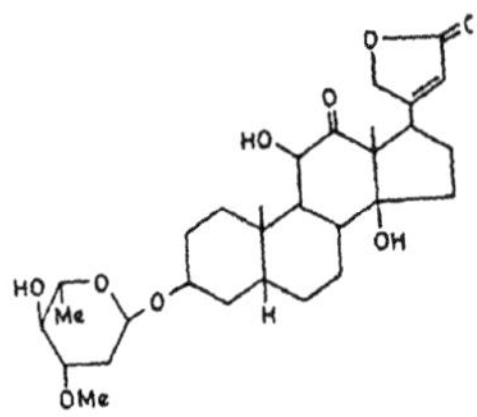

(151) $C_{30}H_{44}O_{9}$ *(80)*

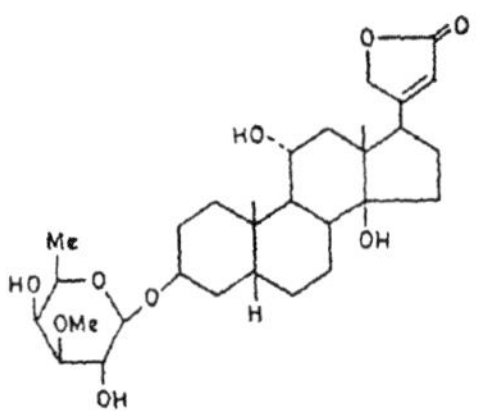

(152) $C_{30}H_{46}O_{9}$ *(80)*

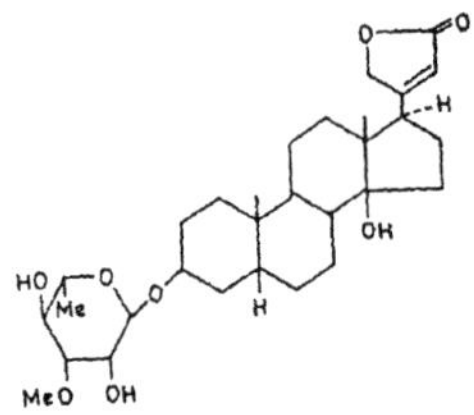

(153) $C_{30}H_{46}O_{8}$ *(81)*

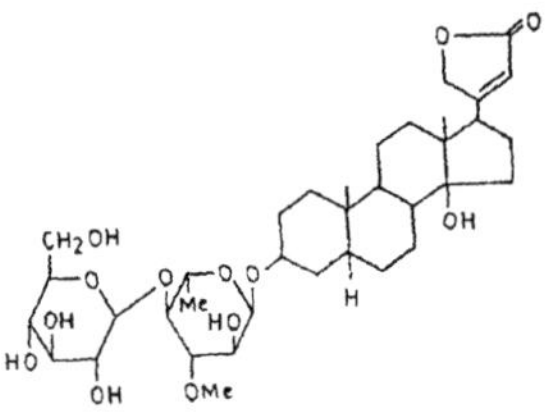

(154) $C_{36}H_{56}O_{13}$ *(82)*

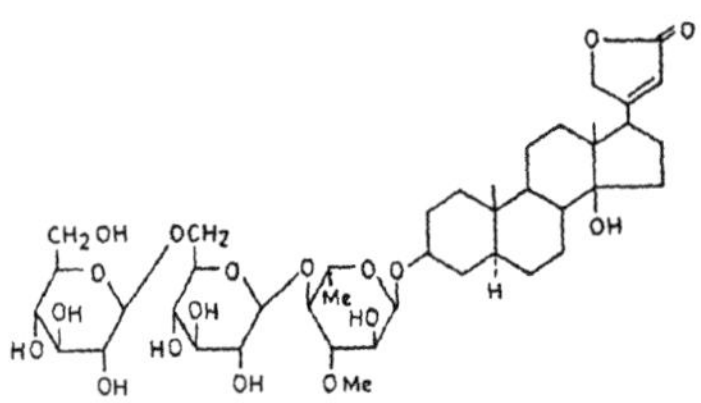

(155) $C_{42}H_{66}O_{18}$ *(82)*

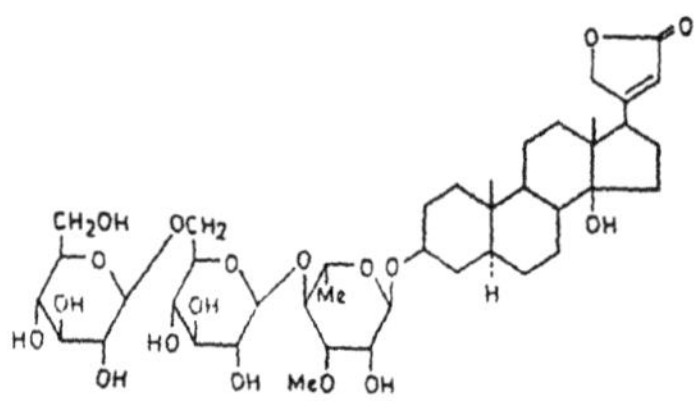

(156) $C_{42}H_{66}O_{18}$ *(82)*

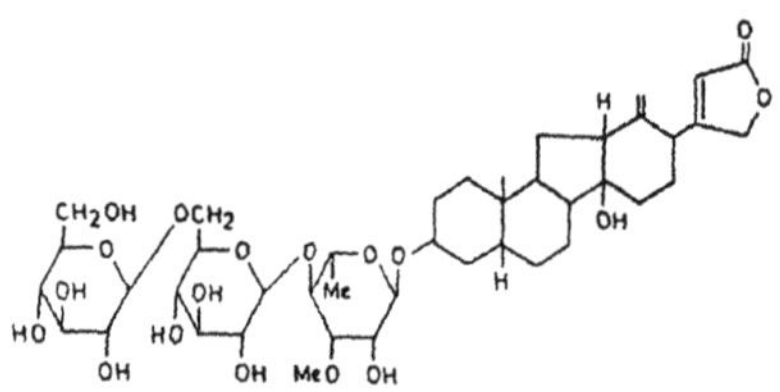

(157) $C_{42}H_{64}O_{18}$ *(82)*

Table 3 (*continued*)

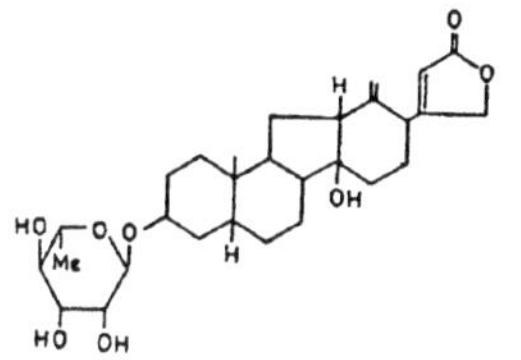

(158) $C_{29}H_{42}O_{8}$ (*82*)

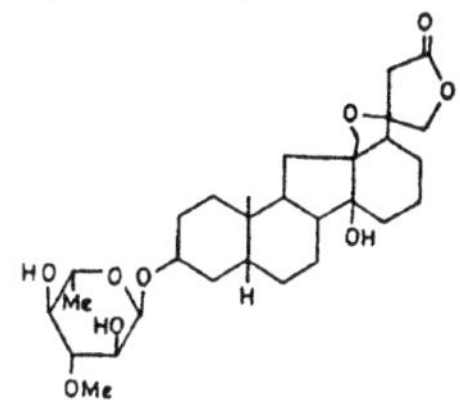

(159) $C_{30}H_{46}O_{9}$ (*82*)

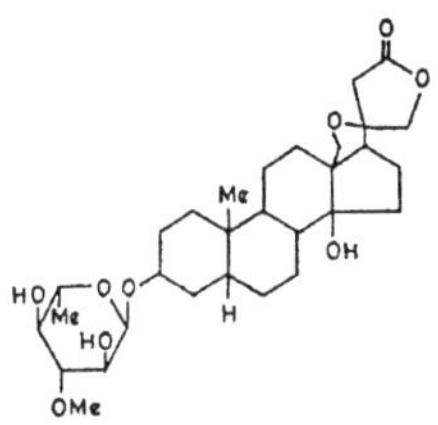

(160) $C_{30}H_{46}O_{9}$ (*82*)

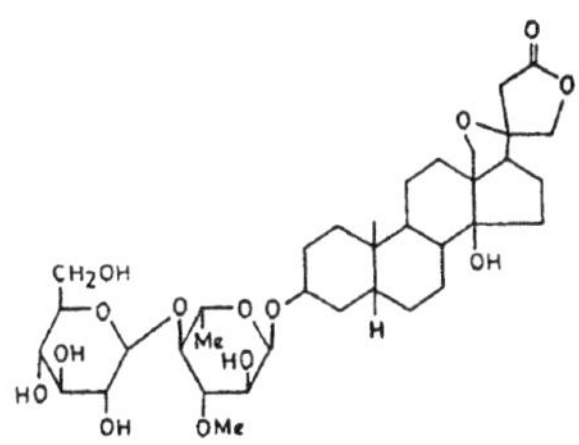

(161) $C_{36}H_{56}O_{14}$ (*82*)

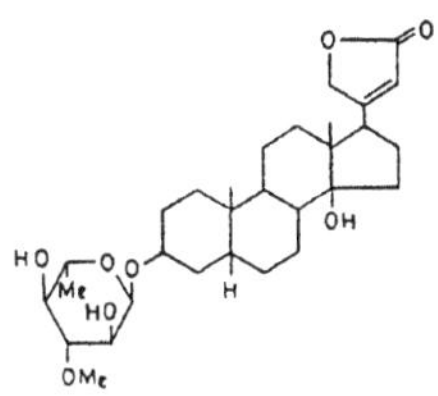

(162) $C_{30}H_{46}O_{8}$ (*10*)

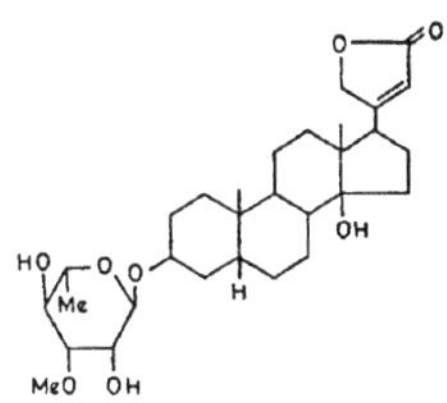

(163) $C_{30}H_{46}O_{8}$ (*10*)

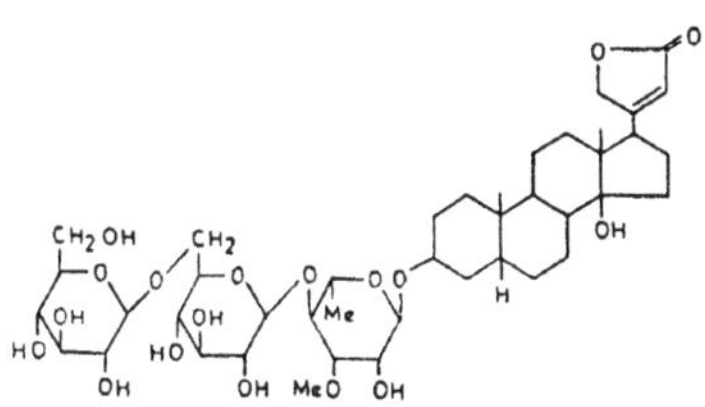

(164) $C_{42}H_{66}O_{18}$ (*10*)

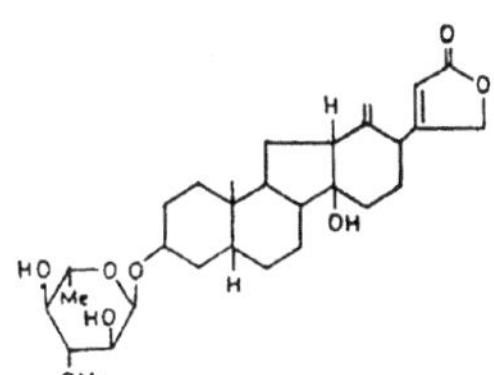

(165) $C_{30}H_{44}O_{8}$ (*10*)

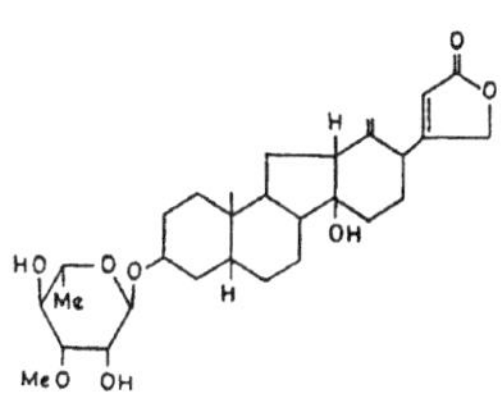

(166) $C_{30}H_{44}O_{8}$ (*10*)

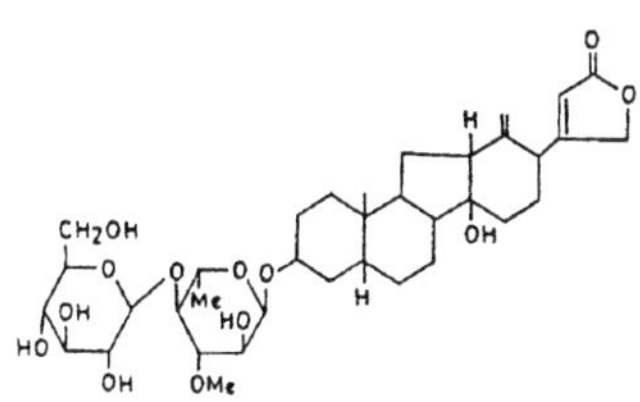

(167) $C_{36}H_{54}O_{13}$ (*10*)

Table 3 (*continued*)

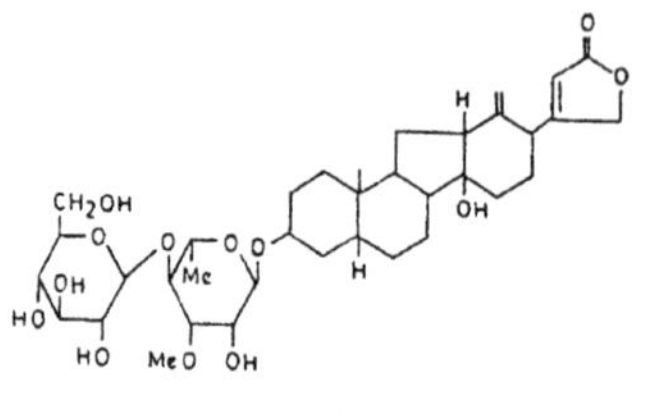

(168) $C_{36}H_{54}O_{13}$ (*10*)

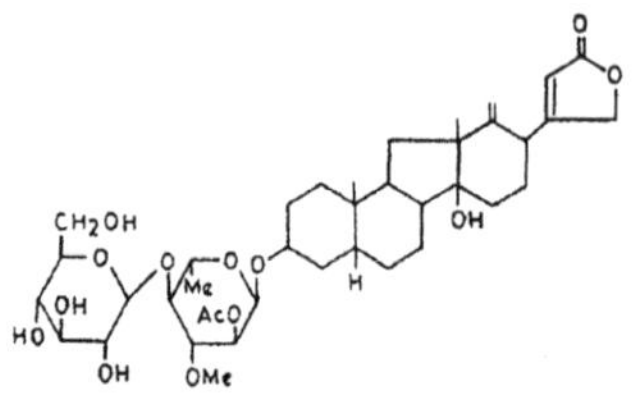

(169) $C_{38}H_{56}O_{14}$ (*10*)

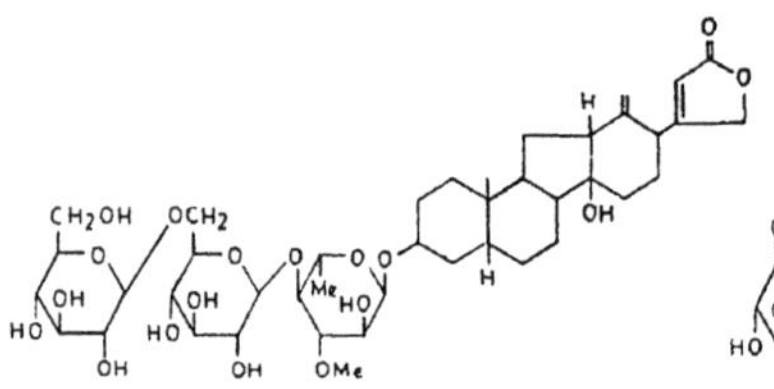

(170) $C_{42}H_{64}O_{18}$ (*10*)

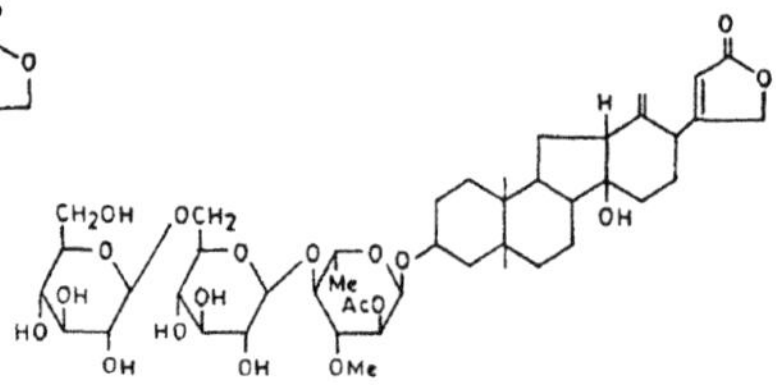

(171) $C_{44}H_{66}O_{19}$ (*10*)

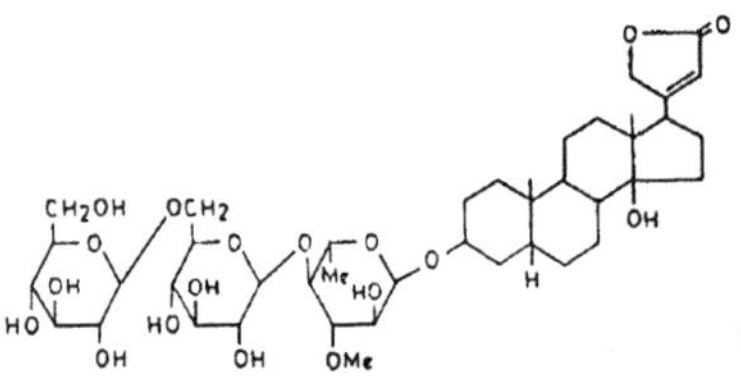

(172) $C_{42}H_{66}O_{18}$ (*10*)

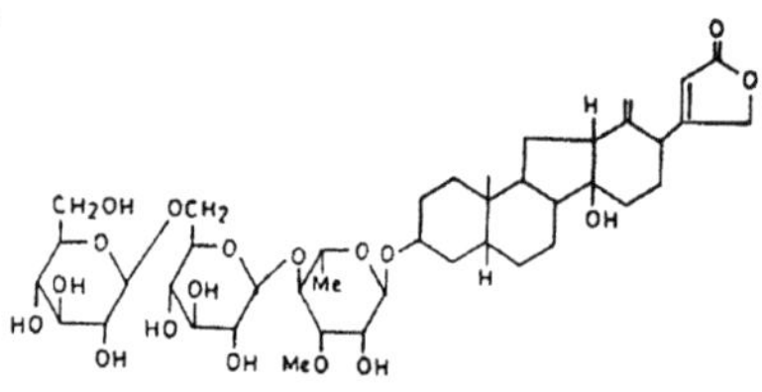

(173) $C_{42}H_{64}O_{18}$ (*10*)

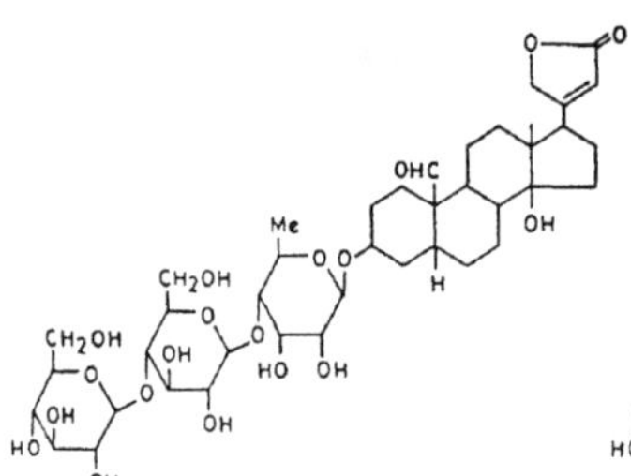

(174) $C_{41}H_{62}O_{19}$ (*83*)

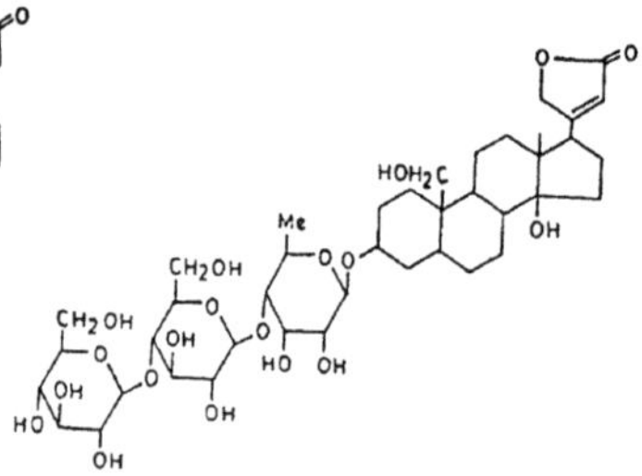

(175) $C_{41}H_{64}O_{19}$ (*83*)

References, pp. 148–155

Table 3 (*continued*)

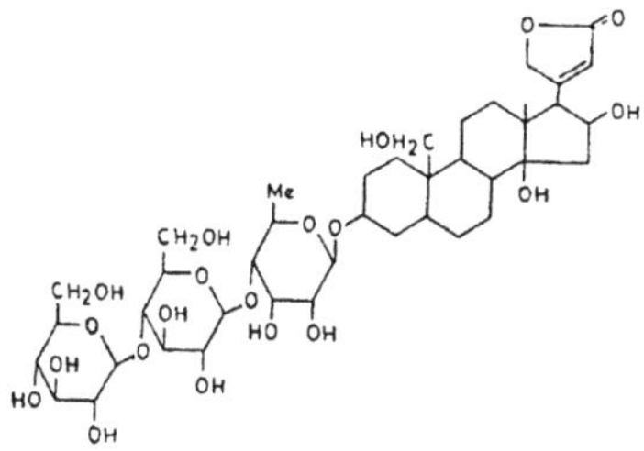

(176) $C_{41}H_{64}O_{20}$ (*83*)

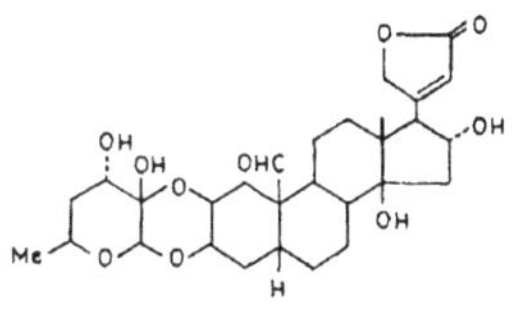

(177) $C_{29}H_{40}O_{10}$ (*83*)

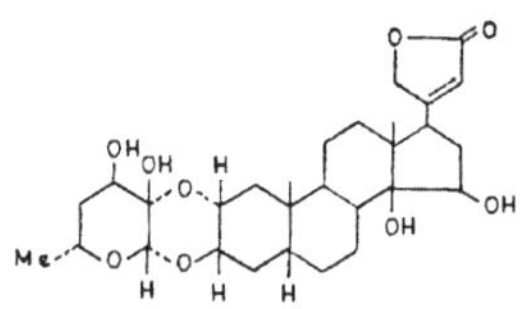

(178) $C_{29}H_{42}O_{9}$ (*84*)

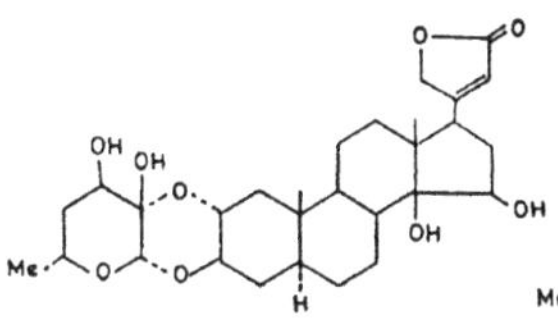

(179) $C_{29}H_{42}O_{9}$ (*85*)

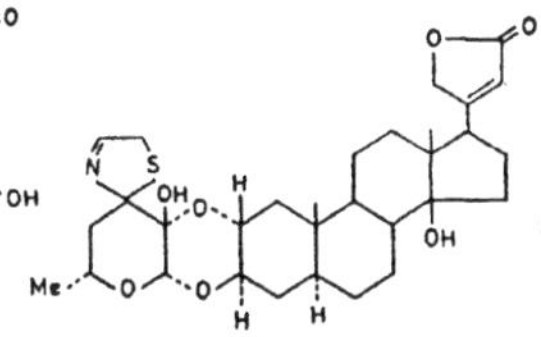

(180) $C_{31}H_{43}O_{6}NS$ (*86*)

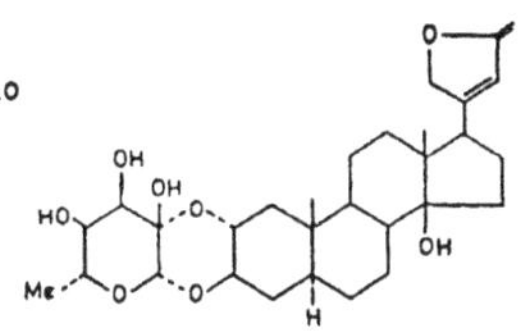

(181) $C_{29}H_{42}O_{9}$ (*86*)

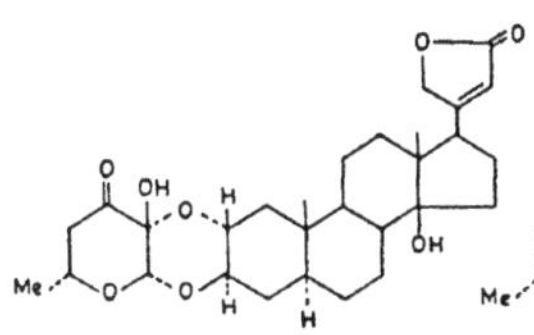

(182) $C_{29}H_{40}O_{8}$ (*86*)

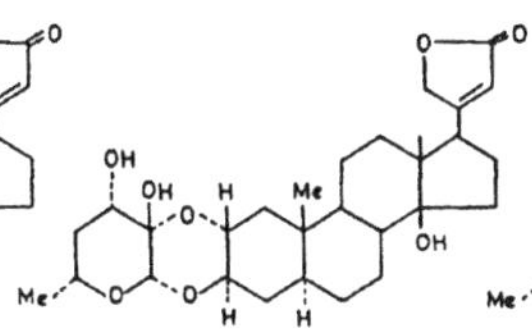

(183) $C_{29}H_{42}O_{8}$ (*86*)

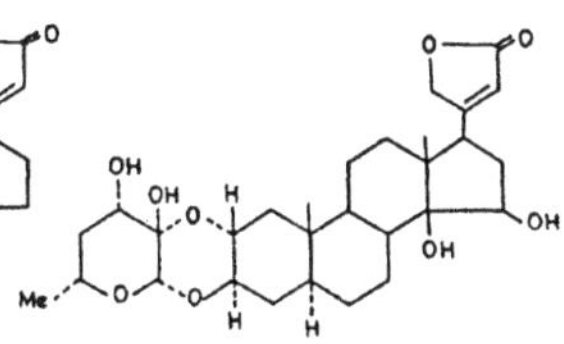

(184) $C_{29}H_{42}O_{9}$ (*86*)

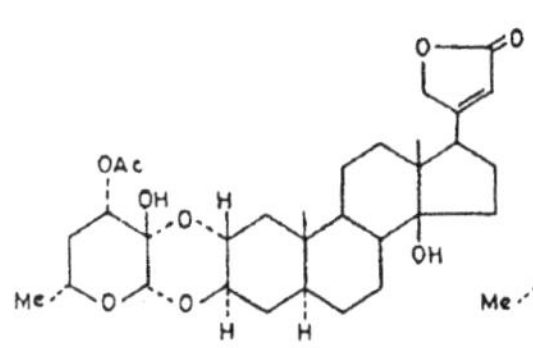

(185) $C_{31}H_{44}O_{9}$ (*86*)

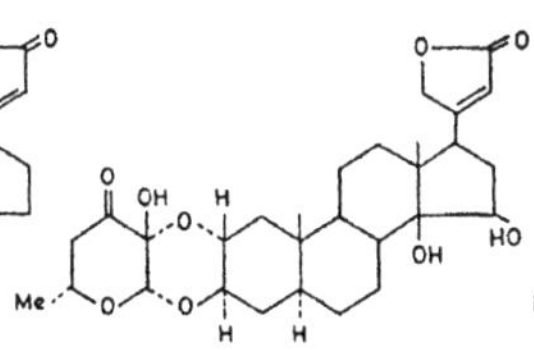

(186) $C_{29}H_{40}O_{9}$ (*86*)

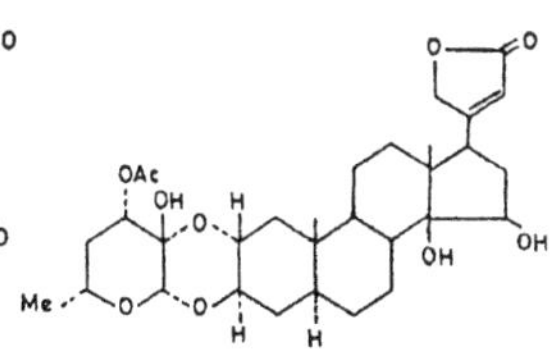

(187) $C_{31}H_{44}O_{10}$ (*86*)

Table 3 (*continued*)

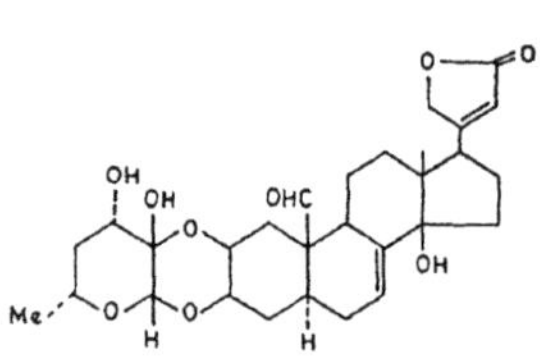

(188) $C_{29}H_{38}O_9$ *(87)*

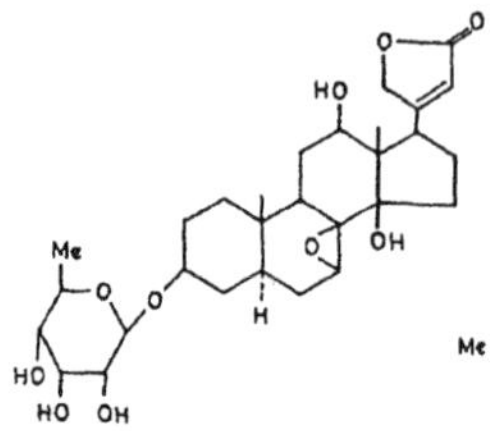

(189) $C_{29}H_{42}O_{10}$ *(88)*

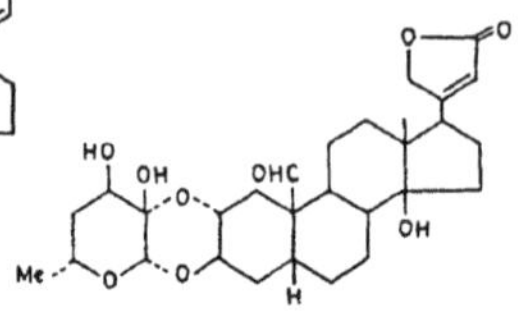

(190) $C_{29}H_{40}O_9$ *(89)*

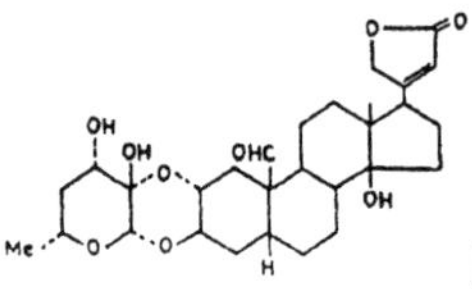

(191) $C_{29}H_{40}O_9$ *(89)*

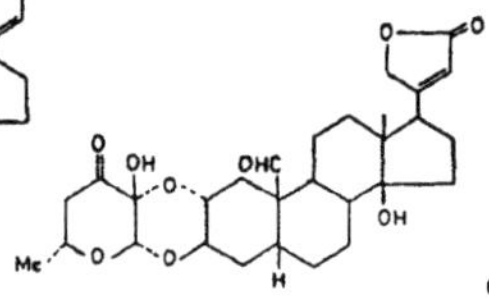

(192) $C_{29}H_{38}O_9$ *(89)*

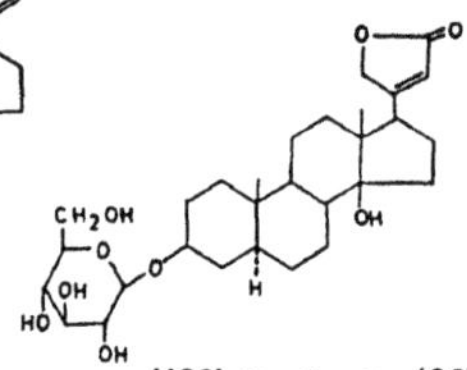

(193) $C_{29}H_{44}O_9$ *(89)*

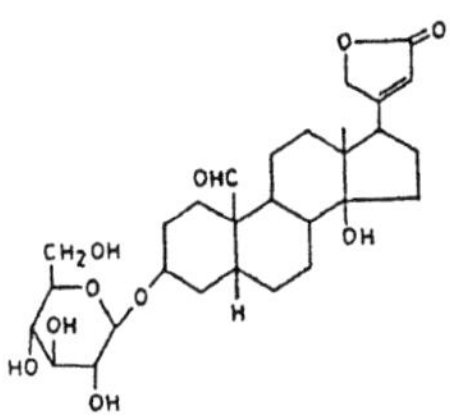

(194) $C_{29}H_{42}O_{10}$ *(89)*

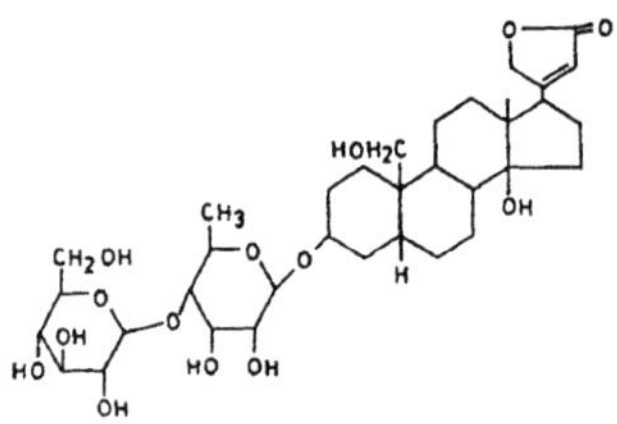

(195) $C_{35}H_{54}O_{14}$ *(89)*

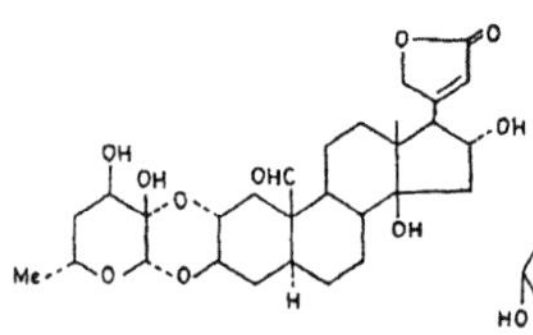

(196) $C_{29}H_{40}O_{10}$ *(89)*

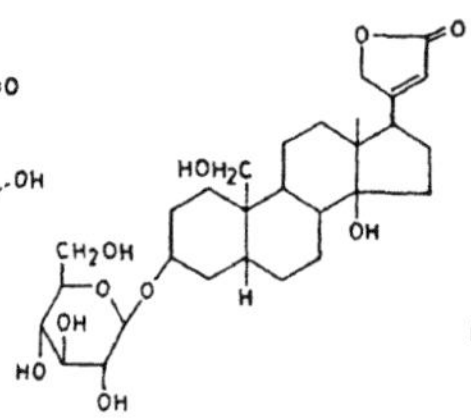

(197) $C_{29}H_{44}O_{10}$ *(89)*

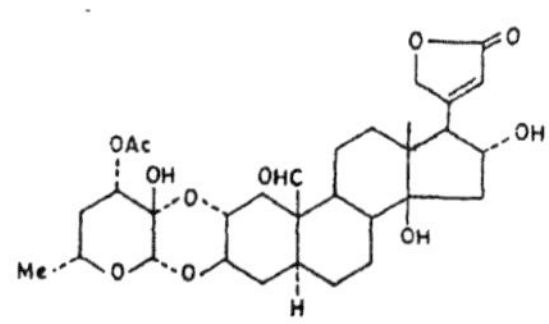

(198) $C_{31}H_{42}O_{11}$ *(90)*

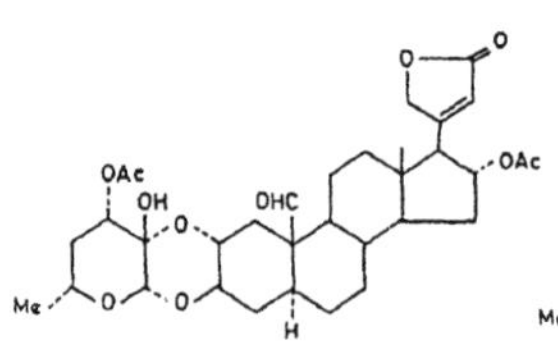

(199) $C_{33}H_{44}O_{11}$ *(90)*

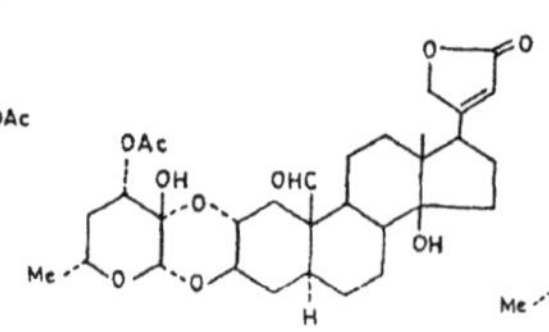

(200) $C_{31}H_{42}O_{10}$ *(90)*

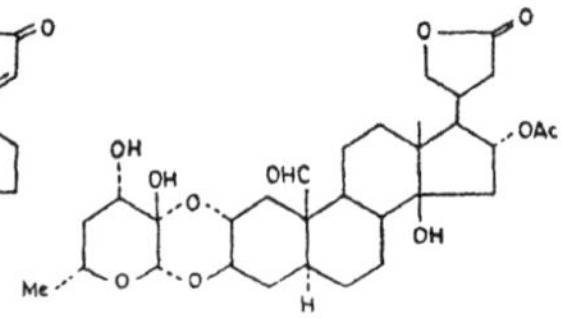

(201) $C_{31}H_{42}O_{11}$ *(90)*

Table 3 (*continued*)

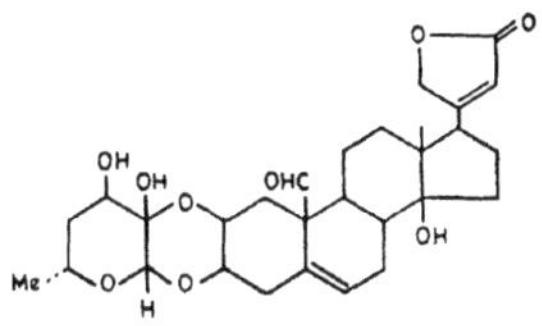

(202) $C_{29}H_{38}O_9$ *(91)*

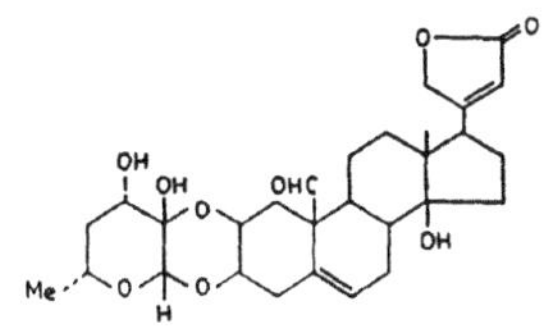

(203) $C_{29}H_{38}O_9$ *(91)*

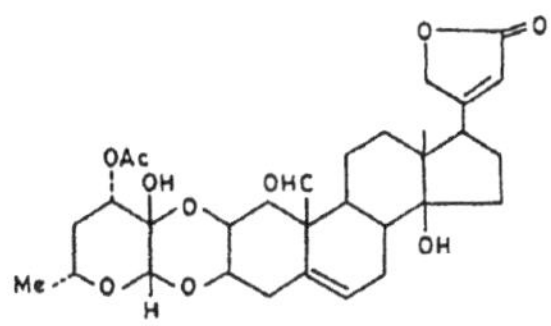

(204) $C_{31}H_{40}O_{10}$ *(91)*

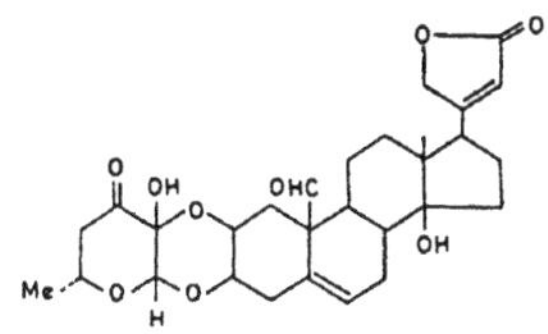

(205) $C_{29}H_{36}O_9$ *(91)*

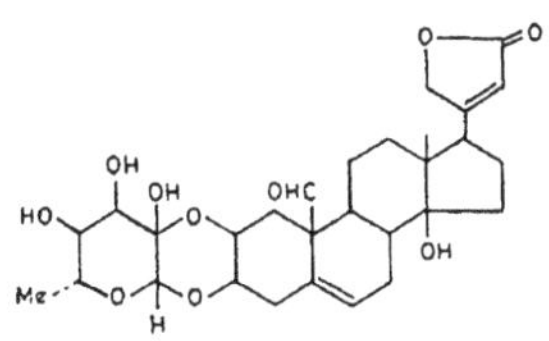

(206) $C_{29}H_{38}O_{10}$ *(91)*

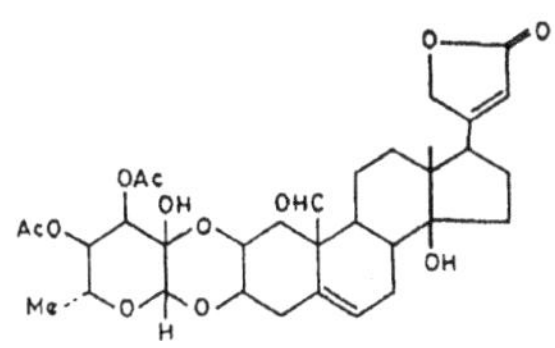

(207) $C_{33}H_{42}O_{12}$ *(91)*

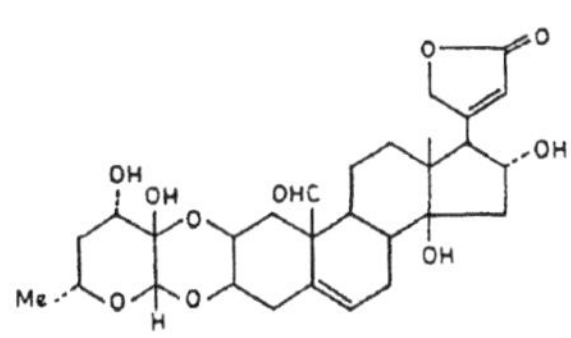

(208) $C_{29}H_{38}O_{10}$ *(91)*

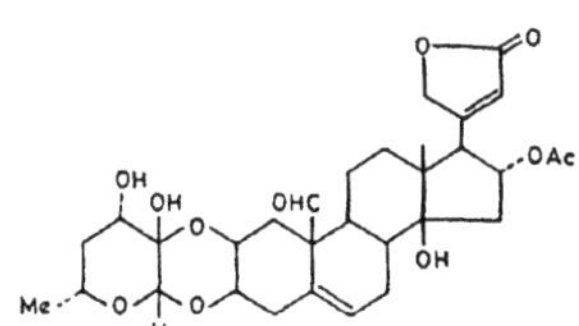

(209) $C_{31}H_{40}O_{11}$ *(91)*

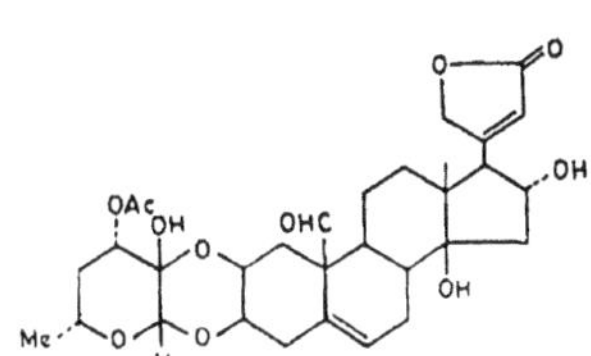

(210) $C_{31}H_{40}O_{11}$ *(91)*

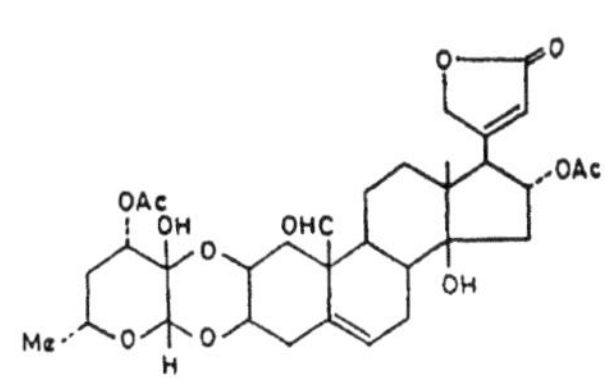

(211) $C_{33}H_{42}O_{12}$ *(91)*

Table 3 (*continued*)

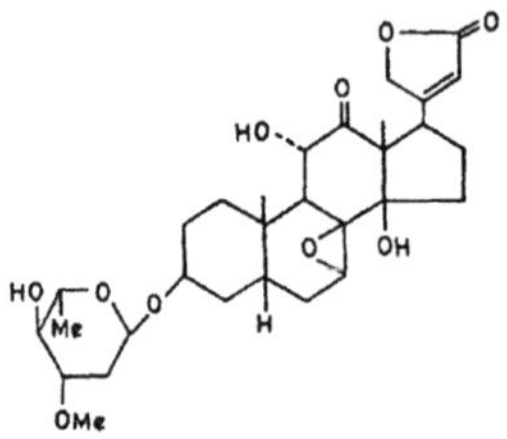

(212) $C_{30}H_{42}O_{10}$ *(92)*

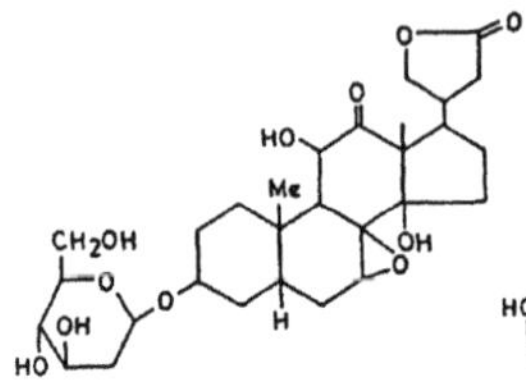

(213) $C_{29}H_{40}O_{11}$ *(93)*

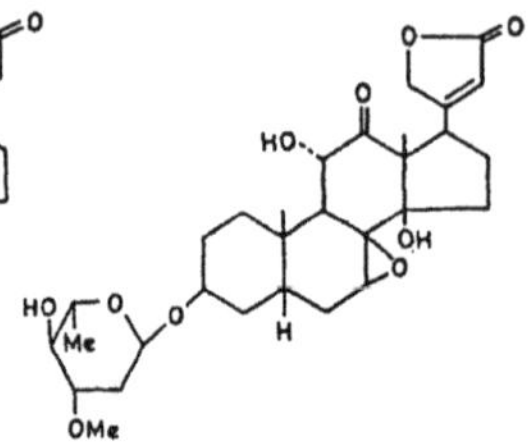

(214) $C_{30}H_{42}O_{10}$ *(94)*

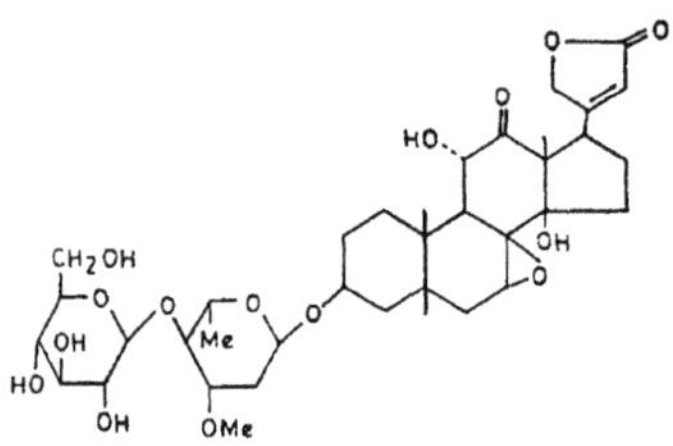

(215) $C_{36}H_{52}O_{15}$ *(94)*

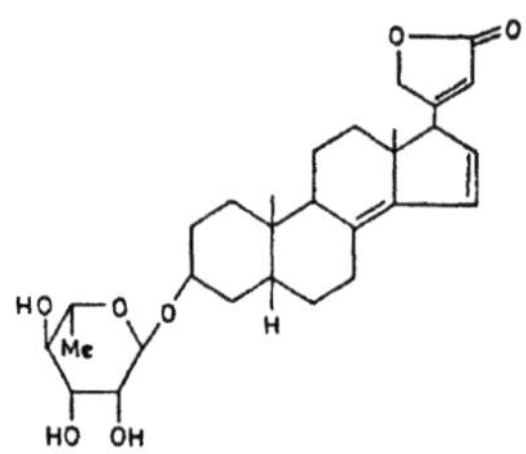

(216) $C_{29}H_{40}O_{7}$ *(95)*

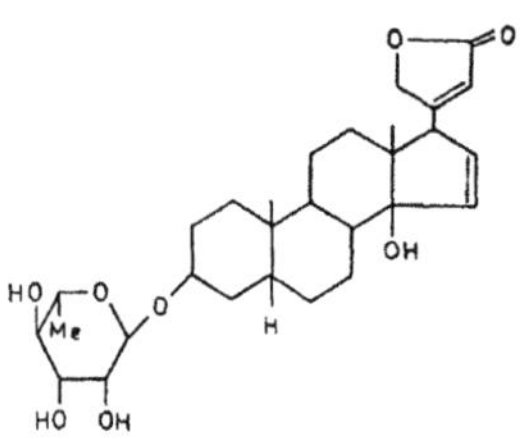

(217) $C_{29}H_{42}O_{8}$ *(95)*

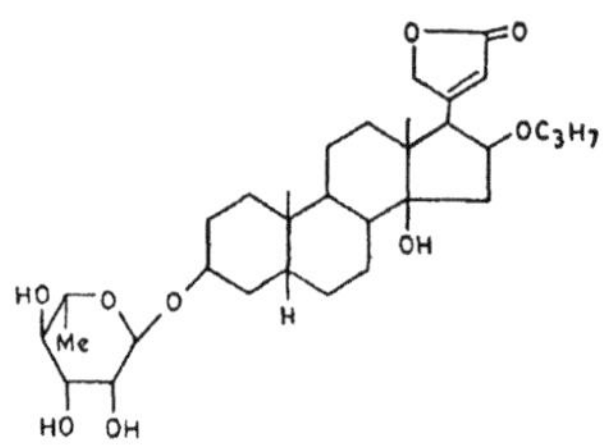

(218) $C_{32}H_{50}O_{9}$ *(95)*

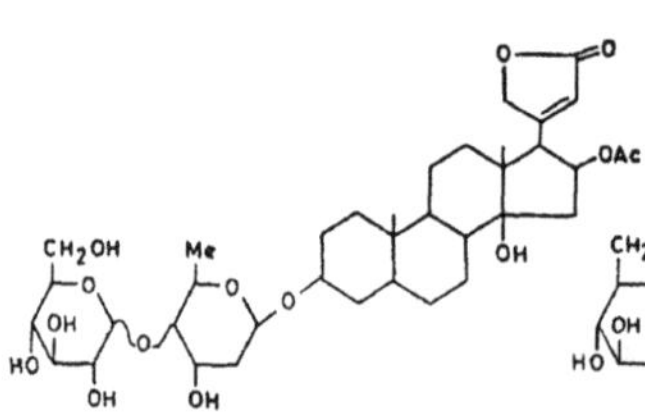

(219) $C_{37}H_{56}O_{14}$ *(96)*

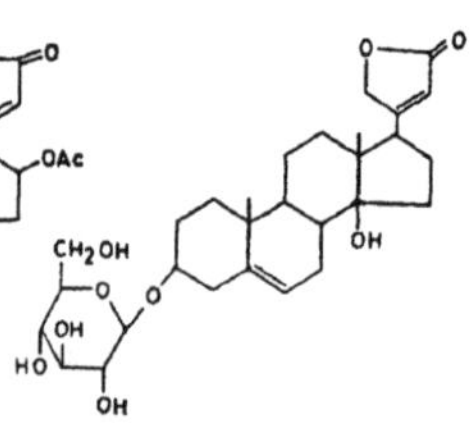

(220) $C_{29}H_{42}O_{9}$ *(97)*

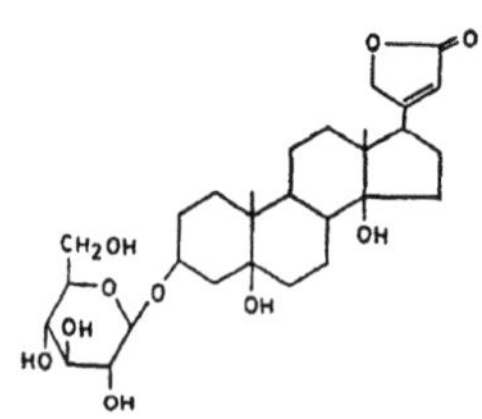

(221) $C_{29}H_{44}O_{10}$ *(97)*

References, pp. 148–155

Table 3 (*continued*)

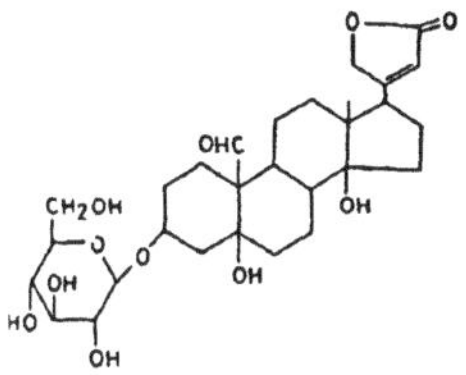

(222) $C_{29}H_{42}O_{11}$ (*97*)

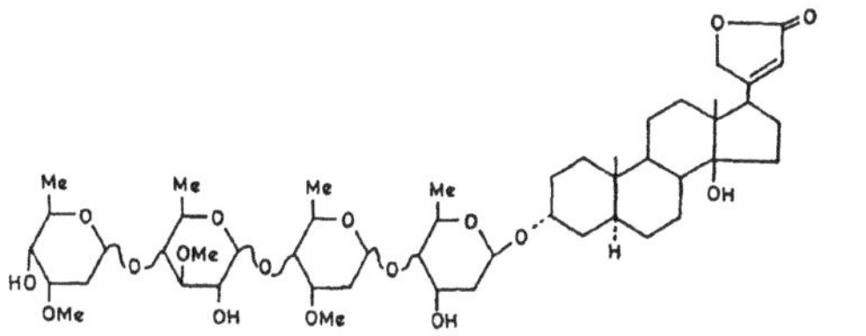

(223) $C_{50}H_{80}O_{17}$ (*22*)

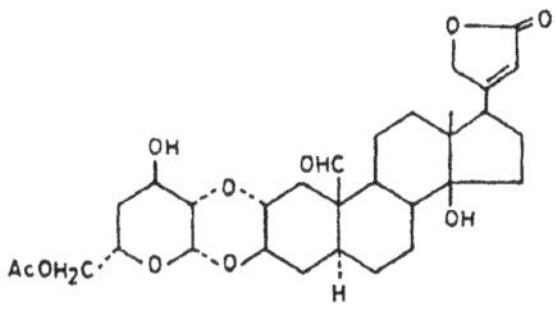

(224) $C_{31}H_{42}O_{10}$ (*98*)

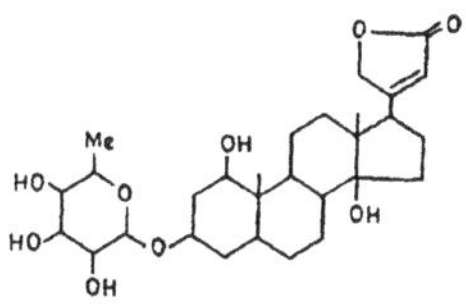

(225) $C_{29}H_{44}O_{9}$ (*99*)

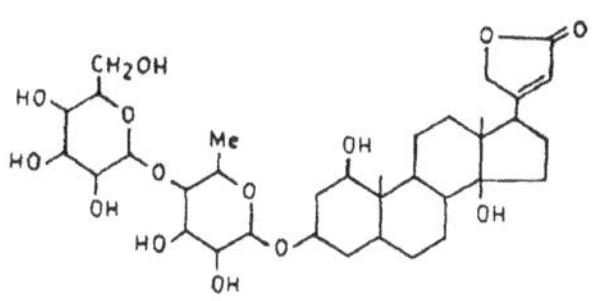

(226) $C_{35}H_{54}O_{14}$ (*99*)

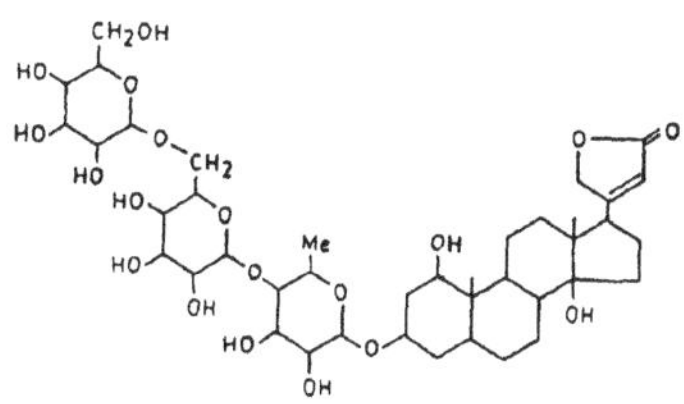

(227) $C_{41}H_{64}O_{19}$ (*99*)

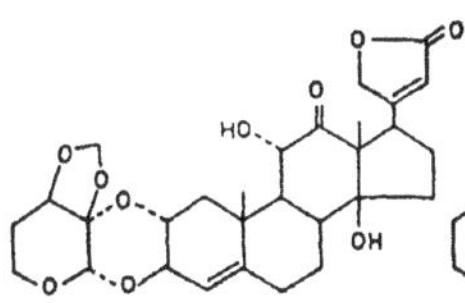

(228) $C_{29}H_{36}O_{10}$ (*100*)

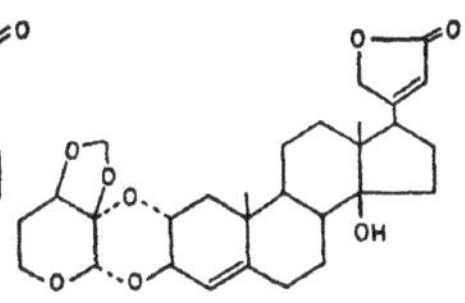

(229) $C_{29}H_{38}O_{8}$ (*100*)

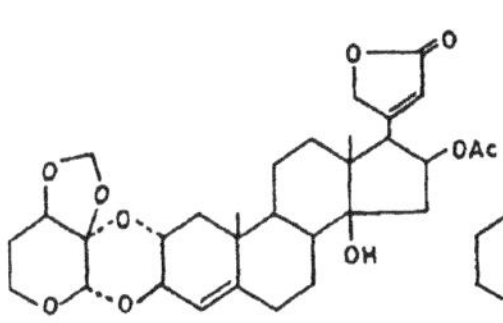

(230) $C_{31}H_{40}O_{10}$ (*100*)

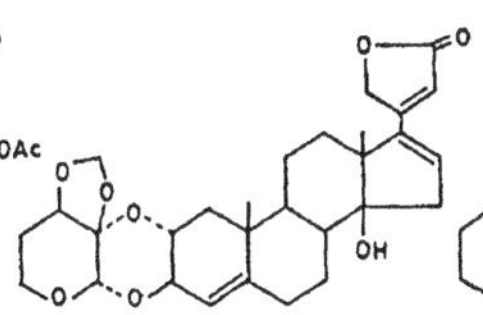

(231) $C_{29}H_{38}O_{8}$ (*100*)

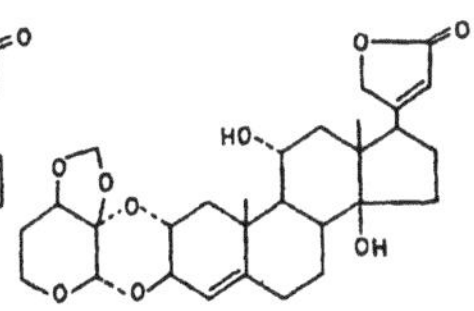

(232) $C_{29}H_{36}O_{9}$ (*100*)

Table 3 (*continued*)

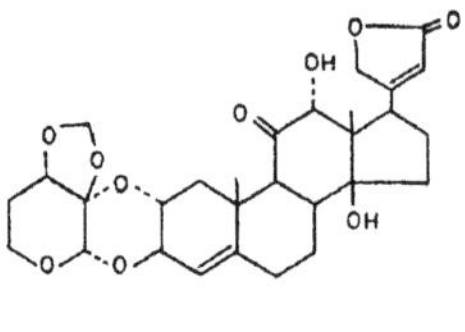

(233) $C_{29}H_{36}O_{10}$ (*100*)

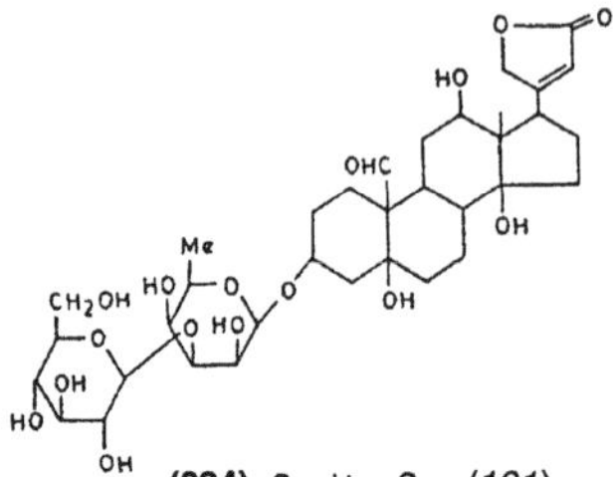

(234) $C_{35}H_{52}O_{16}$ (*101*)

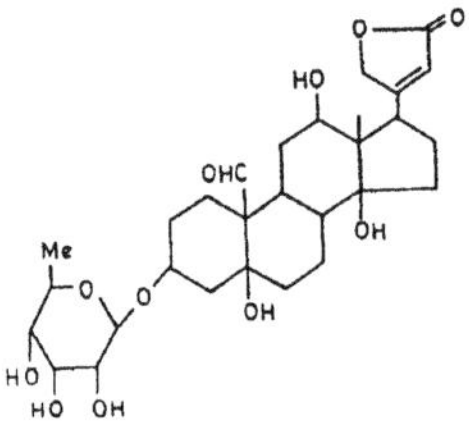

(235) $C_{29}H_{42}O_{11}$ (*101*)

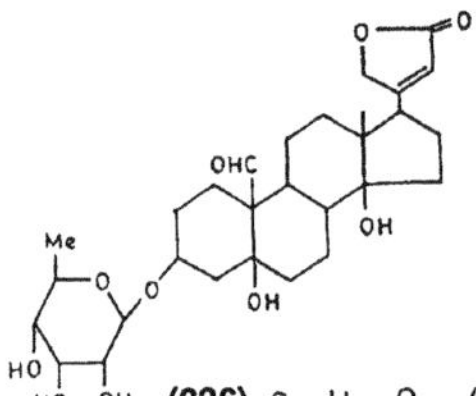

(236) $C_{29}H_{42}O_{10}$ (*101*)

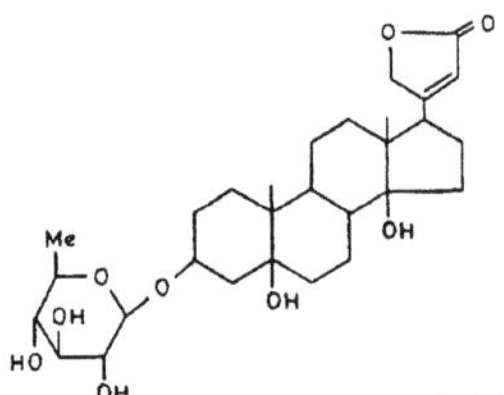

(237) $C_{29}H_{4}O_{9}$ (*101*)

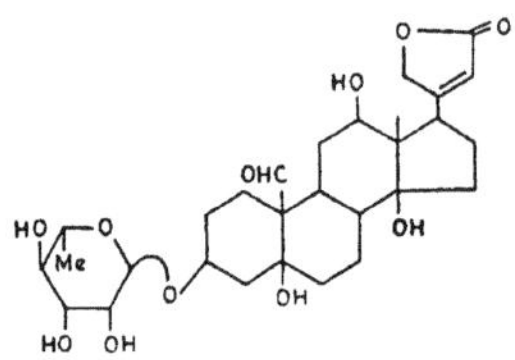

(238) $C_{29}H_{42}O_{11}$ (*101*)

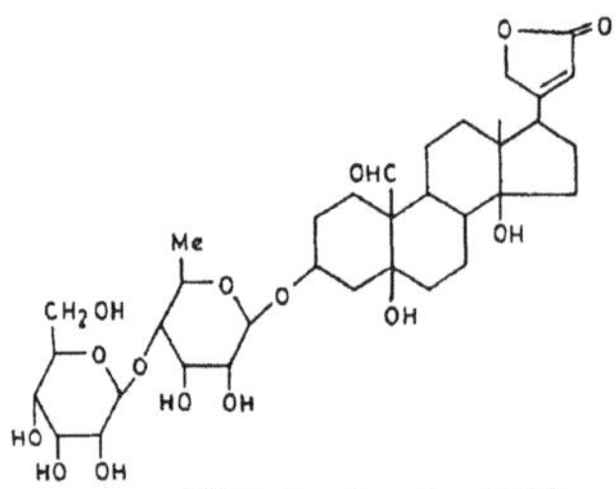

(239) $C_{35}H_{52}O_{15}$ (*101*)

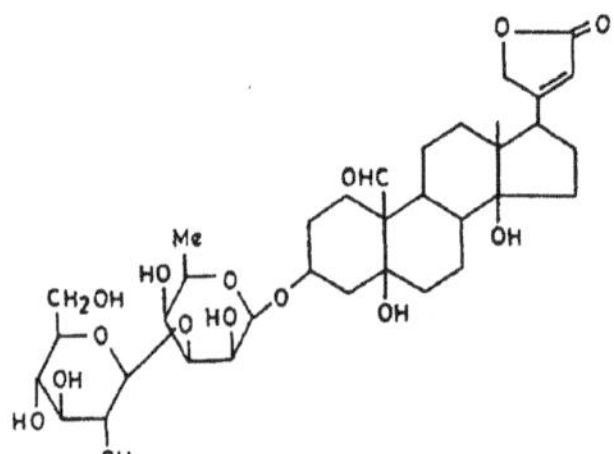

(240) $C_{35}H_{52}O_{15}$ (*101*)

References, pp. 148–155

Table 3 (*continued*)

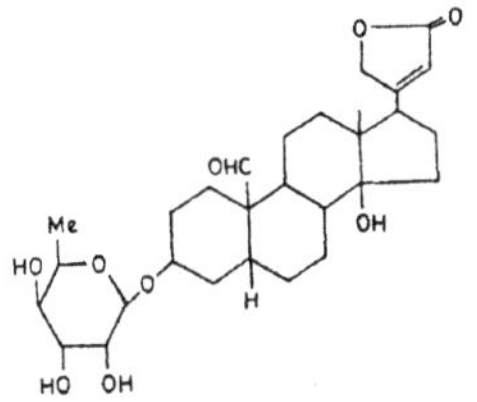

(241) $C_{29}H_{42}O_9$ (*102*)

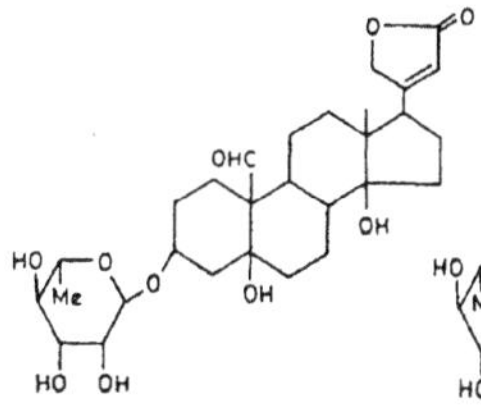

(242) $C_{29}H_{42}O_{10}$ (*103*)

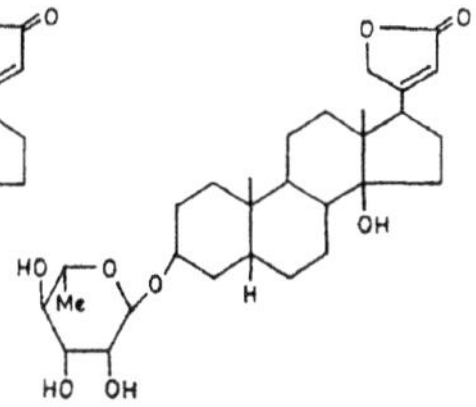

(243) $C_{29}H_{44}O_8$ (*103*)

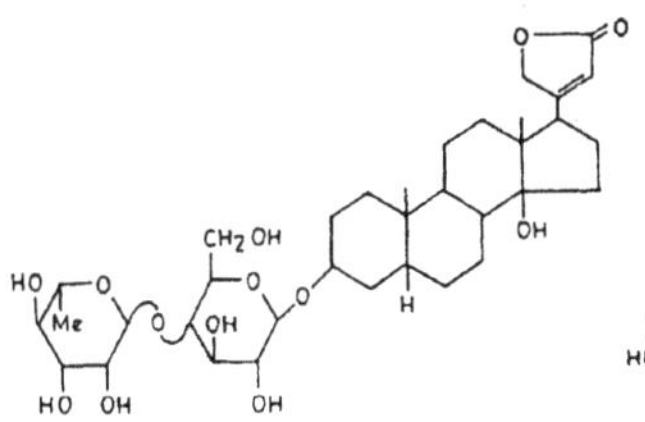

(244) $C_{35}H_{54}O_{13}$ (*103*)

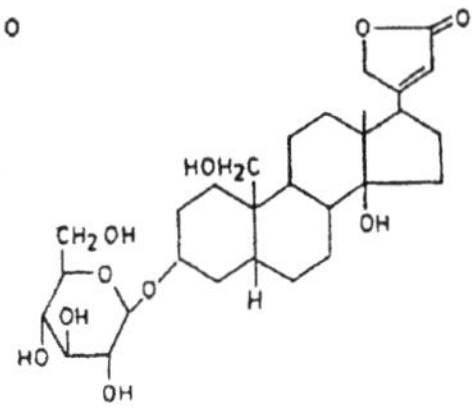

(245) $C_{29}H_{44}O_{10}$ (*103*)

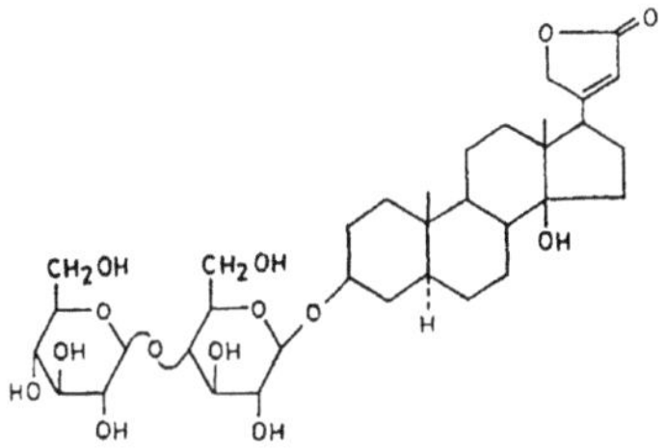

(246) $C_{35}H_{54}O_{14}$ (*104*)

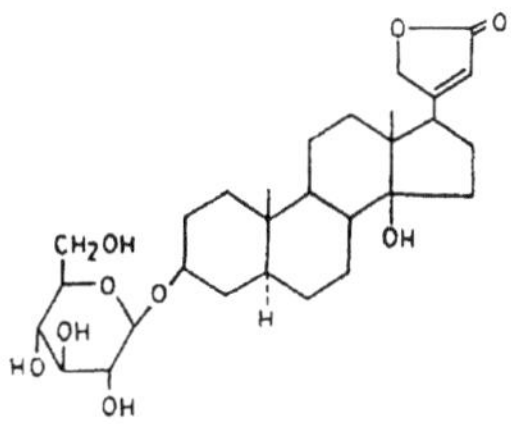

(247) $C_{29}H_{44}O_9$ (*104*)

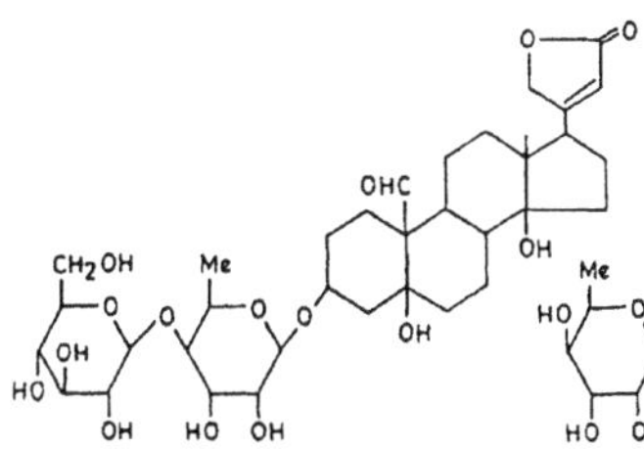

(248) $C_{35}H_{52}O_{15}$ (*105*)

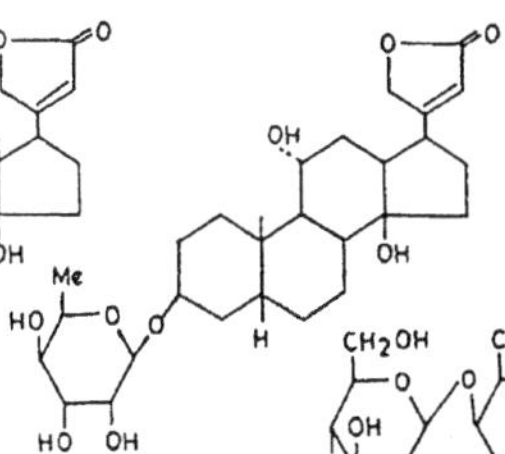

(249) $C_{29}H_{44}O_9$ (*105*)

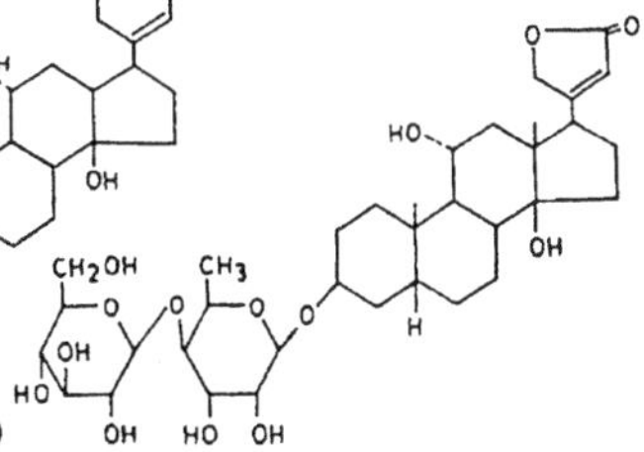

(250) $C_{35}H_{54}O_{14}$ (*105*)

Table 3 (*continued*)

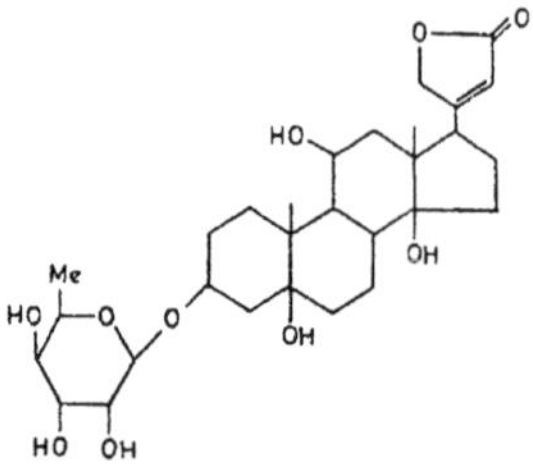

(251) $C_{29}H_{44}O_{10}$ (*106*)

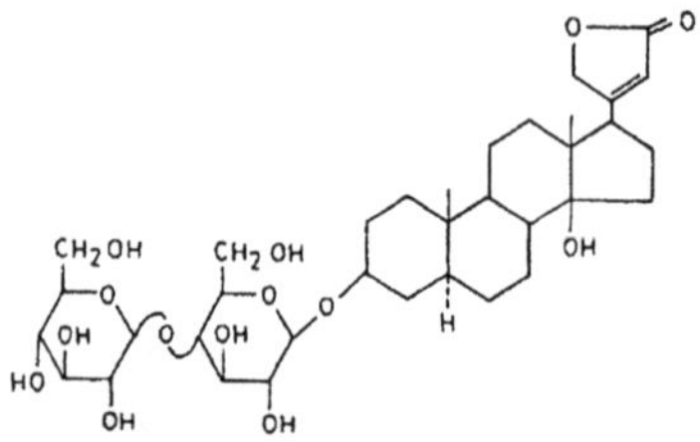

(252) $C_{35}H_{54}O_{14}$ (*106*)

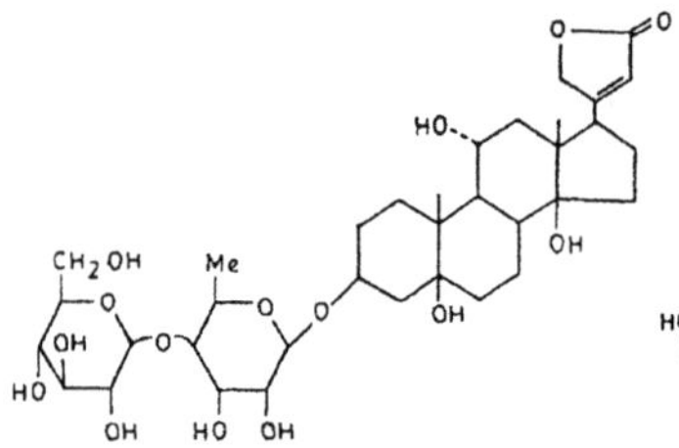

(253) $C_{35}H_{54}O_{15}$ (*106*)

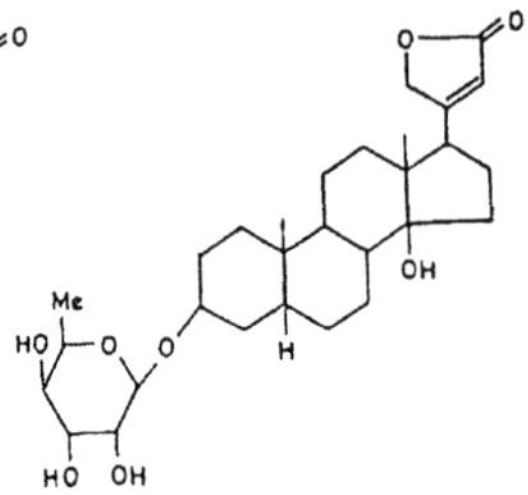

(254) $C_{29}H_{44}O_{8}$ (*107*)

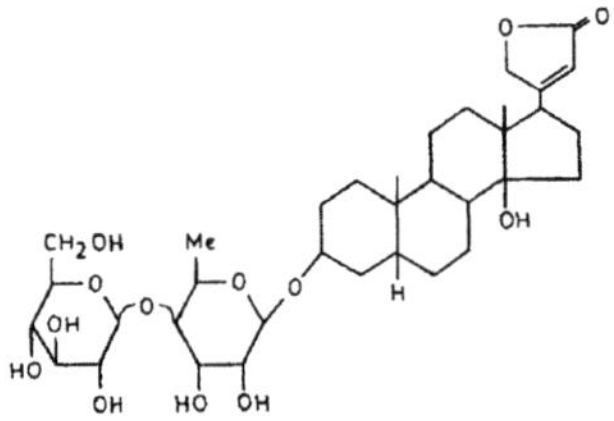

(255) $C_{35}H_{54}O_{13}$ (*107*)

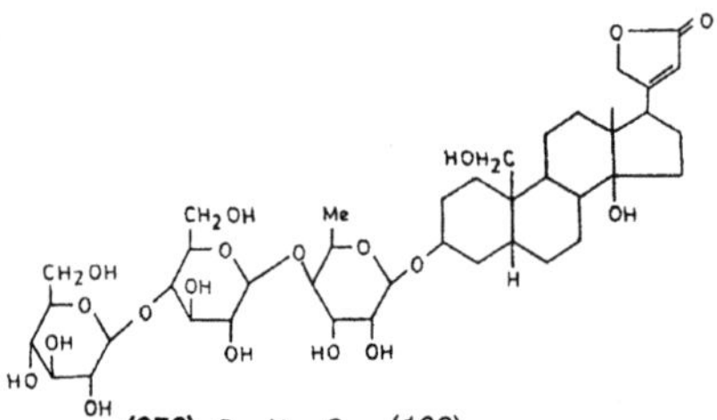

(256) $C_{41}H_{64}O_{19}$ (*108*)

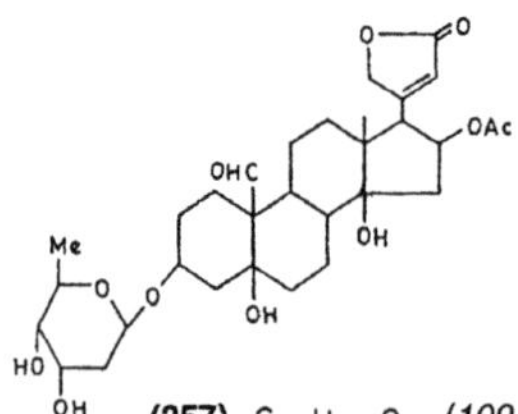

(257) $C_{31}H_{44}O_{11}$ (*109,110*)

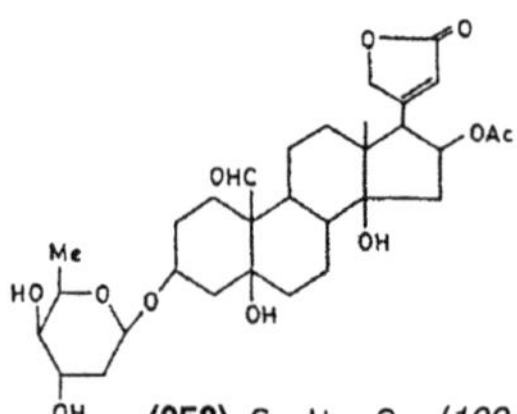

(258) $C_{31}H_{44}O_{11}$ (*109,110*)

References, pp. 148–155

Table 3 (*continued*)

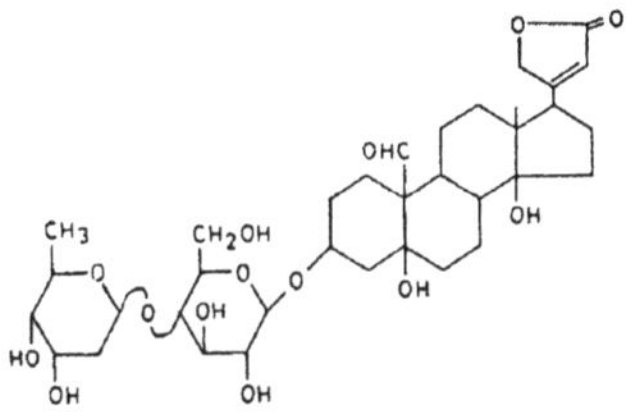

(259) $C_{35}H_{52}O_{14}$ (*111*)

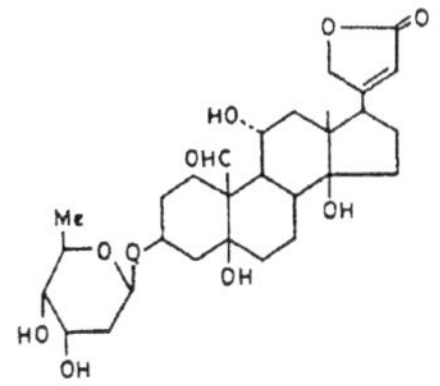

(260) $C_{29}H_{42}O_{10}$ (*112*)

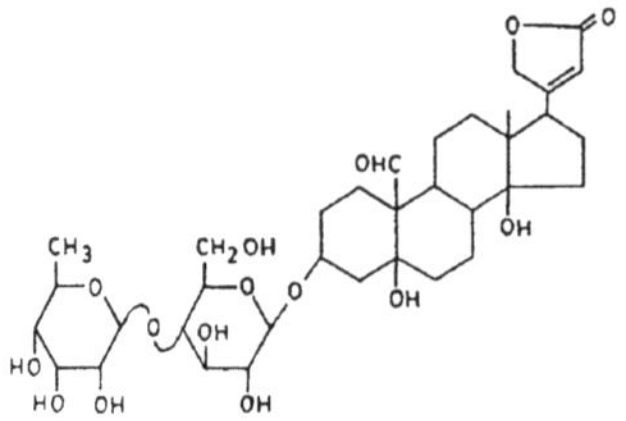

(261) $C_{35}H_{52}O_{15}$ (*113*)

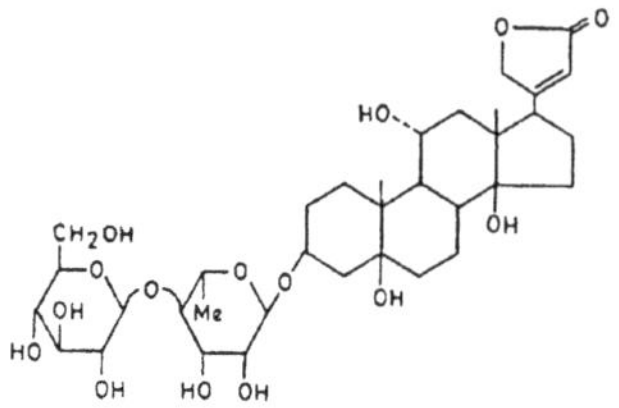

(262) $C_{35}H_{54}O_{15}$ (*114*)

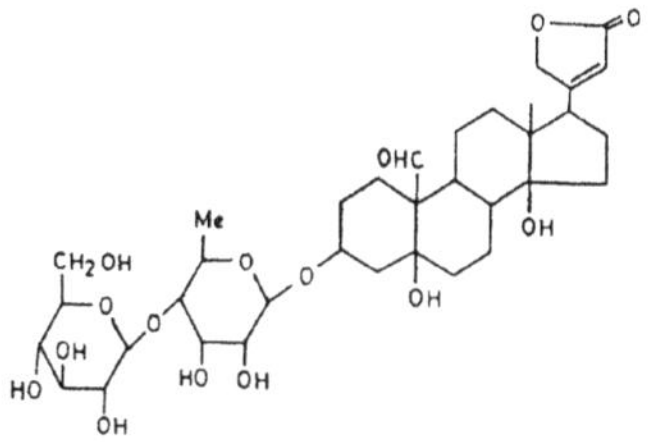

(263) $C_{35}H_{54}O_{15}$ (*115*)

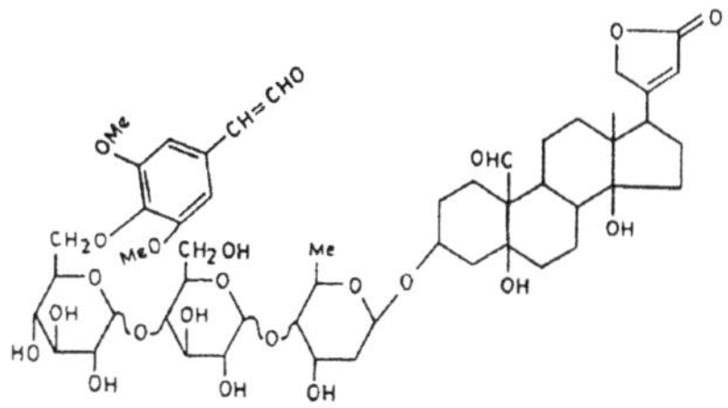

(264) $C_{51}H_{71}O_{22}$ (*116*)

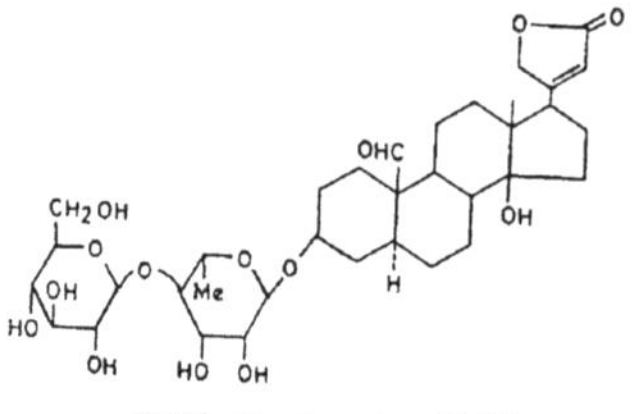

(265) $C_{35}H_{52}O_{14}$ (*117*)

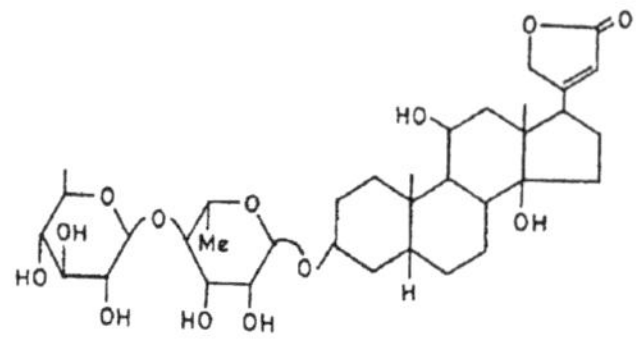

(266) $C_{35}H_{54}O_{13}$ (*117*)

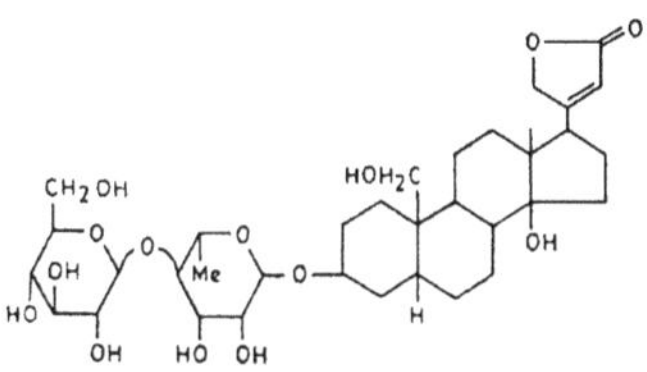

(267) $C_{35}H_{54}O_{14}$ (*117*)

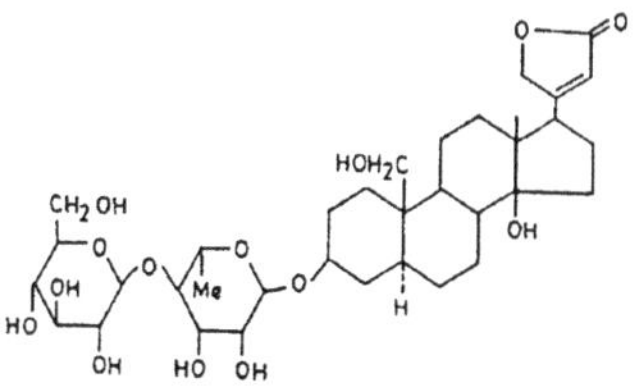

(268) $C_{35}H_{54}O_{14}$ (*117*)

Table 3 (*continued*)

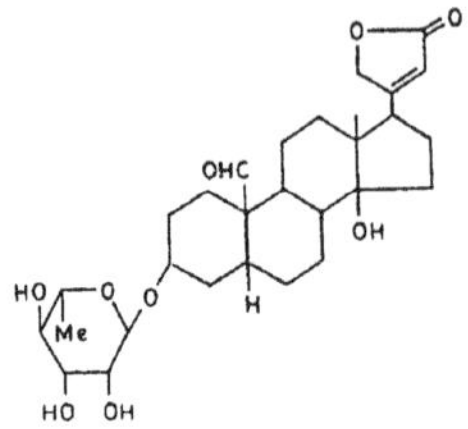

(269) $C_{29}H_{42}O_9$ (*117*)

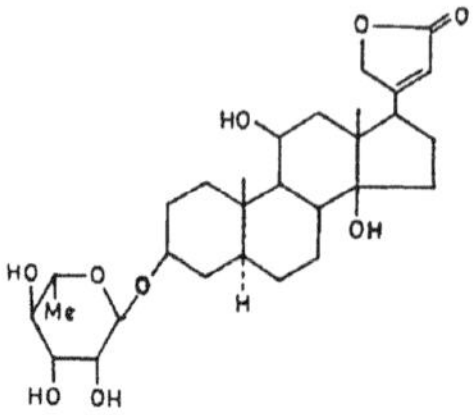

(270) $C_{29}H_{44}O_9$ (*117*)

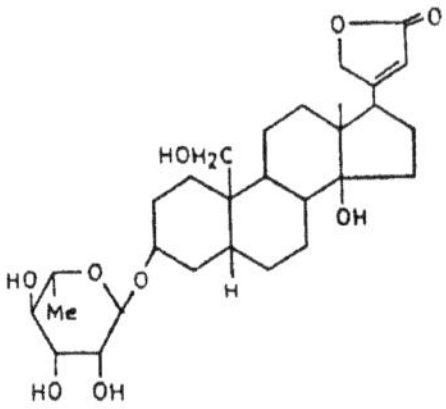

(271) $C_{29}H_{44}O_9$ (*117*)

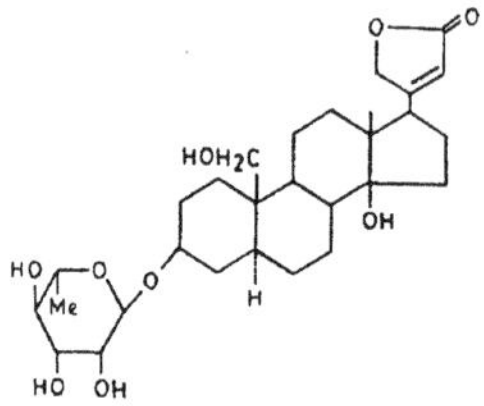

(272) $C_{29}H_{44}O_9$ (*117*)

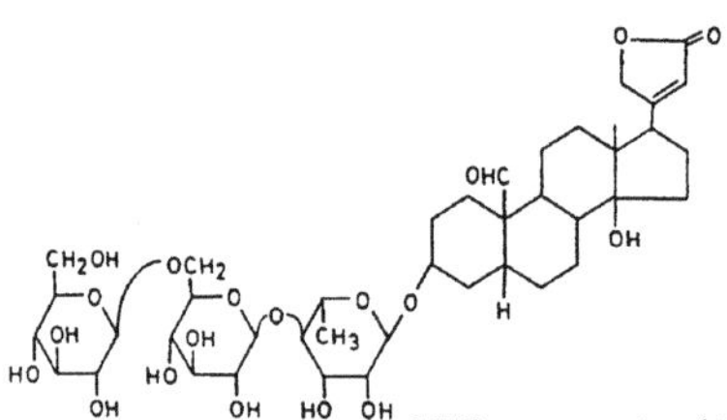

(273) $C_{41}H_{64}O_{19}$ (*118*)

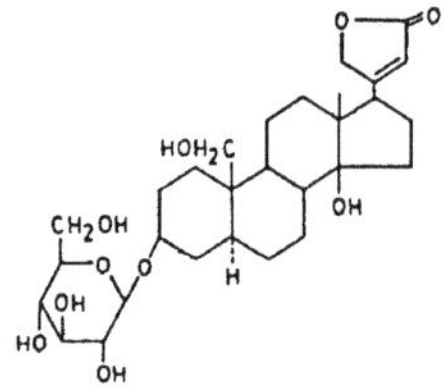

(274) $C_{29}H_{44}O_{10}$ (*119*)

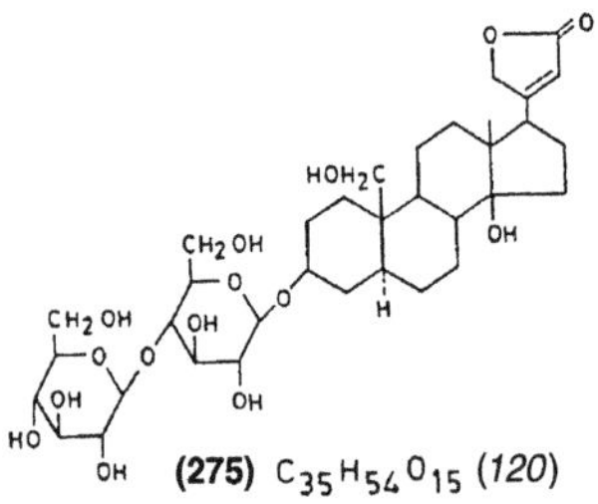

(275) $C_{35}H_{54}O_{15}$ (*120*)

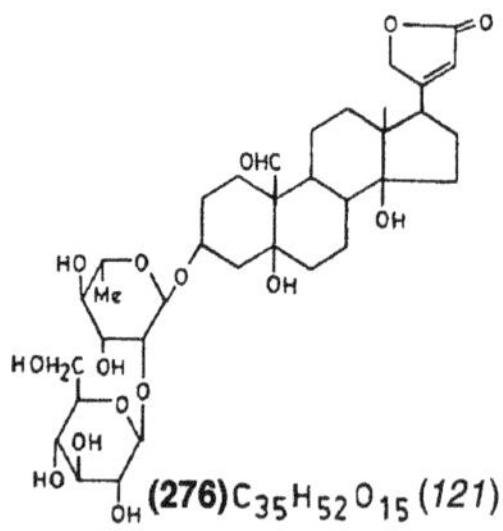

(276) $C_{35}H_{52}O_{15}$ (*121*)

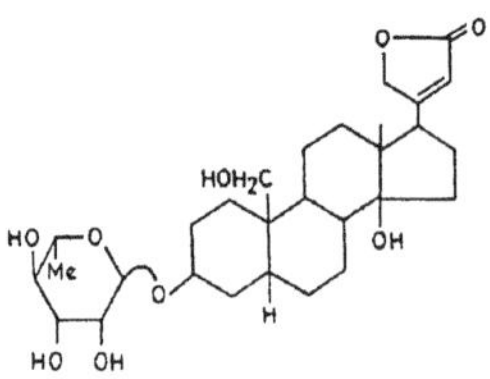

(277) $C_{29}H_{44}O_9$ (*122*)

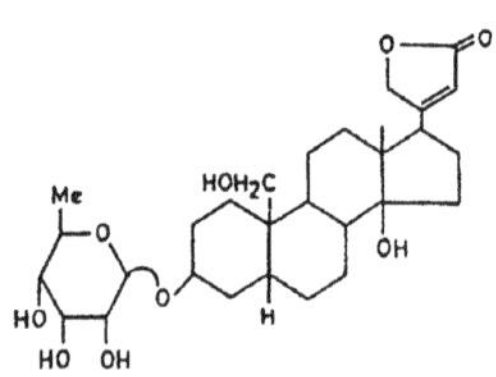

(278) $C_{29}H_{44}O_9$ (*122*)

Table 3 (*continued*)

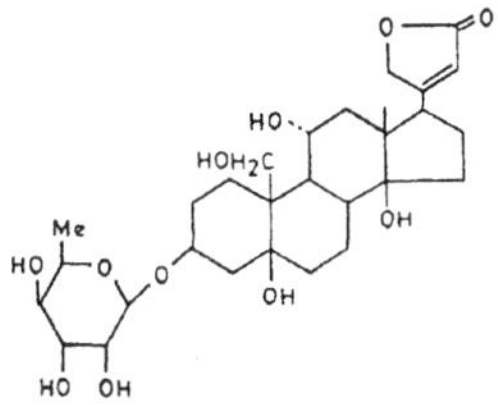

(279) $C_{29}H_{44}O_{11}$ (*123*)

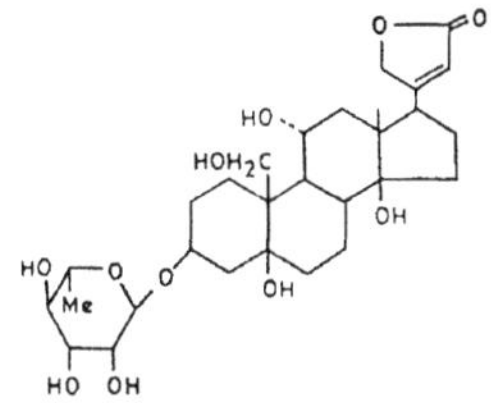

(280) $C_{29}H_{44}O_{11}$ (*123*)

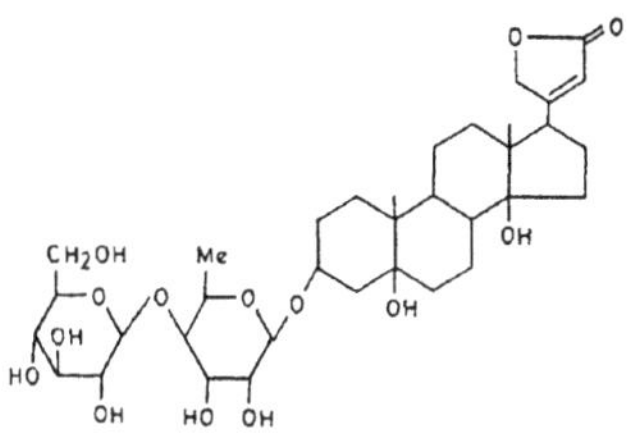

(281) $C_{35}H_{54}O_{14}$ (*124*)

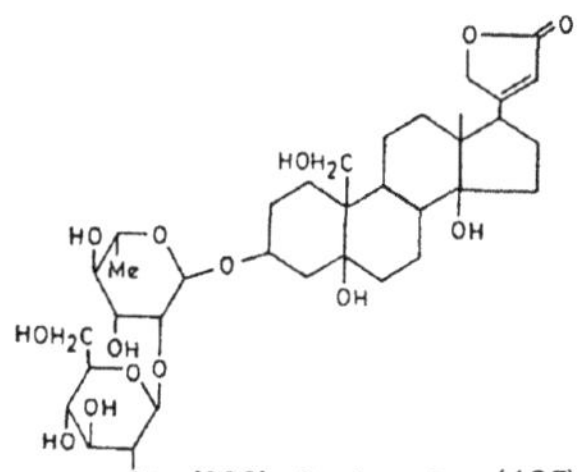

(282) $C_{35}H_{54}O_{15}$ (*125*)

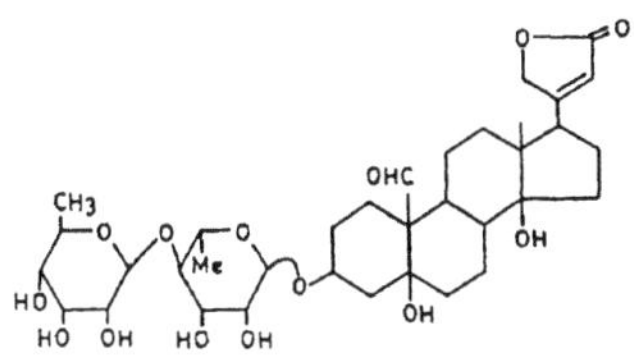

(283) $C_{35}H_{52}O_{14}$ (*126*)

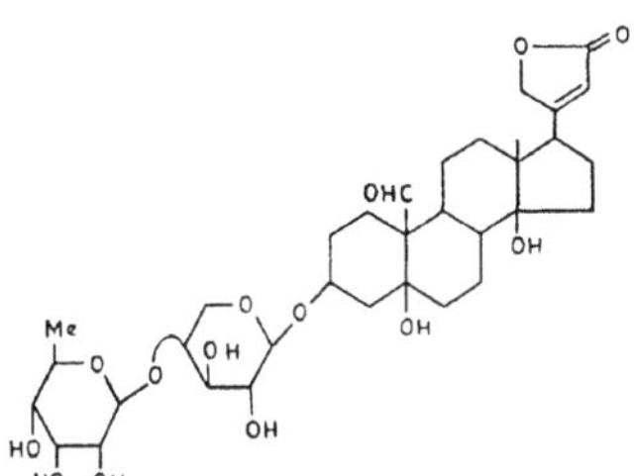

(284) $C_{34}H_{50}O_{14}$ (*126*)

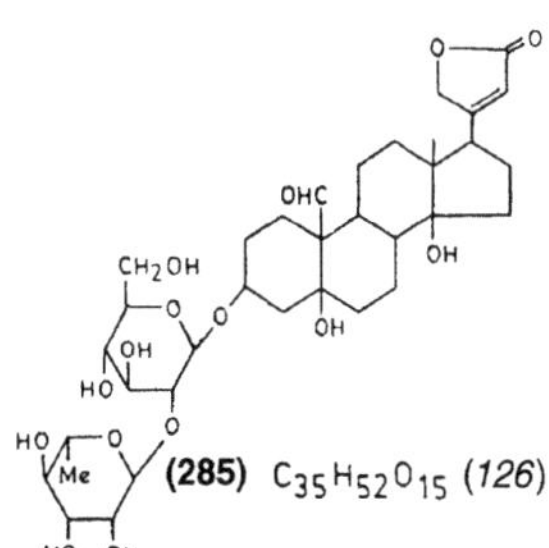

(285) $C_{35}H_{52}O_{15}$ (*126*)

Table 3 (*continued*)

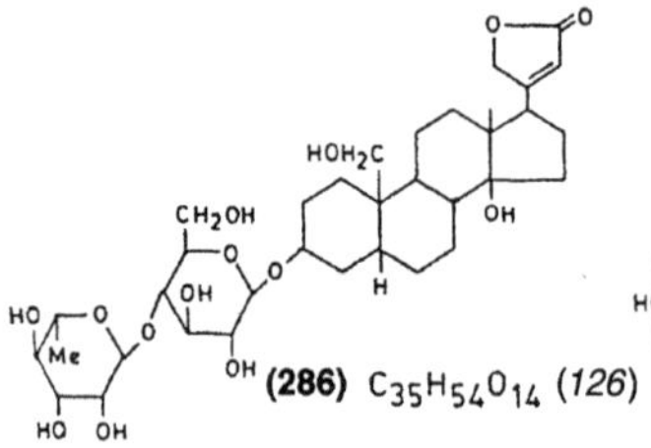

(286) $C_{35}H_{54}O_{14}$ (*126*)

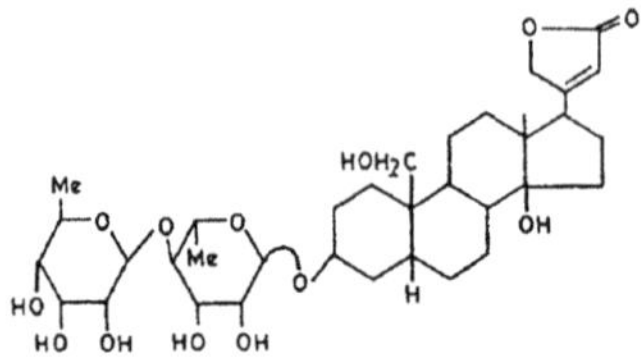

(287) $C_{35}H_{54}O_{13}$ (*126*)

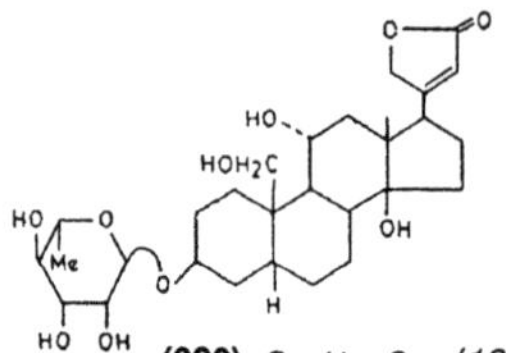

(288) $C_{29}H_{44}O_{10}$ (*126*)

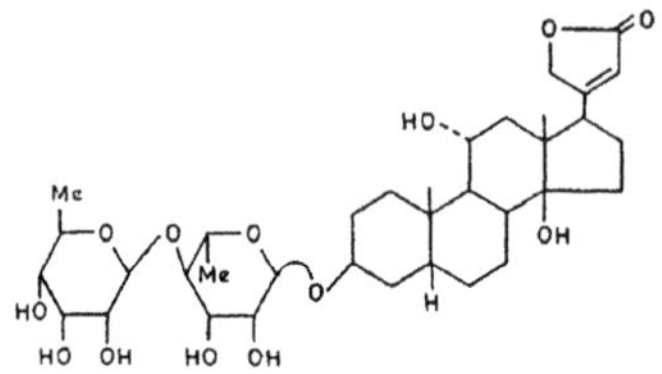

(289) $C_{35}H_{54}O_{13}$ (*126*)

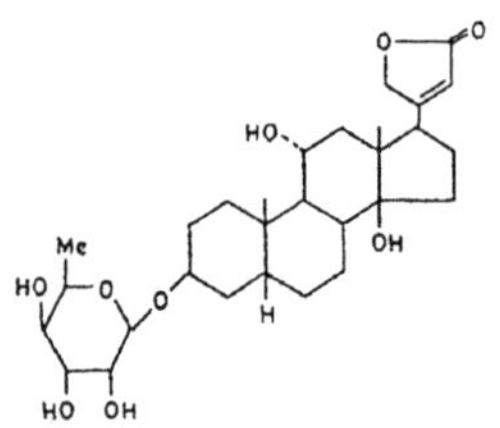

(290) $C_{29}H_{44}O_{9}$ (*126*)

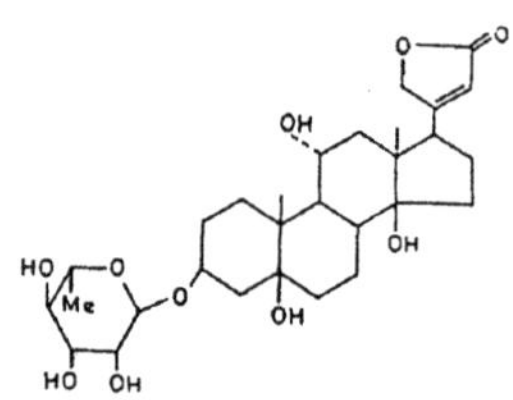

(291) $C_{29}H_{44}O_{10}$ (*127*)

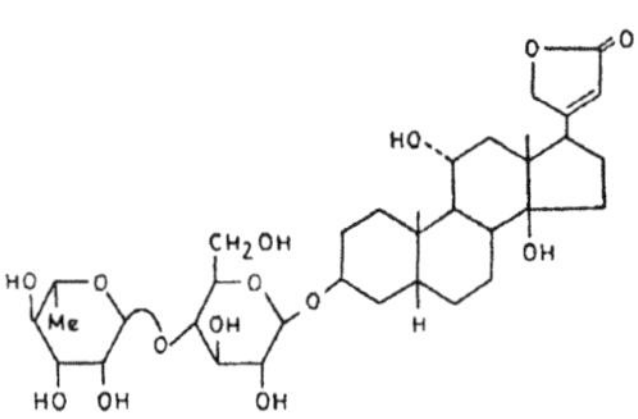

(292) $C_{35}H_{54}O_{14}$ (*128*)

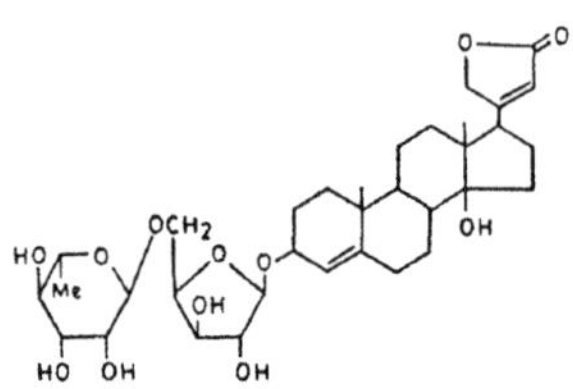

(293) $C_{34}H_{50}O_{12}$ (*129*)

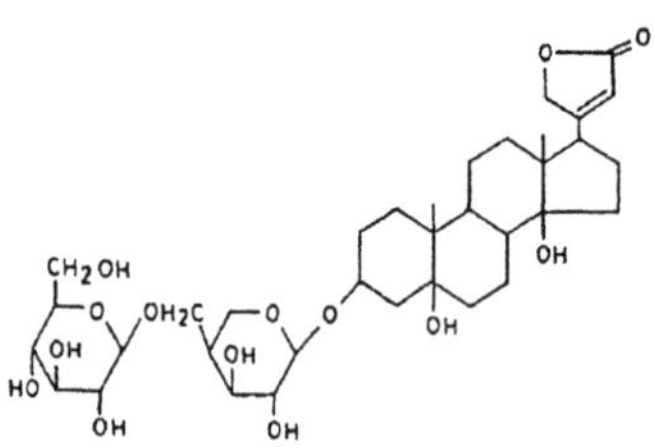

(294) $C_{34}H_{52}O_{14}$ (*130*)

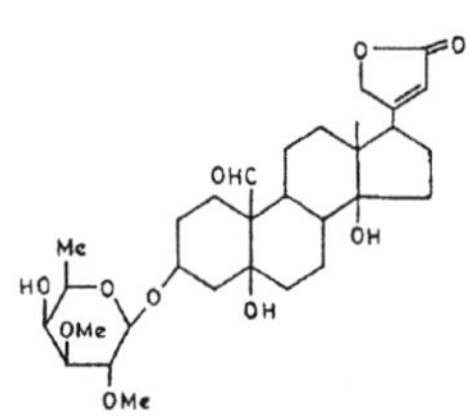

(295) $C_{31}H_{46}O_{10}$ (*131*)

Table 3 (*continued*)

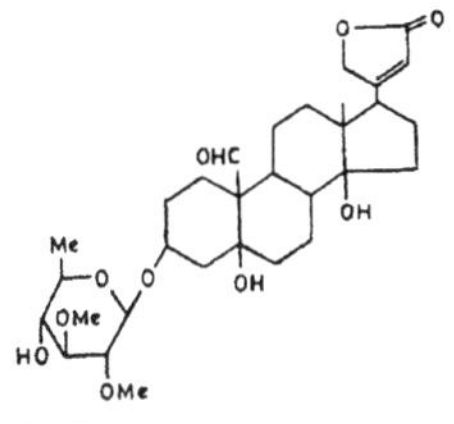

(296) $C_{31}H_{46}O_{10}$ (*131*)

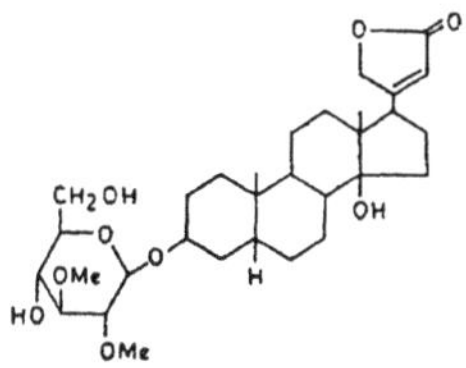

(297) $C_{31}H_{50}O_{9}$ (*132*)

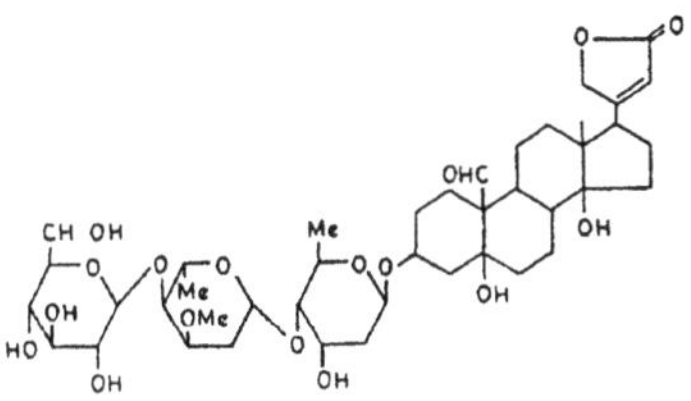

(298) $C_{42}H_{64}O_{17}$ (*133*)

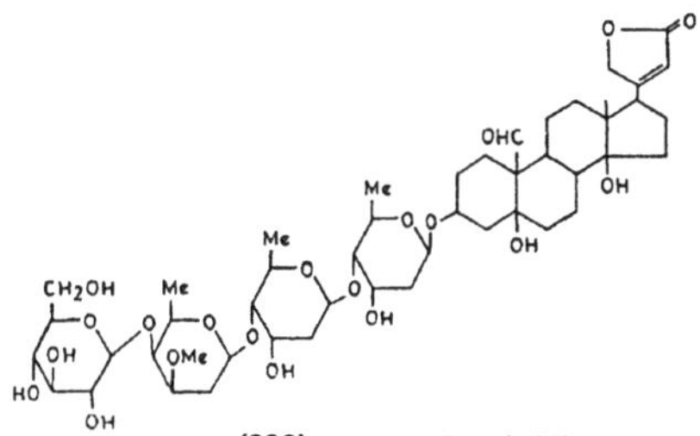

(299) $C_{48}H_{74}O_{20}$ (*133*)

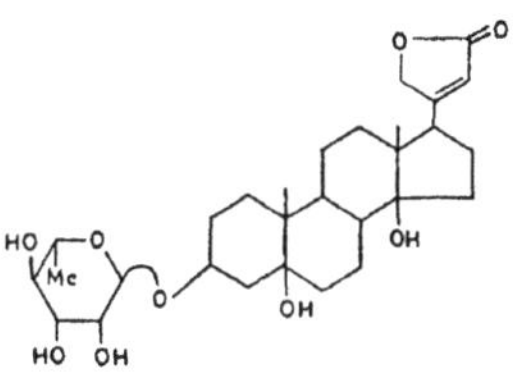

(300) $C_{29}H_{44}O_{9}$ (*134*)

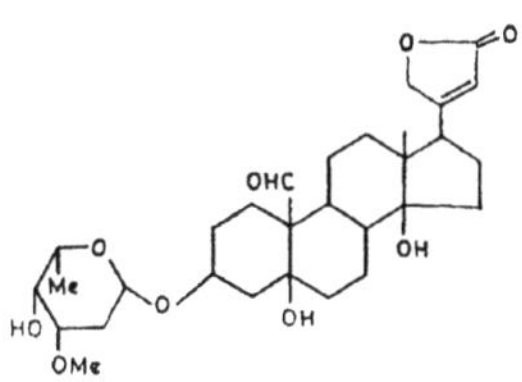

(301) $C_{30}H_{44}O_{9}$ (*134*)

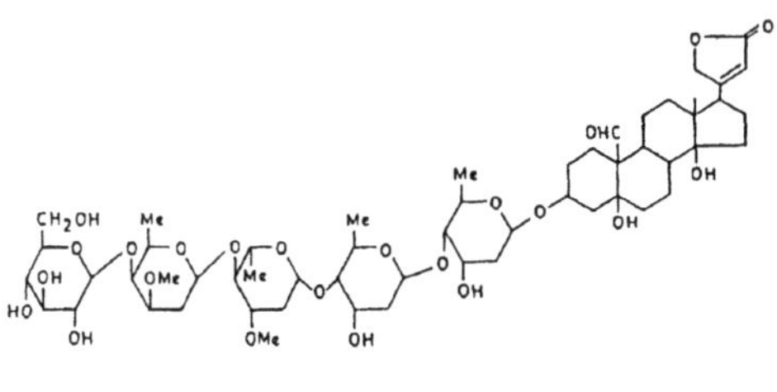

(302) $C_{55}H_{86}O_{23}$ (*135*)

Table 3 (*continued*)

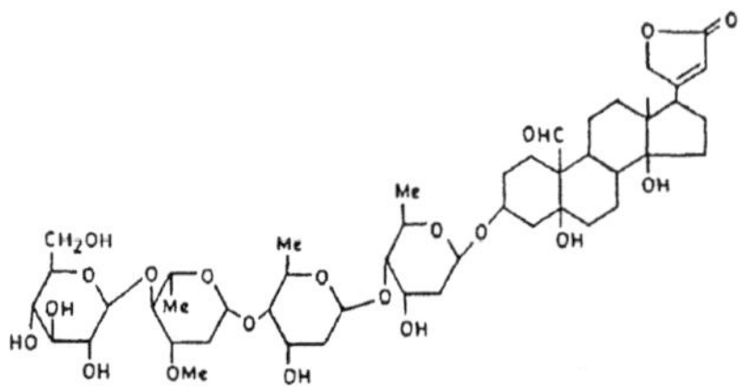

(303) $C_{48}H_{74}O_{20}$ (*135*)

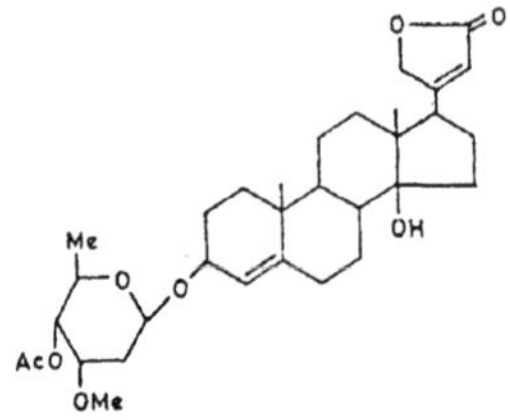

(304) $C_{32}H_{46}O_8$ (*136*)

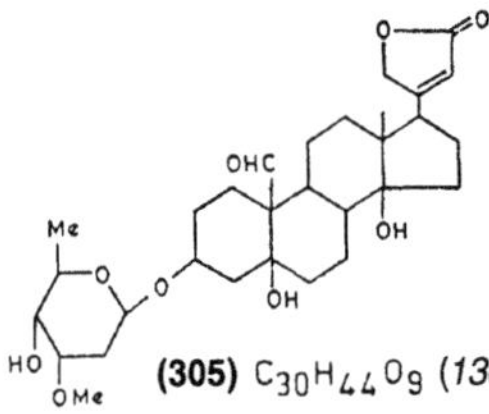

(305) $C_{30}H_{44}O_9$ (*136*)

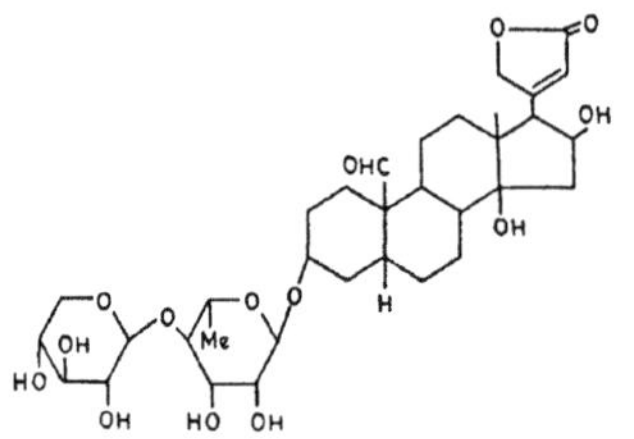

(306) $C_{34}H_{50}O_{14}$ (*137*)

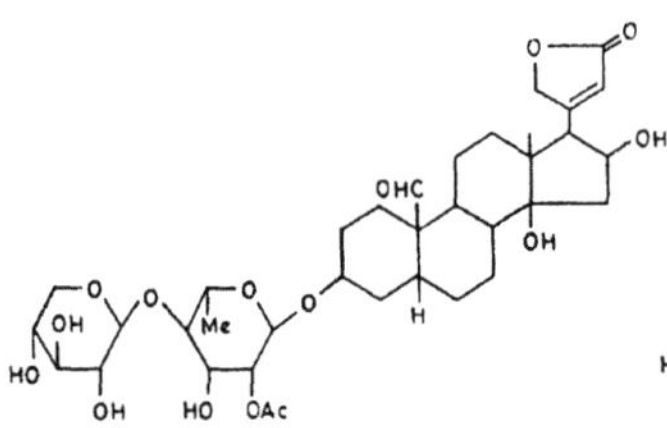

(307) $C_{36}H_{52}O_{15}$ (*137*)

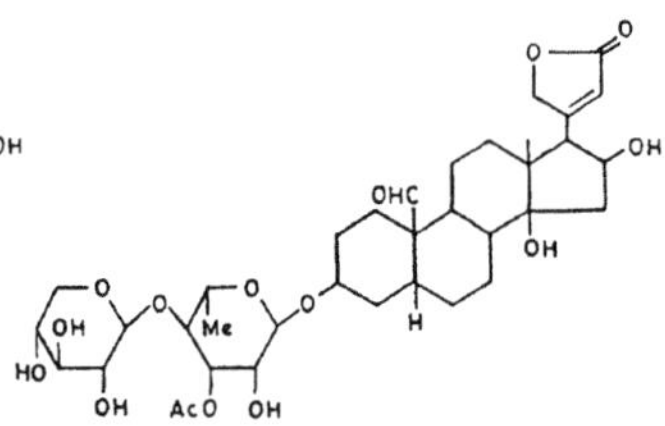

(308) $C_{36}H_{52}O_{15}$ (*137*)

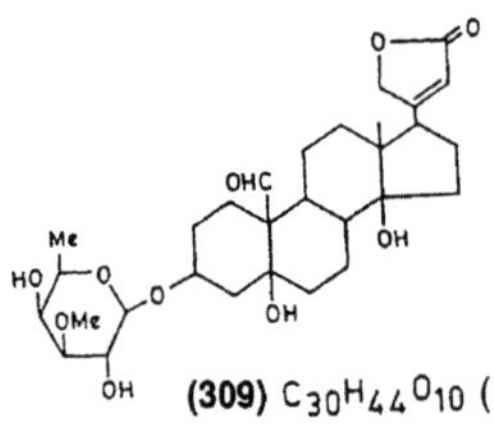

(309) $C_{30}H_{44}O_{10}$ (*138*)

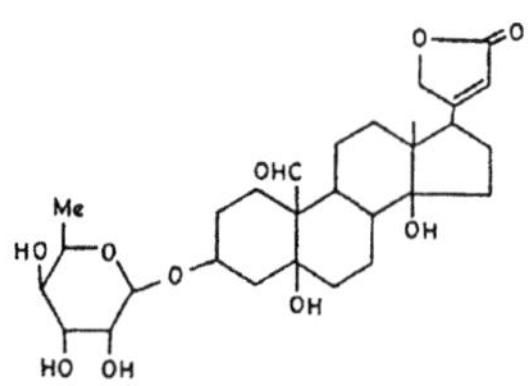

(310) $C_{29}H_{42}O_{10}$ (*138*)

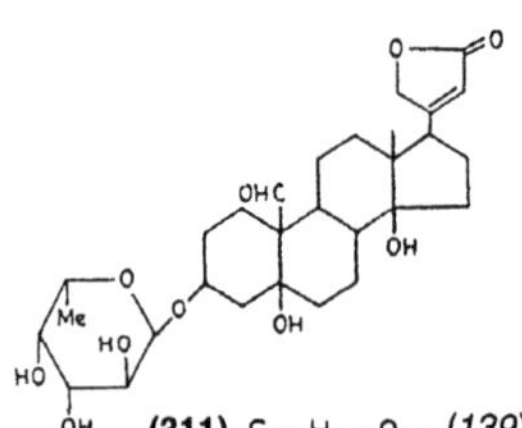

(311) $C_{29}H_{42}O_{10}$ (*139*)

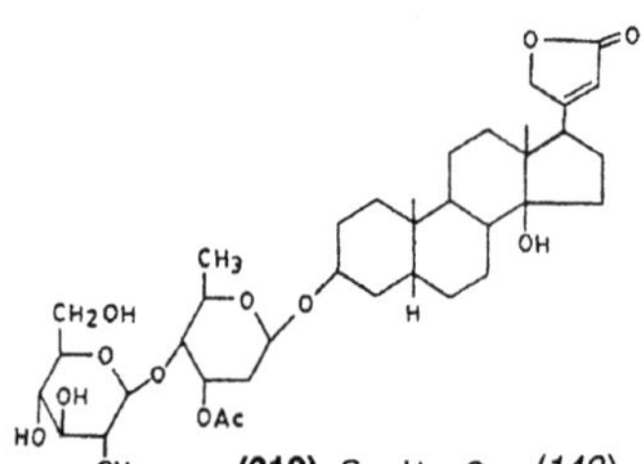

(312) $C_{37}H_{56}O_{13}$ (*140*)

Table 3 (*continued*)

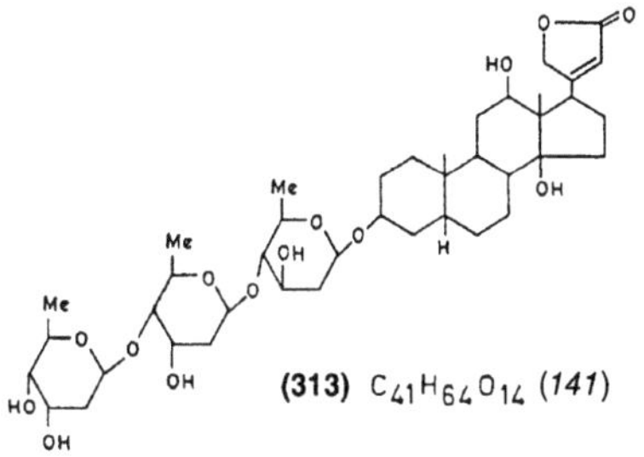

(313) $C_{41}H_{64}O_{14}$ (*141*)

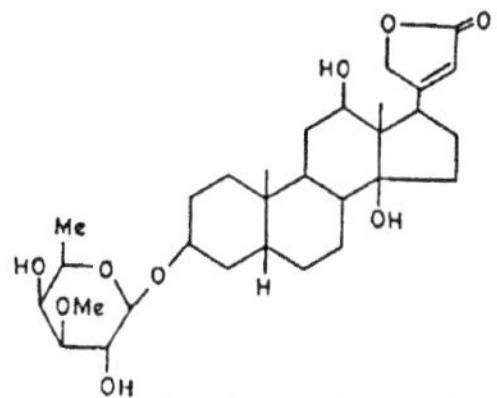

(314) $C_{30}H_{44}O_9$ (*142*)

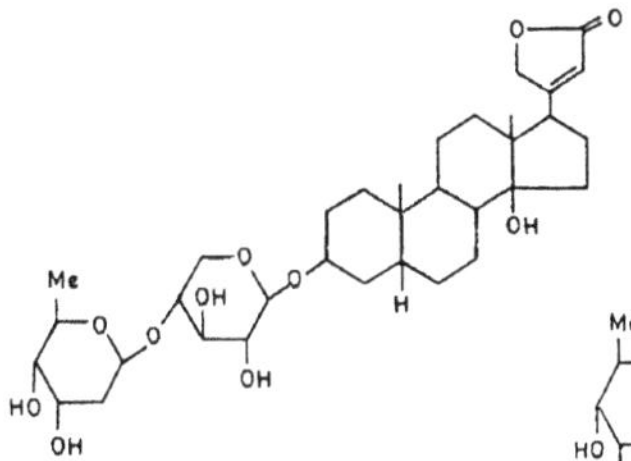

(315) $C_{34}H_{52}O_{11}$ (*143*)

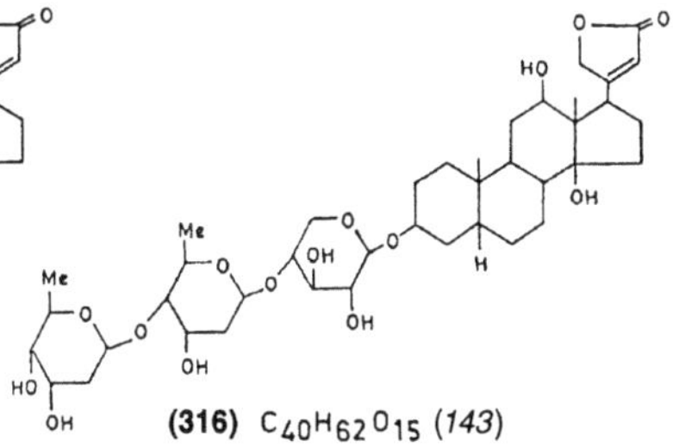

(316) $C_{40}H_{62}O_{15}$ (*143*)

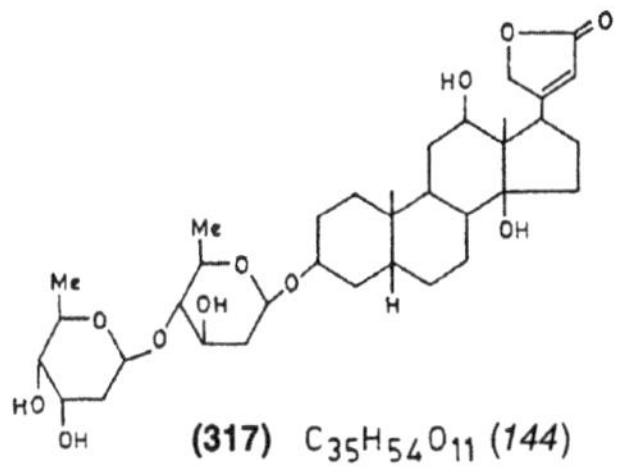

(317) $C_{35}H_{54}O_{11}$ (*144*)

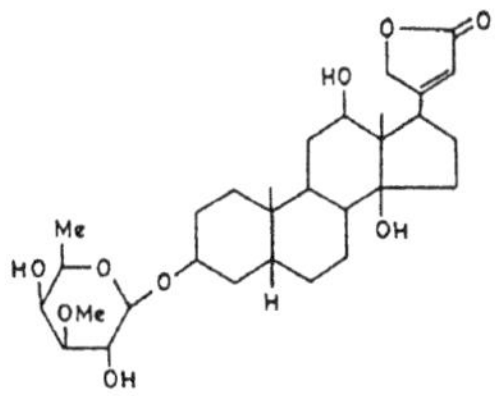

(318) $C_{30}H_{46}O_9$ (*144*)

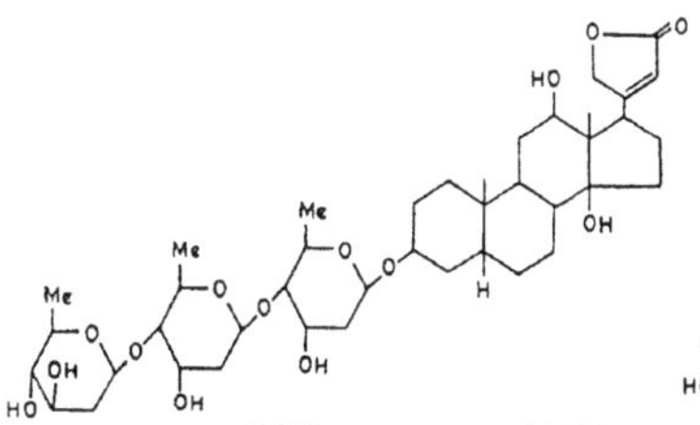

(319) $C_{41}H_{64}O_{14}$ (*145*)

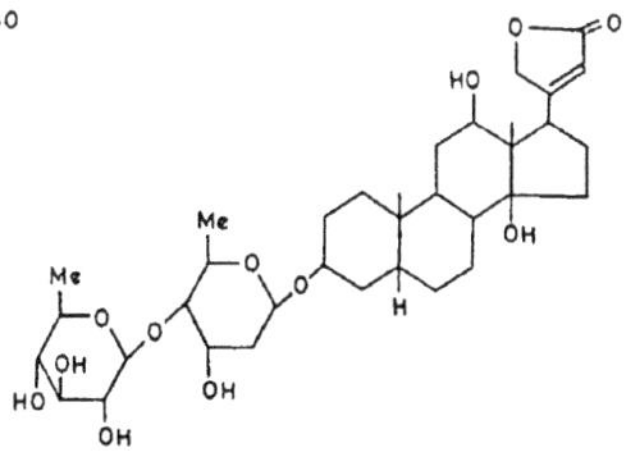

(320) $C_{35}H_{54}O_{12}$ (*145*)

Table 3 (*continued*)

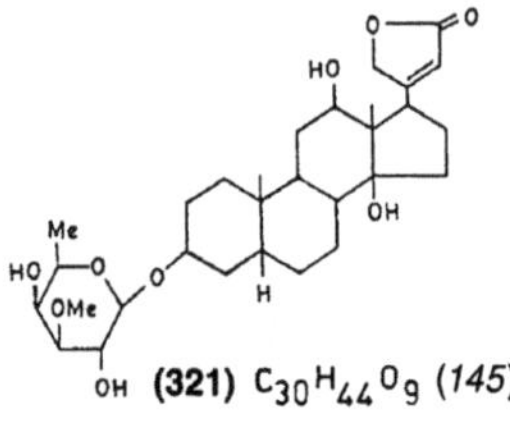

(321) $C_{30}H_{44}O_9$ (*145*)

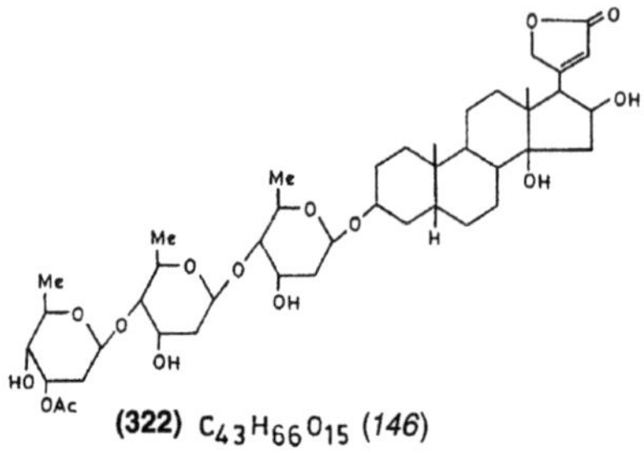

(322) $C_{43}H_{66}O_{15}$ (*146*)

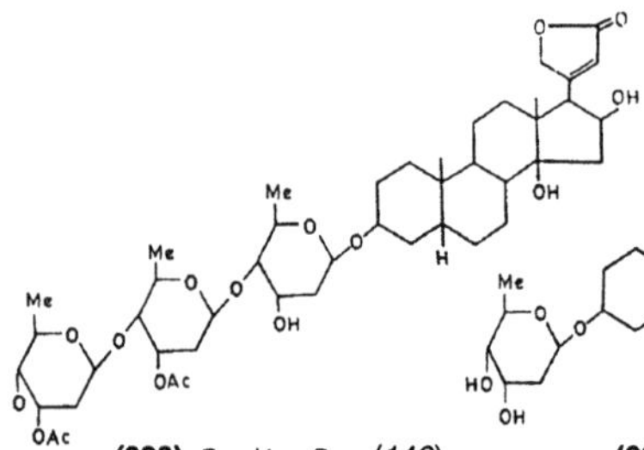

(323) $C_{45}H_{68}O_{16}$ (*146*)

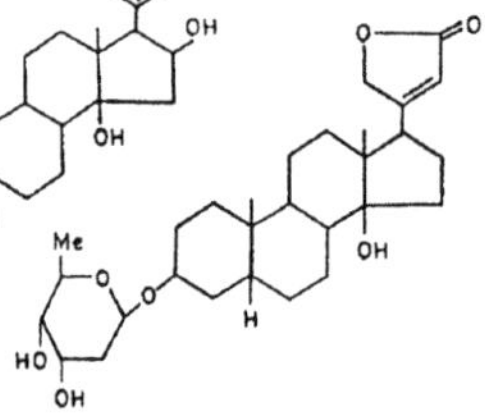

(324) $C_{29}H_{44}O_7$ (*34*)

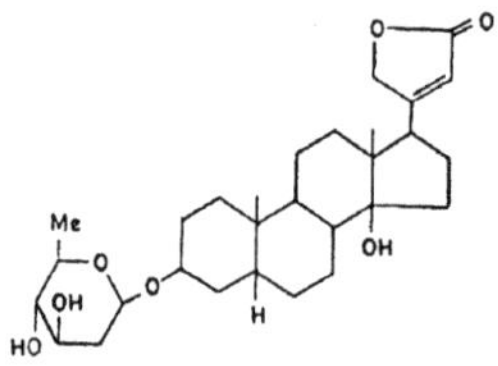

(325) $C_{29}H_{44}O_7$ (*34*)

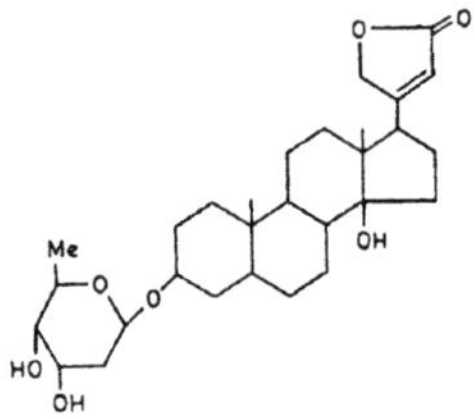

(326) $C_{29}H_{44}O_7$ (*34*)

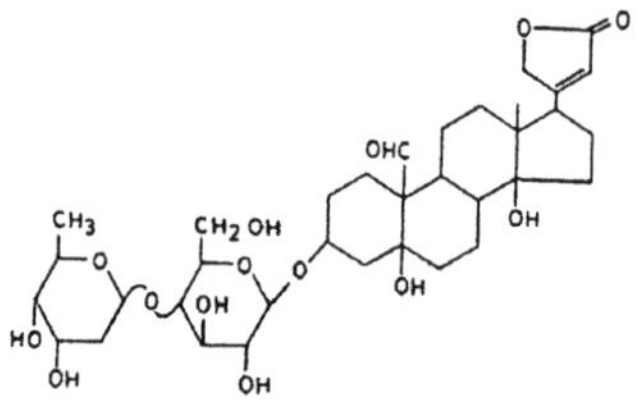

(327) $C_{35}H_{52}O_{14}$ (*147*)

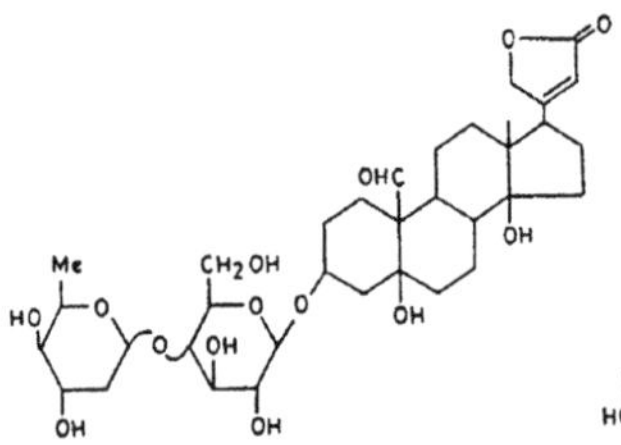

(328) $C_{35}H_{52}O_{14}$ (*147*)

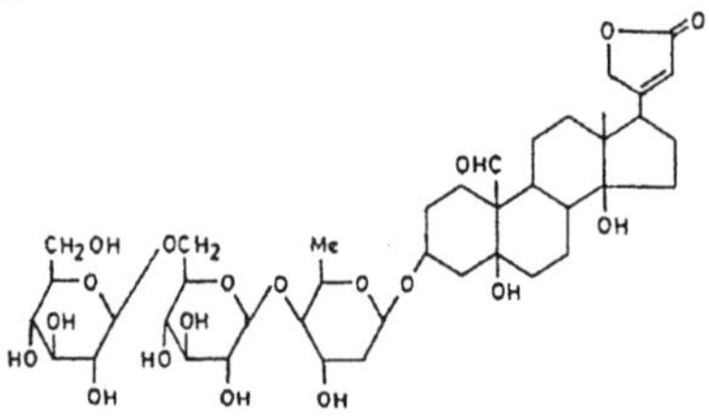

(329) $C_{41}H_{62}O_{19}$ (*148*)

Table 3 (*continued*)

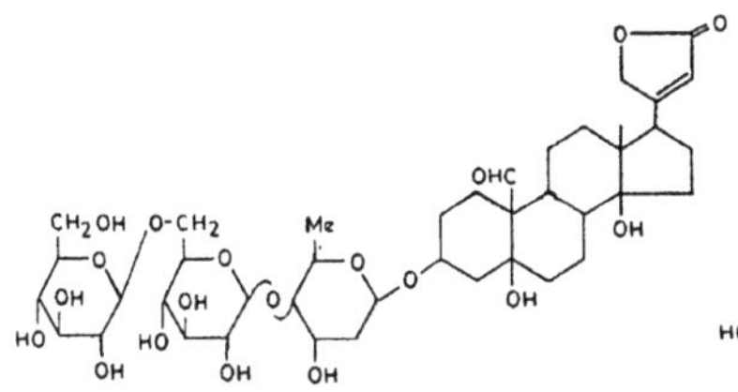

(330) $C_{41}H_{62}O_{19}$ (*149*)

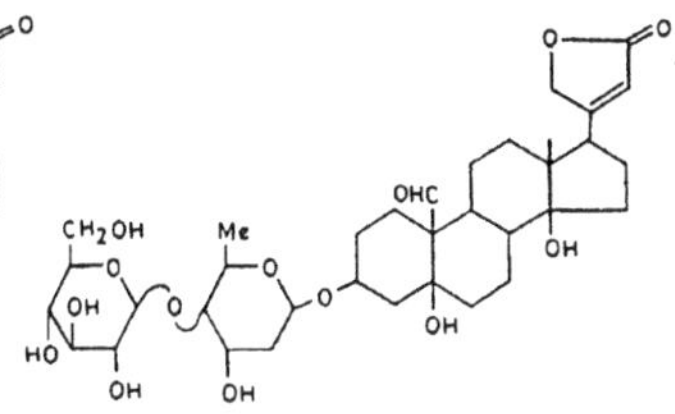

(331) $C_{35}H_{52}O_{14}$ (*149*)

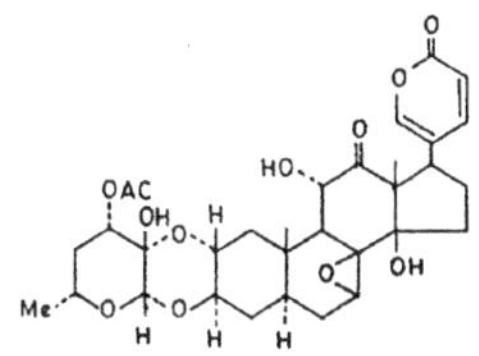

(332) $C_{32}H_{40}O_{12}$ (*150*)

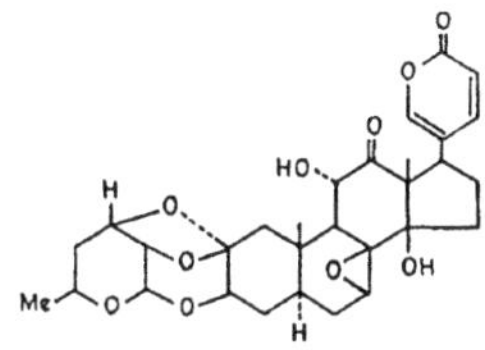

(333) $C_{30}H_{36}O_{10}$ (*150*)

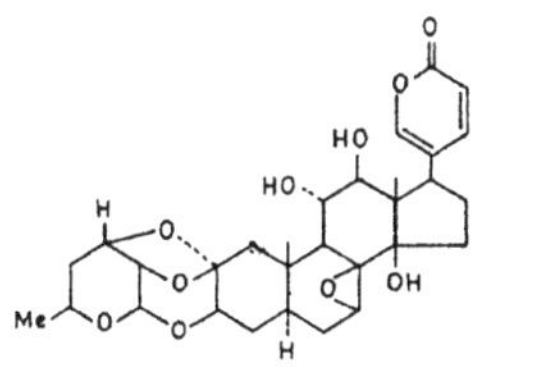

(334) $C_{30}H_{38}O_{10}$ (*150*)

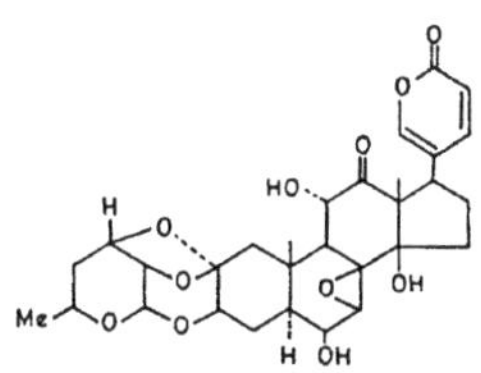

(335) $C_{30}H_{36}O_{11}$ (*150*)

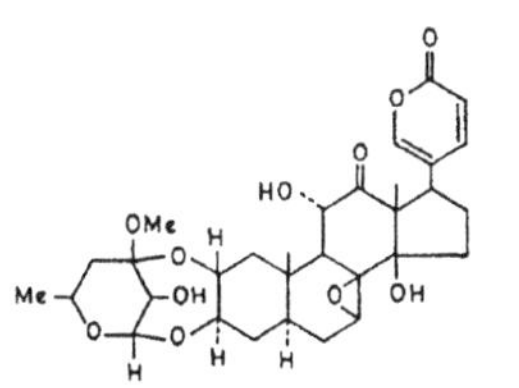

(336) $C_{31}H_{40}O_{11}$ (*150*)

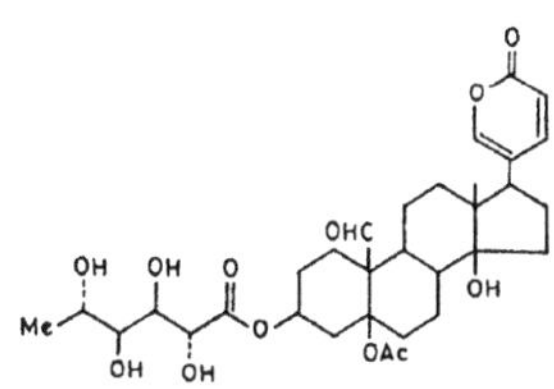

(337) $C_{32}H_{44}O_{12}$ (*151*)

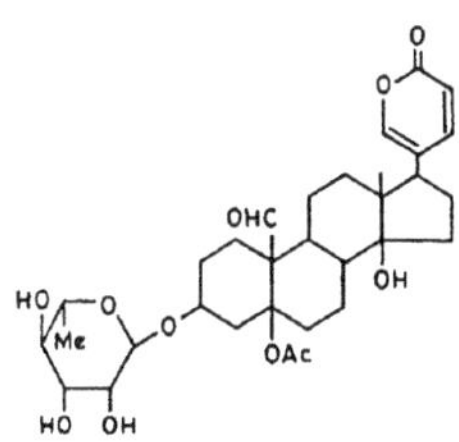

(338) $C_{32}H_{44}O_{11}$ (*151*)

Table 3 (*continued*)

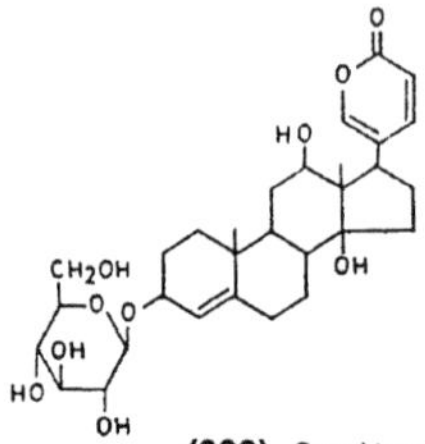

(339) $C_{30}H_{42}O_{10}$ (*152*)

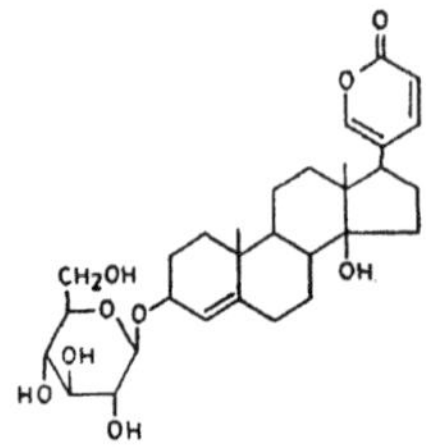

(340) $C_{30}H_{42}O_{9}$ (*152*)

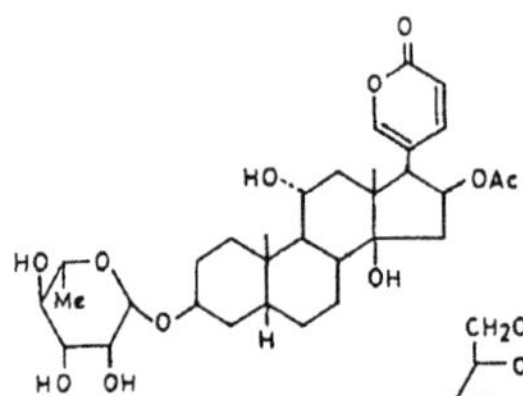

(341) $C_{32}H_{46}O_{11}$ (*153*)

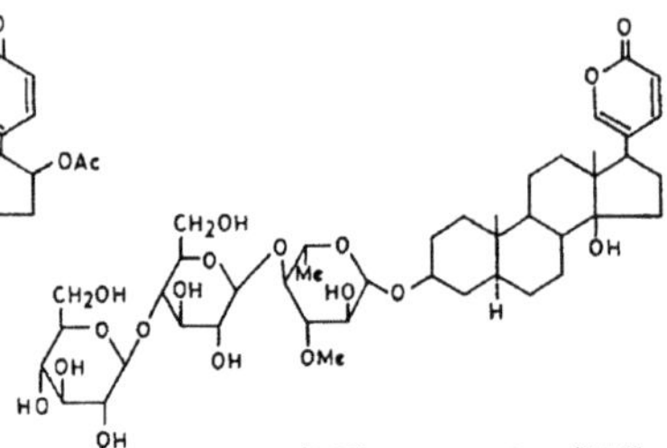

(342) $C_{43}H_{66}O_{18}$ (*152*)

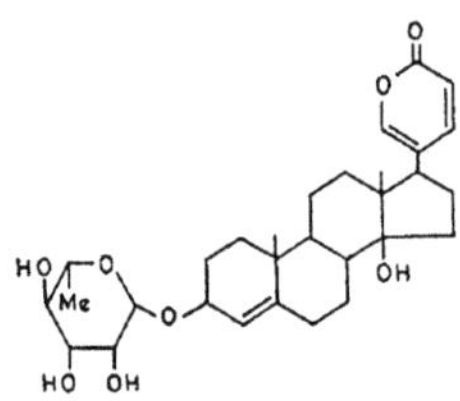

(343) $C_{30}H_{42}O_{8}$ (*153*)

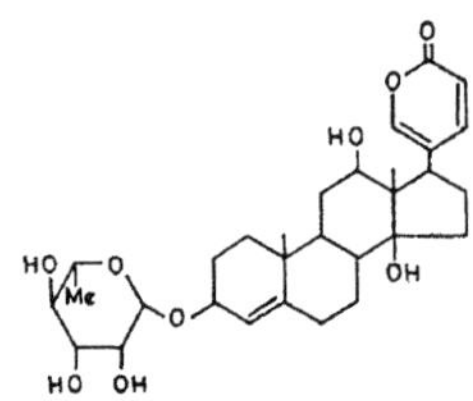

(344) $C_{30}H_{42}O_{9}$ (*153*)

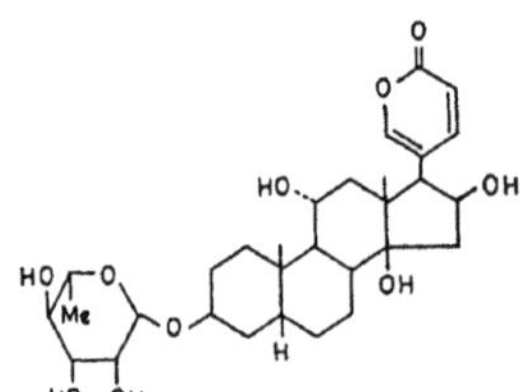

(345) $C_{30}H_{44}O_{10}$ (*153*)

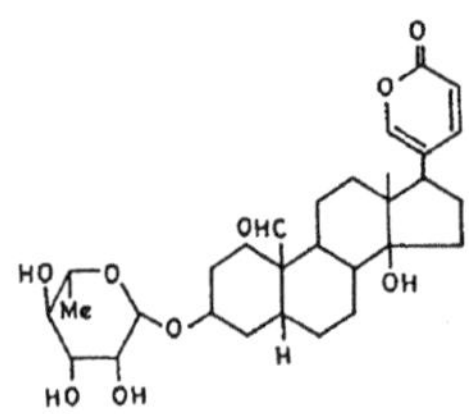

(346) $C_{30}H_{42}O_{9}$ (*153*)

References, pp. 148–155

Table 3 (*continued*)

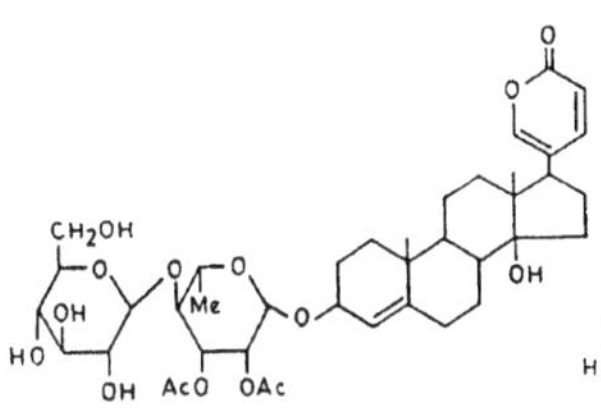

(347) $C_{40}H_{56}O_{15}$ (*153*)

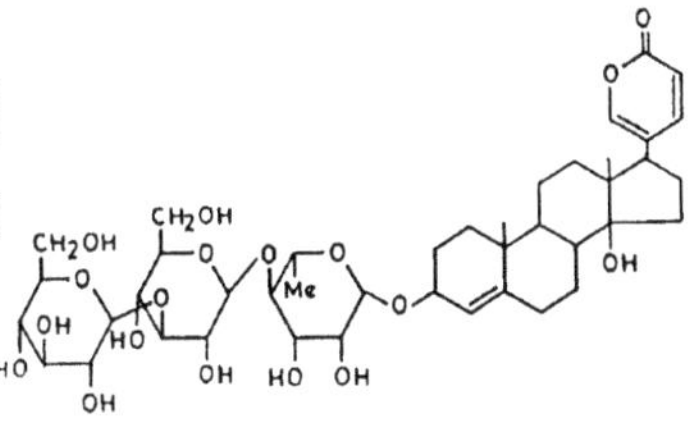

(348) $C_{42}H_{62}O_{18}$ (*153*)

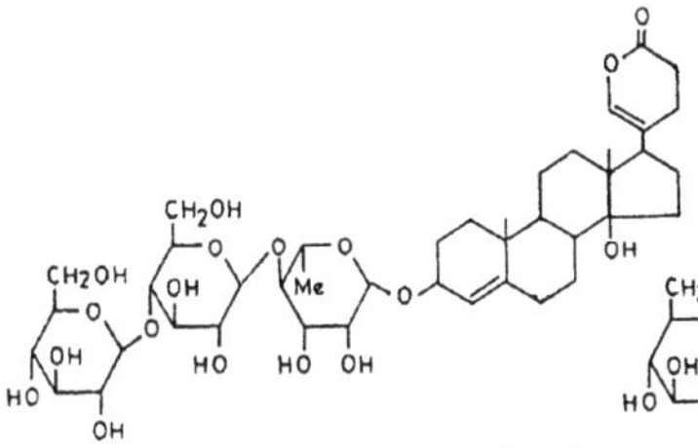

(349) $C_{42}H_{62}O_{18}$ (*153*)

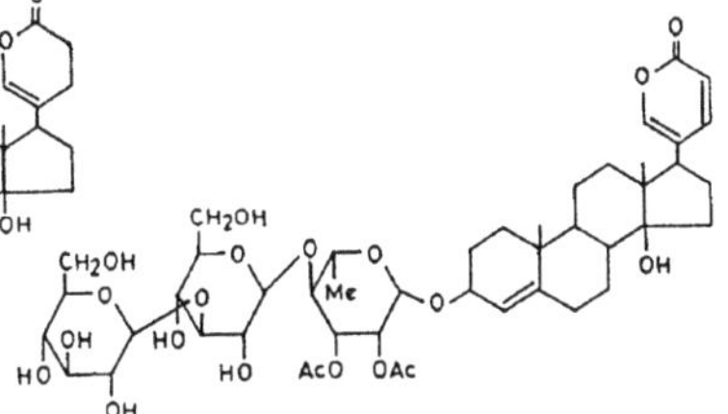

(350) $C_{46}H_{66}O_{20}$ (*153*)

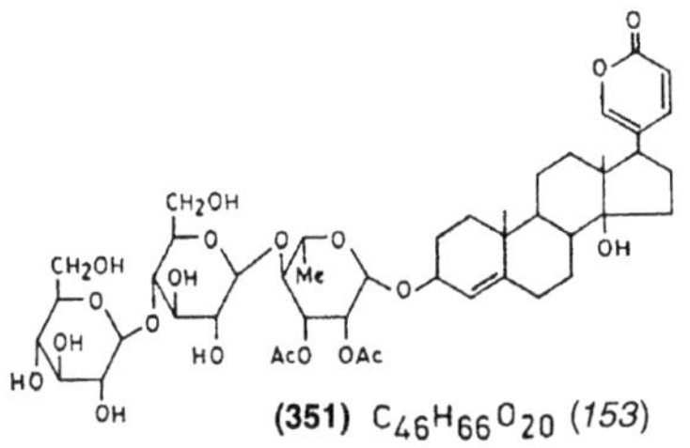

(351) $C_{46}H_{66}O_{20}$ (*153*)

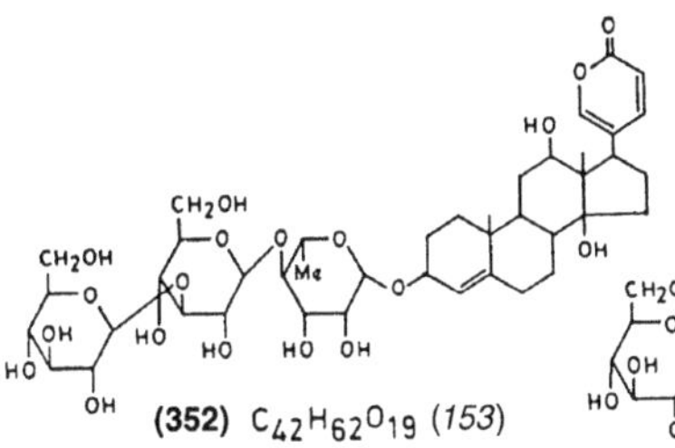

(352) $C_{42}H_{62}O_{19}$ (*153*)

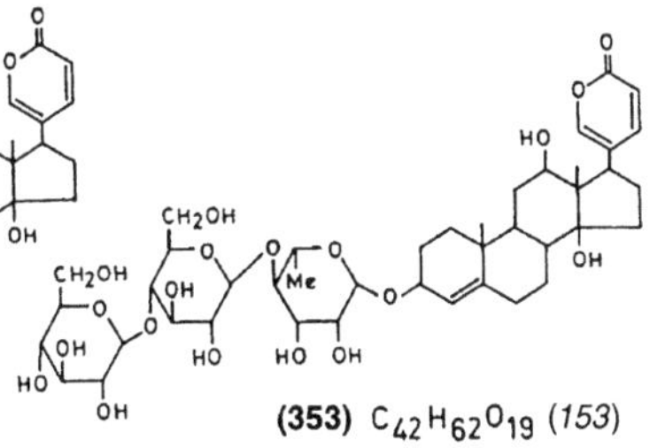

(353) $C_{42}H_{62}O_{19}$ (*153*)

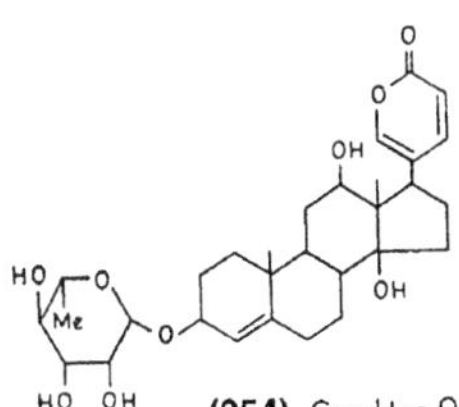

(354) $C_{30}H_{42}O_{9}$ (*154*)

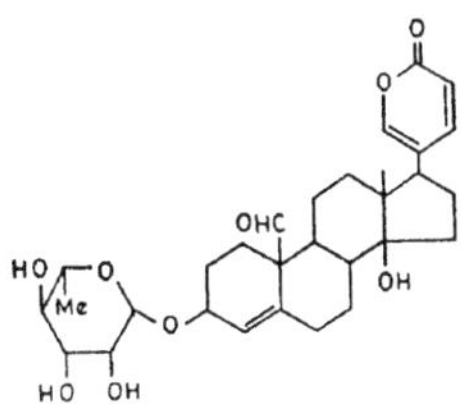

(355) $C_{30}H_{40}O_{9}$ (*154*)

Table 3 (*continued*)

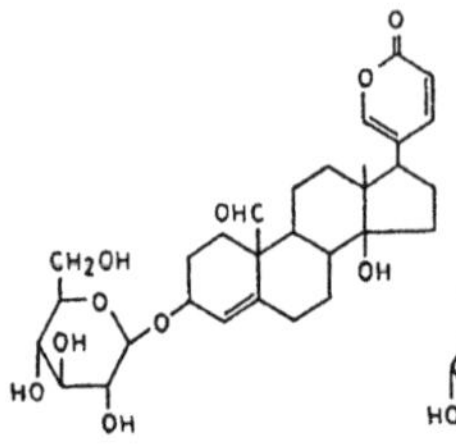

(356) $C_{30}H_{40}O_{10}$ (*154*)

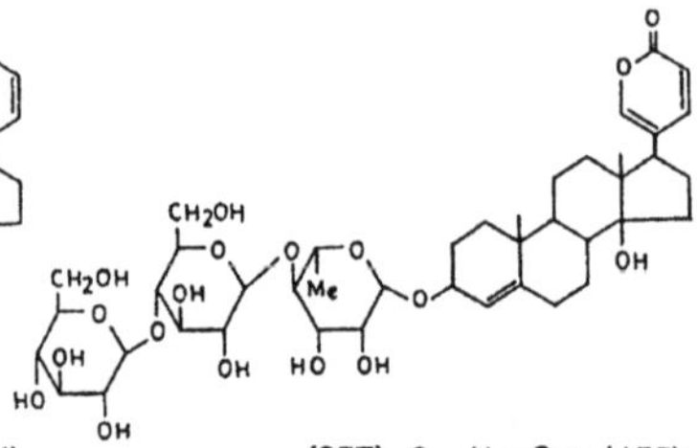

(357) $C_{42}H_{62}O_{18}$ (*155*)

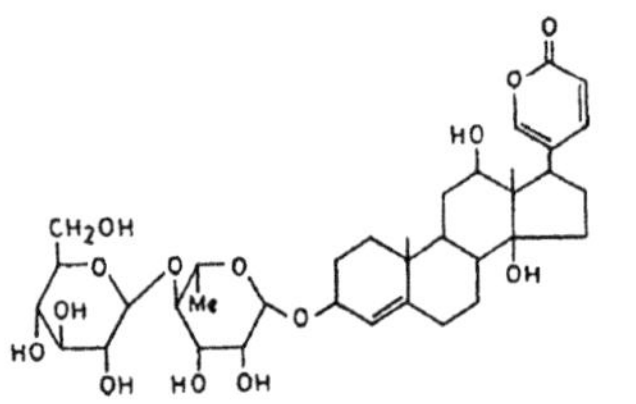

(358) $C_{36}H_{52}O_{14}$ (*155*)

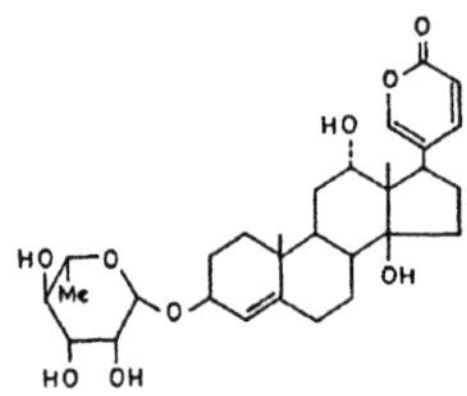

(359) $C_{30}H_{42}O_{9}$ (*155*)

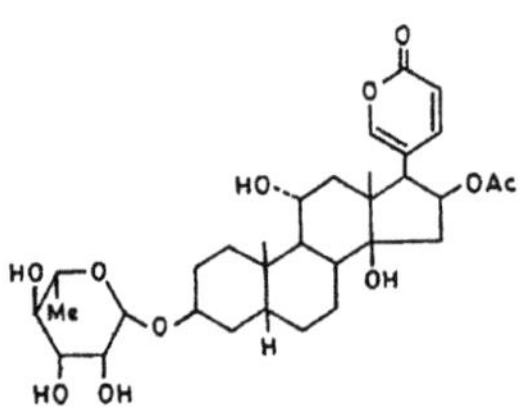

(360) $C_{32}H_{44}O_{11}$ (*155*)

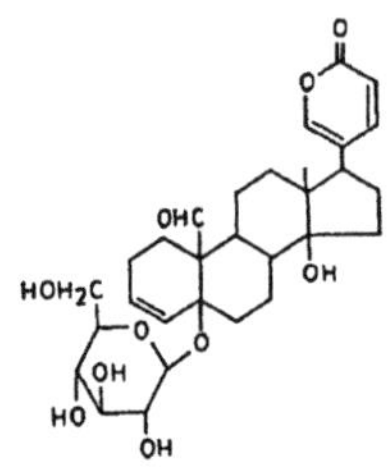

(361) $C_{30}H_{40}O_{10}$ (*155*)

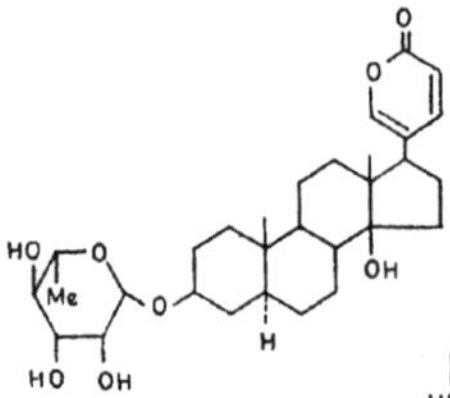

(362) $C_{30}H_{44}O_{8}$ (*155*)

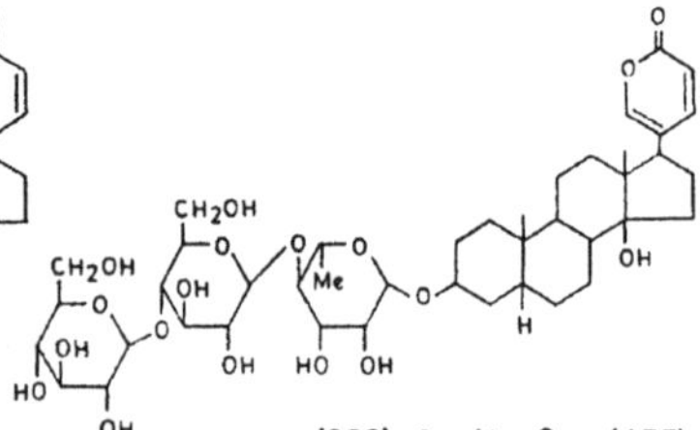

(363) $C_{42}H_{64}O_{18}$ (*155*)

References, pp. 148–155

Table 3 (*continued*)

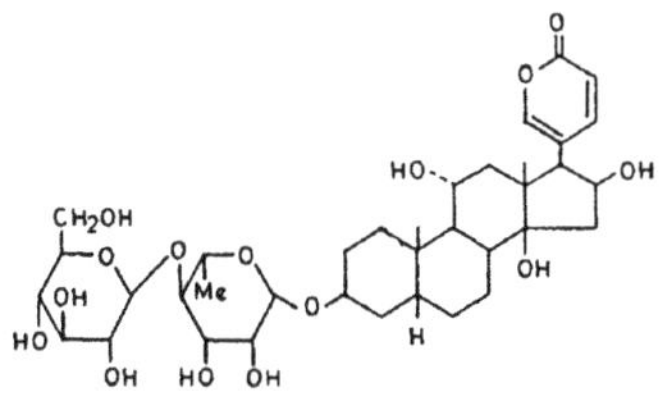

(364) $C_{36}H_{54}O_{15}$ (*155*)

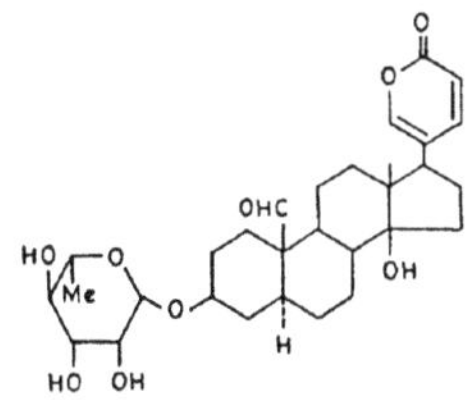

(365) $C_{30}H_{42}O_{9}$ (*155*)

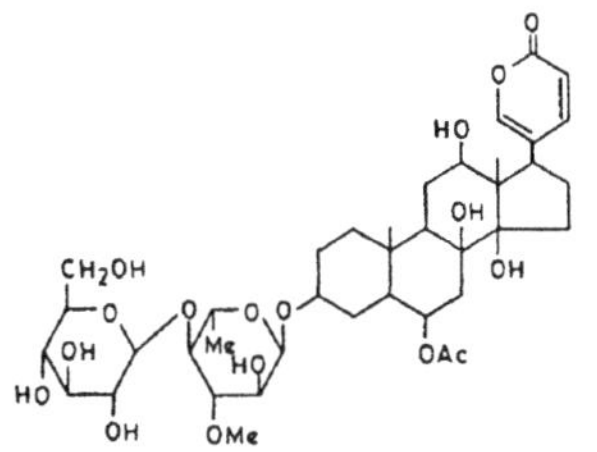

(366) $C_{39}H_{58}O_{17}$ (*156*)

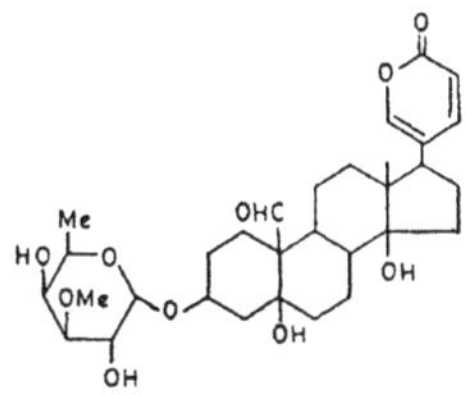

(367) $C_{31}H_{44}O_{10}$ (*157*)

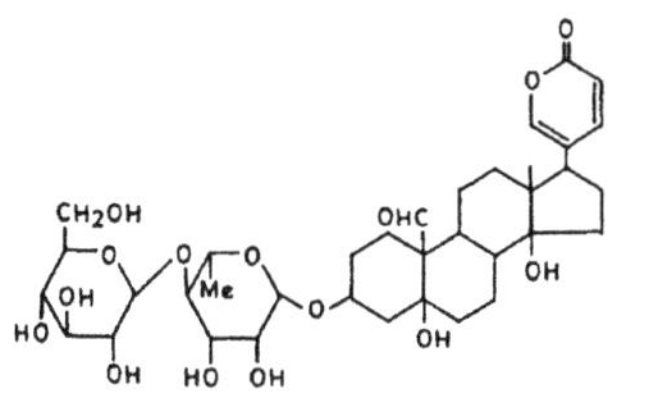

(368) $C_{36}H_{52}O_{15}$ (*158*)

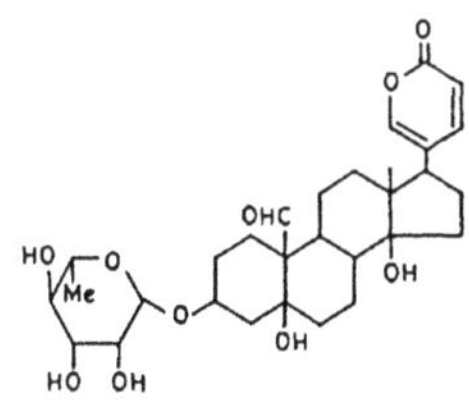

(369) $C_{30}H_{42}O_{10}$ (*158*)

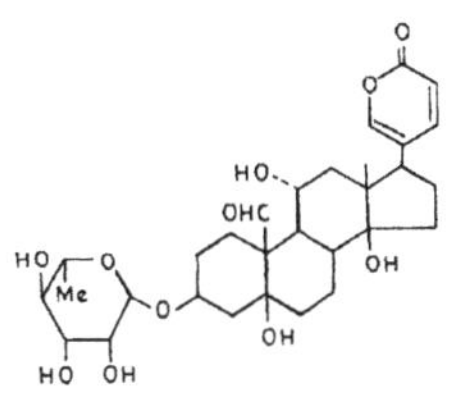

(370) $C_{30}H_{42}O_{11}$ (*158*)

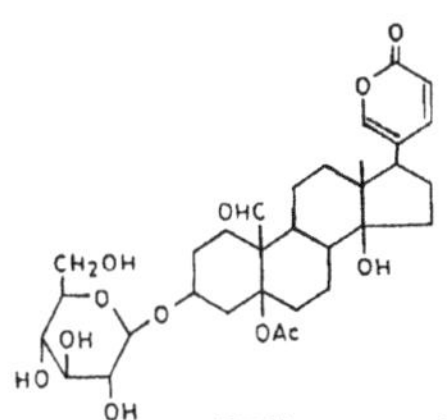

(371) $C_{32}H_{44}O_{12}$ (*159*)

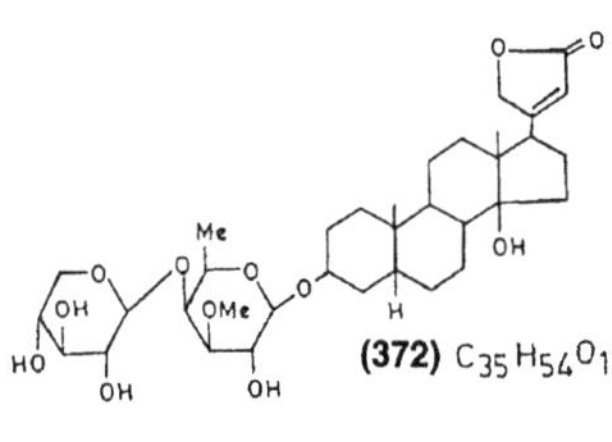

(372) $C_{35}H_{54}O_{12}$ (*160*)

Table 3 (*continued*)

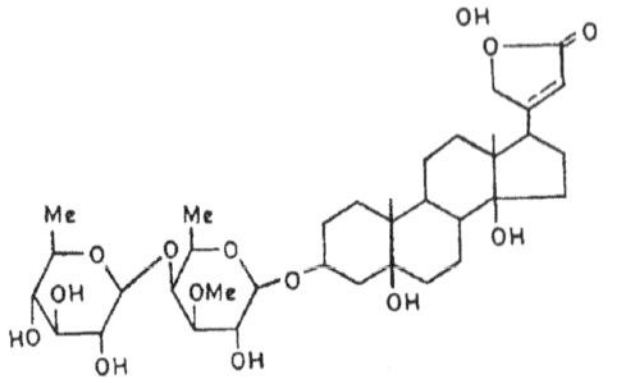

(373) $C_{36}H_{56}O_{13}$ (*160*)

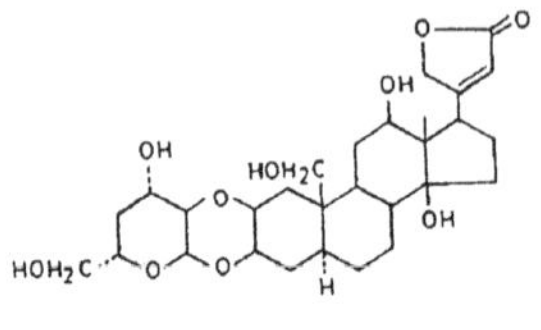

(374) $C_{29}H_{42}O_{9}$ (*98*)

References

1. SINGH, B., and R.P. RASTOGI: Cardenolide Glycosides and Genins. Phytochem., **9**, 315 (1970).

2. REICHSTEIN, T.: Cardenolid und Pregnanglykoside. Naturwiss., **54**, 53 (1967).

3. HOCH, J.H.: A Survey of Cardiac Glycosides and Genins. Columbia: University of South Carolina Press. 1961.

4. WILSON, J.W.: Medicinal Chemistry (A. BURGER, ed.), p. 626. New York: Interscience. 1960.

5. REDONNET, T.A.: Critical and Comparative Study from the Pharmacological View Point of the 16th Revision of the Pharmacopeia of the United States, I: The Selection of the Monograph Chapters Where the Drugs Are Described. Farmacognosia, **25**, 1 (1965).

6. LIBIZOV, N.I., and I.A. GUBANOV: Plants Containing Cardiac Glycosides. Farmatsiya, **16**, 29 (1967).

7. STOLL, A., and A. RENZ: Die Überführung von Scillaren A in Epi-allo-lithocholsäure. Helv. Chim. Acta, **24**, 1380 (1941).

8. CHEU, K.K.: Pharmacological Evaluation of Sedative and Hypnotic Drugs. J. Med. Pharm. Chem., **3**, 111 (1961).

9. CABRERA, G.M., M. E. DELUCA, and G.M.L. SELDES: Cardenolide Glycosides from the Roots of *Mandevilla pentlandiana*. Phytochem., **32**, 1253 (1993).

10. ABE, F., T. YAMAUCHI, and T. NOHARA: C-Nor-D-homo-cardenolide Glycosides from *Thevetia neriifolia*. Phytochem., **31**, 251 (1992).

11. REICHSTEIN, T., and S. RANGASWAMI: Konstitution von Odorosid A und Odorosid B. Helv. Chim. Acta, **32**, 939 (1949).

12. MABRY, T.J., K.R. MARKHAM, and M.B. THOMAS: The Systematic Identification of Flavonoids. New York: Springer. 1970.

13. KRASSO, A.F., EK. WEISS, and T. REICHSTEIN: Die Cardenolide von *Beaumontia grandiflora* Wallich. Helv. Chim. Acta, **46**, 1691 (1963).

14. ZELNIK, R.U., and O. SCHINDLER: Die Glykoside der Samen von *Strophanthus eminii*. Helv. Chim. Acta, **40**, 2110 (1957).

15. REICHSTEIN, T., K.D. ROBERTS, and EK. WEISS: Die Cardenolide der Samen von *Mallotus paniculatus*. Helv. Chim. Acta, **49**, 316 (1966).

16. YING, H., and N.C. SUN: Cardiac Glycosides of *Thevetia peruviana* (II): Isolation and Identification of Cerberin, Ruvoside and a New Cardiac Glycoside Perusitin. Yas Hsueh Pao, **11**, 464 (1964).

17. Reichstein, T., and O. Schindler: Die Reduktion des Isodigitoxigenins nach Wolff-Kishner. Helv. Chim. Acta, **39**, 1876 (1956).
18. Manzetti, A.R., and T. Reichstein: Die Glykoside von *Streblus asper*. Helv. Chim. Acta, **47**, 2320 (1964).
19. Sanlewiz, L., Ek. Weiss, and T. Reichstein: Die Pregnanderivate der Wurzeln von *Asclepias lilacina*. Helv. Chim. Acta, **50**, 530 (1967).
20. Kiliani, H.: Über *Digitalinum verum*. Ber. Deutsch. Chem. Ges., **63**, 2866 (1930).
21. Mannich, C., and G. Siewert: Über g-Strophanthin (Ouabain) und g-Strophanthidin. Ber. Deutsch. Chem. Ges., **75**, 737 (1942).
22. Srivastava, S., A. Khare, and M.P. Khare: A Cardenolide Tetraglycoside from *Oxystelma esculentum*. Phytochem., **30**, 301 (1991).
23. Horiger, N., D. Zivanov, H.H.A. Linde, and K. Meyer: Weitere Cardenolide aus Ch' an Su. Helv. Chim. Acta, **53**, 2051 (1970).
24. Singh, B., and R.P. Rastogi: Chemical Investigation of *Asclepias curassavica* Linn. Indian J. Chem., **7**, 1105 (1969).
25. Brown, P., F. Bruschweiler, and G.R. Pettit: Field Ionization Mass Spectrometry, III: Cardenolides. Org. Mass. Spectrom., **5**, 573 (1971).
26. Brown, P., F. Bruschweiler, and G.R. Pettit: Massenspektrometrische Untersuchungen von Naturprodukten, Cardenolide. Helv. Chim. Acta, **55**, 531 (1972).
27. Bosso, C., F. Taravel, J. Ulrich, and M. Vignon: Utilisation du ^{13}C en Spectrometrie de Masse: Etude de la Fragmentation de Disaccharides. Org. Mass. Spectrom., **13**, 477 (1978).
28. Wood, G.W.: Some Recent Application of Field Ionization/Field Desorption Mass Spectrometry to Organic Chemistry. Tetrahedron, **32**, 1125 (1982).
29. Latimer, R.P., and H.R. Schulten: Anal. Chem., **61**, 1201A (1989).
30. Sethi, A., D. Deepak, M.P. Khare, and A. Khare: A Novel Pregnane Glycoside from *Periploca calophylla*. J. Nat. Prod., **51**, 787 (1988).
31. Jin, Q.D., Q.L. Zhou, and Q.Z. Mu: Two New Pregnane Oligoglycosides from *Dregea sinensis* var. *corrugata*. J. Nat. Prod., **52**, 1214 (1989).
32. Bose, A.K., and P.R. Srinivasan: NMR Spectral Studies, XII: Trichloroacetyl Isocyanate as *in situ* Derivatizing Reagent for ^{13}C NMR Spectroscopy of Alcohols, Phenols and Amines. Tetrahedron, **31**, 3025 (1975).
33. Yamauchi, T., F. Abe, and T. Santisuk: Cardiac Glycosides of *Beaumontia brevituba* and *B. murtonii*. Phytochem., **29**, 1961 (1990).
34. Trujillo, J.M., O. Hernandez, and E. Navarro: Mono-Glycosides from *Isoplexis chalcantha*. J. Nat. Prod., **53**, 167 (1990).
35. Junior, P., and M. Wichtl: 3-Epi-periplogenin; ein neues Cardenolid aus *Adonis vernalis*. Phytochem., **19**, 2193 (1980).
36. Yamauchi, T., F. Abe, and M. Nishi: Carbon-13 NMR of 5-Cardenolides. Chem. Pharm. Bull., **26**, 2894 (1978).
37. Tori, K., H. Ishii, Z.W. Walkowski, C. Chachaty, M. Sangare, F. Piriou, and G. Lukas: Carbon-13 Nuclear Magnetic Resonance Spectra of Cardenolides. Tetrahedron Lett., 1077 (1973).
38. Yoshimura, S.I., H. Narita, K. Hayashi, and H. Mitsuhashi: Studies on the Constituents of Asclepiadaceae Plants, LIX: The Structures of Five New Glycosides from *Dregea volubilis*. Chem. Pharm. Bull., **33**, 2287 (1985).
39. Bock, K., and C. Pedersen: Carbon-13 Nuclear Magnetic Resonance Spectroscopy of Monosaccharides. Adv. in Carb. Chem. and Biochem., **41**, 27 (1983).
40. Agrawal, P.K., and M.C. Bansal: Carbon-13 NMR of Flavonoids, p. 283. Amsterdam: Elsevier. 1989.

41. ALLERHAND, A., and D. DODDRELL: Strategies in the Application of Partially Relaxed Fourier Transform Nuclear Magnetic Resonance Spectroscopy in Assignments of Carbon-13 Resonances of Compex Molecules: Stachyose. J. Am. Chem. Soc., **93**, 2777 (1971).

42. MORRIS, G.A.: Modern NMR Techniques for Structure Elucidation. Magn. Res. in Chem., **24**, 371 (1986).

43. HOFFMANN, J.F., and B. FORBUSH: Structure and Function of the Na/K Pump, III, **19**. New York: Academic Press. 1983.

44. GUNTERT, TH.W., and H.H.A. LINDE: Handbook of Experimental Pharmacology (K. GEEFF, ed.), p. 56. Berlin: Springer. 1981.

45. FRANCIS, C.K.C., and R.W. THOMAS: Conformational Factors in Cardiac Glycoside Activity. J. Med. Chem., **28**, 509 (1985).

46. GUNTERT, TH.W., and H.H.A. LINDE: Cardiac Glycosides: Prerequisites for the Development of New Cardiotonic Compounds. Experientia, **33**, 697 (1977).

47. ZÜRCHER, W., E.W. BERG, and CH. TAMM: 14-Epi-digitoxigenin und 3-Desoxydigitoxigenin. Helv. Chim. Acta, **52**, 2449 (1969).

48. ROHRER, D.C., M. KIHARA, T. DEFFO, H. RATHORE, K. AHMED, A.H.L. FROM, and D.S. FULLERTON: Cardiac Glycosides, 4: A Structural and Biological Analysis of β-D-Digitoxosides, β-D-Digitoxose Acetonides and Their Genins. J. Am. Chem. Soc., **106**, 8269 (1984).

49. POLONIA, J., M.H. ARANJO, and M.A. POLONIA: An. Fac. Farm. Porto., **26**, 83 (1966).

50. COOMBE, R.G., and T.R. WATSON: The Cardiac Glycosides of *Gomphocarpus fruticosus* R. Br., III: Gomphoside. Aust. J. Chem., **17**, 92 (1964).

51. KUPCHAN, S.M., J.R. KNOX, J.E. KELSEY, and J.A.S. RENAULD: Calotropin, a Cytotoxic Principle Isolated from *Asclepias curassavica*. Science, 1685 (1964).

52. GERAN, R.I., N.H. GREENBERG, M.N. MACDONALD, A.M. SCHUMACHER, and B.J. ABBOTT: Protocols for Screening Chemical Agents and Natural Products Against Animal Tumors and Other Biological Systems. J. Cancer Chemother. Rep., Part 3, **3**, 17 (1972).

53. MARKS, W.H., H.H.S. FONG, M. TIN-WA, and N.R. FARNSWORTH: Cytotoxic Principles of *Parquetina nigrescens* Bullock. J. Pharm. Sci., **64**, 1674 (1975).

54. KIHARA, M., K. YOSHIKA, T. DEFFO, D.S. FULLERTON, and D.C. ROHRER: Cardiac Glycosides, 3: Synthesis of β-D-Digitoxose Analogues. Tetrahedron, **40**, 1121 (1984).

55. GILMAN, G., L. GOODMAN, and A. GILMAN (eds.): The Pharmacological Basis of Therapeutics, 6th Ed. New York: Macmillan. 1980.

56. FERTH, R., A. BAUMANN, W. ROBIEN, and B. KOPP: Cardenolides from *Ornithogalum mutans*, Part 1. Z. Naturforsch., **B47**, 1444 (1992).

57. FERTH, R., A. BAUMANN, K.K. MAYER, W. ROBIEN, and B. KOPP: Cardenolides from *Ornithogalum mutans*, Part 2. Z. Naturforsch., **B47**, 1459 (1992).

58. VENKATESWARA, R.R., and C.S. VAIDYANATHAN: Chemistry and Biochemical Pharmacology of Cardiac Glycosides. J. Ind. Inst. Sci., **71**, 329 (1991).

59. ERDMANN, E., and K. GREEFF (eds.): Extra Cardiac Effects of Cardiac Glycosides, Cardiac Glycosides 1785–1985, p. 347. Fed. Rep. Ger. 1986.

60. ZUEVA, N.A.: Action Mechanism of Cardiac Steroids. Biochim. Zhivotn. Chel., **16**, 66 (1992).

61. PIERI, F., G.M.C. ARNOULD, and E.H. SEFRAUI: Cardiotonic Glycosides from *Acokanthera spectabilis*. Fitoterapia, **63**, 333 (1992).

62. YAMAUCHI, T., and F. ABE: Cardiac Glycosides and Pregnanes from *Adenium obesum*. Chem. Pharm. Bull., **38**, 669 (1990).

63. ABE, F., and T. YAMAUCHI: Affinoside A and Companion Glycosides from the Stem and Bark of *Anodendron affine*. Chem. Pharm. Bull., **30**, 1183 (1982).
64. ABE, F., and T. YAMAUCHI: Anodendron, IV: Affinosides S-1/S-VII. Chem. Pharm. Bull., **31**, 1199 (1983).
65. ABE, F., and T. YAMAUCHI: Anodendron, V: Affinosides M and K Cardenolide from the Seeds of *Anodendron affine*. Chem. Pharm. Bull., **33**, 847 (1985).
66. ABE, F., and T. YAMAUCHI: Affinosides L_a–L_e, Major Cardenolide Glycosides from the Leaves of *Anodendron affine*. Chem. Pharm. Bull., **33**, 3662 (1985).
67. ABE, F., T. YAMAUCHI, T. FUJIOKA, and K. MIHASHI: Anodendron, VIII: Minor Cardenolide Glycosides and a Cardenolide from Leaves of *Anodendron affine*. Chem. Pharm. Bull., **34**, 2774 (1986).
68. ABE, F., and T. YAMAUCHI: Re-Investigation of Cardenolides from Fresh Leaves of *Anodendron affine*. Phytochem., **33**, 457 (1993).
69. HANADA, R., F. ABE, Y. MORI, and T. YAMAUCHI: Studies on Anodendron, Part 10: Reinvestigation of Cardenolide Glycosides from Seeds of *Anodendron affine*. Phytochem., **31**, 3547 (1992).
70. FUKUYAMA, Y., M. OCHI, H. KASSAI, and M. KODAMA: Insect Growth Inhibitory Cardenolide Glycosides from *Anodendron affine*. Phytochem., **32**, 297 (1993).
71. KULSHRESHTHA, D.K., and R.P. RASTOGI: Cardioactive Constituents from *Carissa spinarum*. Indian J. Chem., **7**, 1102 (1969).
72. ABE, F., and T. YAMAUCHI: Studies on *Cerbera*, I: Cardiac Glycosides in the Seeds, Bark and Leaves of *Cerbera manghus*. Chem. Pharm. Bull., **25**, 2744 (1977).
73. YAMAUCHI, T., F. ABE, and A.S.C. WAN: Cardenolide Monoglycosides from the Leaves of *Cerbera odollum* and *Cerbera manghus*. Chem. Pharm. Bull., **35**, 2744 (1987).
74. YAMAUCHI, T., F. ABE, and A.S.C. WAN: Studies on *Cerbera*, IV: Polar Carenolide Glycosides from the Leaves of *Cerbera odollum* and *Cerbera manghus*. Chem. Pharm. Bull., **35**, 4813 (1987).
75. KISLICHENKO, S.G., I.P. MAKAREVICH, and D.G. KOLESNIKOV: Glucoevonogenin and Glucoevonoloside. Khim. Prir. Soedin., **5**, 193 (1969).
76. YAMAUCHI, T.: Cardiokinetic and Diuretic Gluconerigoside in *Nerium* Leaves. Japan Kokai, 74061131 (1974).
77. YAMAUCHI, T., N. TAKATA, and T. MIMURA: *Nerium*, 5: Cardiac Glycosides of the Leaves of *Nerium odorum*. Phytochem., **14**, 1379 (1975).
78. SIDDIQUI, S., F. HAFFEZ, S. BEGUM, and B.S. SIDDIQUI: Isolation and Structure of Two Cardiac Glycosides from the Leaves of *Nerium oleander*. Phytochem., **26**, 237 (1986).
79. HANADA, R., F. ABE, and T. YAMAUCHI: *Nerium*, Part 14: Steroid Glycosides from the Roots of *Nerium odorum*. Phytochem., **31**, 3183 (1992).
80. CHEN, R., F. ABE, T. YAMAUCHI, and M. TAKI: Cardenolide Glycosides of *Strophanthus divaricatus*. Phytochem., **26**, 2351 (1987).
81. JOLAD, S.D., J.J. HOFMANN, J.R. COLE, M.S. TEMPESTA, and R.B. BATES: 3′-O-Methylevomonoside, a New Cytotoxic Cardiac Glycoside from *Thevetia ahonia*. J. Org. Chem., **46**, 1946 (1981).
82. ABE, F., T. YAMAUCHI, and A.S.C. WAN: *Thevetia*, Part 2: Cardiac Glycosides from the Leaves of *Thevetia neriifolia*. Phytochem., **31**, 3189 (1992).
83. ABE, F., Y. MORI, and T. YAMAUCHI: Cardenolide Glycosides from the Seeds of *Asclepias curassavica*. Chem. Pharm. Bull., **40**, 2917 (1992).
84. ABE, F., Y. MORI, and T. YAMAUCHI: 3′-Epi-19-norafroside and 12-Hydroxycoroglaucigenin from *Asclepias curassavica*. Chem. Pharm. Bull., **39**, 2709 (1991).
85. CHEUNG, H.T.A., R.G. COOMBE, W.T.L. SIDWELL, and T.R. WATSON: Afroside, a 15-Hydroxycardenolide. J. Chem. Soc. Perkin Trans. I, 64 (1981).

86. CHEUNG, H.T.A., C.K.C. FRANCIS, R.W. THOMAS, and J.W. ROBERT: Cardenolide Glycosides of the Asclepiadaceae. New Glycosides from *Asclepias fruticosa* and the Stereochemistry of Uscharin, Voruscharin and Calotoxin. J. Chem. Soc. Perkin I, 2827 (1983).
87. SABUJO, N., S. MURRAY, J.V.S. BLUM, and J.H. ROBERT: Structure of Humistratin, a Novel Cardenolide from the Sandhill Milkweed *Asclepias humistrata*. J. Org. Chem., **47**, 2154 (1982).
88. CHEUNG, H.T.A., T.R. WATSON, S.M. LEE, M.M. MACCHERNEY, and J.N. SEIBER: Structure of Aspecioside from the Monarch Butterfly Larvae Foodplants *Asclepias speciosa* and *syriaca*. J. Chem. Soc. Perkin Trans. I, 61 (1986).
89. SHIVANAND, D.J., B.B. ROBERT, R.C. JACK, J.H. JOSEPH, J.S. TERUNA, and N.T. BARBARA: Cardenolide and a Lignan from *Asclepias subulata*. Phytochem., **25**, 2581 (1986).
90. CHEUNG, H.T.A., C.J. NELSON, and T.R. WATSON: New Cardenolide Glycosides with Doubly Linked Sugars from *Asclepias vestita*. J. Chem. Res. Synop., **6** (1989).
91. CHEUNG, H.T.A., and C.J. NELSON: Cardenolide Glycosides with 5,6-Unsaturation from *Asclepias vestita*. J. Chem. Soc. Perkin I, 1563 (1989).
92. KHARE, M.P., and B.B. SHAH: Structure of Buchanin, a New Cardenolide from *Cryptolepis buchanani*. J. Nep. Chem. Soc., **3**, 21 (1983).
93. VENKATESWARA, R., K.S. RAO, and C.S. VAIDYANATHAN: Cryptosin, a New Cardenolide in Tissue Culture and Intact Plants of *Cryptolepis buchanani*. Planta Cell Rep., **6**, 291 (1987).
94. PURUSHOTHAMAN, K.K., S. VASANTHA, J.D. CONNOLLY, and D.S. RYCROFT: New Sarverogenin and Isosarverogenin Glycosides from *Cryptolepis buchanani*. Rev. Latinoam. Quim., **19**, 28 (1988).
95. SANDUJA, R., W.Y.R. LO, K.L. EULER, and M. ALAM: Cardenolides of *Cryptostegia madagascariensis*. J. Nat. Prod., **47**, 260 (1984).
96. LICHINS, J.J., K.D. EL, and M. WICHTL: Subalpinoside (S1), a New Oleandrigenin Glycoside and Other Cardenolides from *Digitalis subalpina* var. *subalpina*. Planta Med., **57**, 159 (1991).
97. REICHSTEIN, P., H. KAUFMANN, W. STÖCKLIN, and T. REICHSTEIN: Glycosides and Glycons. Glycosides from *Glossostelma carsoni* Roots. Helv. Chim. Acta, **50**, 2114 (1967).
98. MANSOUR, S., M.S. HIFNAWY, A.T. MCPHAIL, and D.R. MCPHAIL: Ghalakinoside, a Cytotoxic Cardiac Glycoside from *Pergularia tomentosa*. Phytochem., **27**, 3245 (1988).
99. TAKIDO, M., S. KITANAKA, K. MIZOGAMI, and S. NAKAIKE: Isolation of Antitumor Cardenolide Glycosides from *Euonymus alata*. Chem. Abstr., **119**, 68402 p (1993).
100. SHIMADA, K., T. KYUNO, T. NAMBARA, and I. UCHIDA: Elaeodendroside D, E, H, I, and J, New Cardiac Steroids from *Elaeodendron glaucum*. Heterocycles, **15**, 355 (1981).
101. WAGNER, H., H. HABERMEIER, and H.R. SCHULTEN: Constituents of Celastraceae, Part B: Cardiac Glycosides of the Arrow Poison from *Lophopetalum toxicum* Loher. Helv. Chim. Acta, **67**, 54 (1984).
102. MAKAREVICH, I.F.: Cardiac Glycosides of *Cheiranthus allionii*, VI. Khim. Prir. Soedin., **6**, 331 (1970).
103. MAKAREVICH, I.F.: Cardiac Glycosides of *Cheiranthus allionii*, VII. Khim. Prir. Soedin., **8**, 188 (1972).
104. MAKAREVICH, I.F.: Cardiac Glycosides of *Cheiranthus allionii*, VIII. Khim. Prir. Soedin., **8**, 681 (1972).

105. MAKAREVICH, I.F., D.G. KOLESNIKOV, and V.F. BELOKON: Cardiac Glycosides of *Cherianthus allionii.* Khim. Prir. Soedin., **10**, 607 (1974).
106. MAKAREVICH, I.F.: Cardiac Glycosides of *Cheiranthus allionii*, X. Khim. Prir. Soedin., **10**, 738 (1974).
107. MAKAREVICH, I.F., D.D. KOLESNIKOV, I.P. KOVALEV, V.G. GORDIENKO, and V.S. KABANOV: Cardiac Glycosides of *Cheiranthus allionii.* Khim. Prir. Soedin., **11**, 754 (1975).
108. MAKAREVICH, I.F., A.I. PARLII, and S.I. MAKAREVICH: Cardiac Glycosides of *Cheiranthus allionii*, XIII: Glucoerycordin. Khim. Prir. Soedin., **25**, 73 (1989).
109. GONZALEZ, A.G., and E.M. LUQUE: New Cardiac Glycosides from *Cheiranthus scoparius.* An. Quim., **71**, 97 (1975).
110. GONZALEZ, A.G., and E.M. LUQUE: New Cardiac Glycosides from *Cheiranthus scoparius.* Farm. Naeva, **41**, 261, 267, 273 (1976).
111. MASLENNIKOVA, V.A., G.L. GINKINA, R.U. UMAROVA, A.M. NAVRUZOVA, and N.K. ABUBAKIROV: Glycosides of Plants of *Erysimum* Species. Khim. Prir. Soedin., **3**, 173 (1967).
112. MAKAREVICH, I.F., K.V. ZHEMOKLEV, T.V. SLYUSARSKAYA, A.D., MAGOMEDOVA, T.N. TEREKHOVA, and G.T. SIRENKO: Cardiac Glycosides of *Erysimum contractum.* Khim. Prir. Soedin., **27**, 58 (1991).
113. MAKAREVICH, I.F., O.I. KHIMENKO, and D.G. KOLESNIKOV: Cardiac Glycosides of *Erysimum crepidifolium.* Khim. Prir. Soedin., **10**, 611 (1974).
114. MASLENNIKOVA, V.A., R.U. UMAROVA, and N.K. ABUBAKIROV: Glucosides of Erysimum, X: Cardenolides of *Erysimum cuspidatum.* Khim. Prir. Soedin., **11**, 166 (1975).
115. KOLOROVA, B., and M. BOYADZHIEVA: Phytotechnological Investigation of *Erysimum repandum.* Probl. Farm., **6**, 39 (1978).
116. MAKSYUTINA, N.P.: Sinapoyl Ester of Glucoerysimoside in *Erysimum marschallianum.* Khim. Prir. Soedin., **11**, 603 (1975).
117. OKABE, H., K.P. INONE, and T. YAMAUCHI: Studies on the Constituents of *Mallotus japonicus* Muell. Arg., I: Cardiac Glycosides from the Seeds. Chem. Pharm. Bull., **24**, 108 (1976).
118. OKABE, H., and T. YAMAUCHI: Studies on the Constituents of *Mallotus japonicus* Muell. Arg., II: A Corotoxigenin Trioside from the Seeds. Chem. Pharm. Bull., **24**, 2886 (1976).
119. KOVOLEV, V.N., and A.N. KOMISSARENKO: Cardenolides of *Coronilla glauca* and *C. scorpioides.* Khim. Prir. Soedin., **21**, 676 (1985).
120. KOMISSARENKO, A.N., and V.N. KOVALEV: Coronillobioside (I), a Cardenolide Glycoside from Seeds of *Coronilla scorpioides.* Khim. Prir. Soedin., **24**, 726 (1988).
121. KOMISSARENKO, N.F., and E.P. STUPAKOVA: Neoconvalloside. A Cardenolide Glycoside from Plants of the Genus *Convallaria.* Khim. Prir. Soedin., **22**, 201 (1986).
122. SCHENK, B., P. JUNIOR, and M. WICHTL: Cannogenol-3-O-α-L-rhamnoside and Cannogenol-3-O-β-D-allomethyloside, Two New Cardiac Glycosides from *Convallaria majalis.* Planta Med., **40**, 1 (1980).
123. BUCHVAROV, YA.: Cardenolides of *Convallaria majalis*, VII: Isolation of Sarmentologenin Cardiac Glycosides. Farmatsiya, **34**, 6 (1984).
124. ATANASUVA, B. (ed.): Cardenolides of *Convallaria majalis.* In: Proc. Int. Conf. Chem. Biotechnol. Bio. Acta Nat. Prod., **3**, 150 (1981) [Bulg. Acta Sci., Sofia, Bulgaria].
125. BUCHAROV, YA., and N.F. KOMISSARENKO: Neoconvallatoxoloside. Cardenolide Glycoside from *Convallaria majalis.* Khim. Prir. Soedin., **13**, 537 (1977).
126. KOPP, B., and W. KUBELKA: New Cardenolide from *Convallaria majalis.* Planta Med., **45**, 195 (1982).

127. KUCHUKHIDZE, D.K., N.F. KOMISSARENKO, and L.I. ERISTAVI: Cardiac Glycosides of *Rhodea japonica*. Soobsch. Akad. Nauk. Gruz. SSR Ser. Khim., 64, 597 (1971).

128. KUCHUKHIDZE, D.K., N.F. KOMISSARENKO, and L.I. ERISTAVI: Rhodexoside from the *Rhodea japonica*. Iz. Akad. Nauk. Gruz. SSR Ser. Khim., **8**, 157 (1982).

129. SAXENA, V.K., and P.K. CHATURVEDI: A Novel Cardenolide Canarigenin-3-O-α-L-rhamnopyranosyl-(1 → 5)-O-β-D-xylofuranoside. J. Nat. Prod., **55**, 39 (1992).

130. SAXENA, V.K., and S.K. CHATURVEDI: Cardiac Glycosides from the Roots of *Streblus asper*. Planta Med., **4**, 343 (1985).

131. FIEBIG, M., C. DUH, and P. YIH: Plant Anticancer Agents, XLI: Cardiac Glycosides from *Streblus asper*. Planta Med., **48**, 981 (1985).

132. CHATTERJEE, R.K., N. FATMA, P.K. MURTHY, P. SINHA, D.K. KULSHRESHTHA, and B.N. DHAWAN: Macrofilaricidal Activity of the Stem Bark of *Streblus asper* and Its Major Active Constituents. Drug. Dev. Res., **26**, 67 (1992).

133. KOPP, B., L. KRENN, E. KUBELKA, and W. KUBELKA: Cardenolides from *Adonis aestivalis*. Phytochem., **31**, 3195 (1992).

134. JUNIOR, P., D. KRUEGER, and C. MARBURG: *Adonis aleppica* Boiss. Phytochemical Studies, Isolation and Structure Elucidation of Cardenolides. Dtsch. Apoth. Ztg., **125**, 1945 (1985).

135. PAULI, F.G., P. JUNIOR, S. BERGER, and U. MATTHIESEN: Alepposides, Cardenolide Oligoglycosides from *Adonis aleppica*. J. Nat. Prod., **56**, 67 (1993).

136. MANNA, F., M.L. STEIN, B. ANZALONE, and E. POSOCCO: Study on the Active Principles of *Adonis distorta* Ten. Fitoterapia, **49**, 56 (1978).

137. WINKLER, C., and M. WICHTL: New Cardenolides from *Adonis vernalis*. Pharm. Acta Helv., **60**, 243 (1985).

138. WICHTL, M., and P. JUNIOR: Strophanthidin Digitaloside (I) and Strophanthidin Gulomethyloside (II), Two New Cardenolide Glycosides from *Adonis vernalis*. Arch. Pharm., **300**, 905 (1977).

139. WICHTL, M., K. JEUTZSCH, and K. TUERK: Strophanthidin Fucoside: A New Cardenolide Glycoside from *Adonis vernalis*. Monatsh. Chem., **103**, 889 (1972).

140. IMRE, Z., and T. VURDUN: Cardiac Glycosides from the Seeds of *Digitalis cariensis*. Planta Med., **54**, 529 (1988).

141. DAGMAR, K., and W. MAX: New Cardiac Glycosides from *Digitalis lanata*, Part 3. Planta Med., **50**, 265 (1984).

142. KAMILIA, F.T.: Isolation of a New Cardenolide from *Digitalis lanata*. J. Drug. Res., **15**, 245 (1984).

143. DAGMAR, K., and W. MAX: New Cardiac Glycosides from *Digitalis lanata*, Part 4. Planta Med., **50**, 267 (1984).

144. DAGMAR, K., and W. MAX: New Cardiac Glycosides from *Digitalis lanata*. Planta Med., **50**, 168 (1984).

145. DAGMAR, K., J. PETER, and W. MAX: New Cardiac Glycosides from *Digitalis lanata*. Planta Med., **49**, 74 (1983).

146. SAM, F.A., D. MINGEL, J.M. CERRAL, P. PUEBLA, M. MEDARDE, and A.F. BARRERO: Digithapsins and Other Components from *Digitalis thapsi*. An. Quim., **C84**, 31 (1988).

147. RAO, E.V., and D.V. RAO: Cardenolides of the Seeds of *Corchorus acutangulus*. Indian J. Chem., **7**, 1276 (1969).

148. GINLIANA, S., and C.C. RICCA: ^{13}C NMR Spectra of Strophanthidin Glycosides of *Corchorus capsularis*. Gazz. Chem. Ital., **112**, 349 (1982).

149. MAHATO, S.B.D., N.P. SAHU, S.K. ROY, and B.N. PRAMANIK: Cardiac Glycosides from *Corchorus olitorius*. J. Chem. Soc. Perkin I, 2065 (1989).

150. STEYN, P.S., H. VAN, R. FANIE, R. VLEGGAAR, and L.A.P. ANDERSON: Bufadienolide Glycosides of the Crassulacea. Structure and Stereochemistry of *Cotyleden orbiculata.* J. Chem. Soc. Perkin Trans. I, 1633 (1986).

151. ANDERSON, L.A.P., P.S. STEYN, H. VAN, and R. FANIE: The Characterization of Two Novel Byfadienolides, Lancetoxins A and B from *Kalanchoe lanceolata* Forsk. J. Chem. Soc. Perkin Trans. I, 1573 (1984).

152. KRENN, L., B. KOPP, C.E. GRIESMAYER, and W. KUBELKA: Bufadienolides from *Urginea aphylla speta.* Sci. Pharm., **60**, 65 (1992).

153. KRENN, L., M. JAMBRITS, and B. KOPP: Bufadienolides from *Urginea hesperia.* Planta Med., **54**, 227 (1988).

154. KUNTH, K.B., and M. DANNER: Bufadienolides, I: Structure of the Bufadienolides of *Urginea indica.* Sci. Pharm., **51**, 227 (1983).

155. KRENN, L., R. FRETH, W. ROBIEN, and B. KOPP: Bufadienolides from *Urginea maritima sensu/strictu.* Planta Med., **57**, 560 (1991).

156. KRENN, L., M. BAMBERGER, and B. KOPP: Bufadienolides, VI: A New Bufadienolide from *Urginea pancration.* Planta Med., **58**, 284 (1992).

157. VAN, H., R. FANIE, R. VLEGGAAR, and L.A.P. ANDERSON: Bufadienolide Glycoside from *Urginea physodes.* First Report of Natural 14-Deoxybufadienolides. S. Afr. J. Chem., **41**, 145 (1988).

158. KISSMER, B., and M. WICHTL: Bufadienolides from the Seeds of *Helleborus odorus.* Planta Med., 152 (1986).

159. VAN, H., R. FANIE, R. VLEGGAAR, and L.A.P. ANDERSON: Structure Elucidation of Thesiuside, a Bufadiendiolide Glycoside from *Thesium lineatum.* S. Afr. J. Chem., **41**, 39 (1988).

160. SRIVASTAVA, S., M.P. KHARE, and A. KHARE: Cardenolide Diglycosides from *Oxystelma esculentum.* Phytochem., **32**, 1019 (1993).

(*Received November 29, 1994*)

Aspects of the Enzymology of the Shikimate Pathway

E. HASLAM, Department of Chemistry, University of Sheffield, Sheffield, S3 7HF, U.K.

Contents

1. Introduction

"C'était un homme de génie, qui a devance son temps à maints égards; ses grands merites seront plus reconnus sans aucun doute par la posterité qu'ils ne l'ont été par ses contemporains."

These are the concluding words of an appreciation of the life and scientific work of JOHAN FREDERIK EYKMAN (1851–1915) published in *Recueil Travaux Chimique des Pays-Bas*, **XXXV**, 365–420 (1916), written by A. F. HOLLEMAN. It is very doubtful if the writer appreciated, at the time, the full significance of his words. Johan Frederik was the second of eight children born to Christiaan Eykman and his wife Johanna Alida in the village of Nijkerk. He qualified in pharmaceutical science in 1874 from the University of Amsterdam and a year later began to prepare to study for his doctorate in physical sciences at the University of Leiden. These studies were interrupted forever by an extraordinary circumstance. EYKMAN was approached by the Japanese government and nominated as director of a laboratory charged with the analysis of medicaments and research into indigenous materials. In 1881 he was elected to a chair in Chemistry in the Faculty of Medicine at the University of Tokyo; the first Dutchman to hold a chair in this University. In 1886 EYKMAN returned to Holland but before doing so he was received in audience by the Emperor who rewarded him for his services to the Empire with "l'ordre du Soleil Levant".

Thus it is that the first publication on shikimic acid, "Sur les Principes Constituants de l'*Illicium religiosum* (Seib)", emanated from the Laboratoire Pharmaceutique, Tokyo in 1884 (*Recueil Travaux Chimique des*

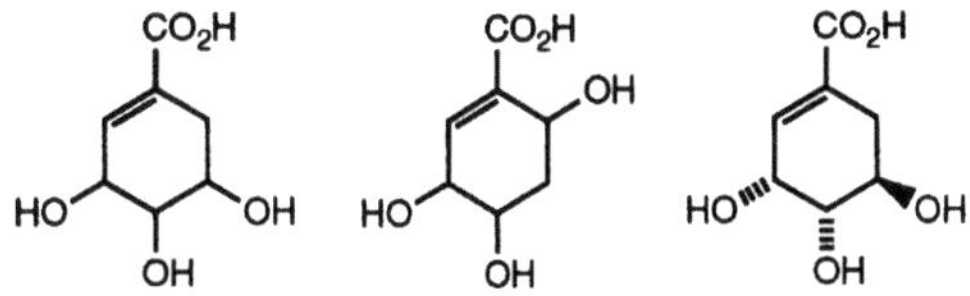

(1) (2) (3)

Pays-Bas, **4**, 32–54 (1885)). Two further publications followed in 1886 (*Recueil Travaux Chimique des Pays-Bas*, **5**, 299–304 (1886)) and in 1891, from his private laboratory in Amsterdam, a full paper – "Über die Shikimisäure", *Ber. Dtsch. Chem. Ges.*, **24**, 1278–1303 (1891) – which gives a comprehensive description of the chemistry of shikimic acid. It is, by any standards, an excellent paper and reveals a man with superb experimental skills. EYKMAN made some remarkably percipient observations on shikimic acid ($C_7H_{10}O_5$; m.p. 184 °C, $[\alpha]_D = -176°$ in ethanol; $[\alpha]_D = -246°$ in water) and gave the acid its name, deriving it from shikimi-no-ki–the Japanese name of the plant *Illicium religiosum*. He underlined the relationship of shikimic acid to quinide, the lactone of quinic acid ($C_7H_{12}O_6$) and described the preparation of triacyl derivatives and the formation of dihydro- and dibromoshikimic acids. He was also able to transform the latter acid to a bromolactone [H_2O, Ag_2O-bromoquinide] and thence [$Ba(OH)_2$] to dihydroxyshikimic acid. He eventually put forward six possible structures for shikimic acid and, left finally with a choice between two of these (**1** and **2**), chose the wrong one! Pioneering studies by H. O. L. FISCHER in the 1930s established not only the vicinal triol structure (**1**) but also the absolute stereochemistry of (–)-shikimic acid as (**3**), (*58–62*). A crucial feature of this work was the degradation of (–)-shikimic acid to 2-deoxy-D-*arabino*-hexono-γ-lactone which established the stereochemical relationship to D-glucose.

(–)-Shikimic acid was isolated by EYKMAN from the fruits of *Illicium religiosum* and *Illicium anisatum* where it occurs in surprisingly large quantities (up to ~ 20%). GREWE and LORENZEN (*76*) introduced a substantial improvement in the method of isolation of (–)-shikimic acid from *Illicium* fruit in 1953 by the application of ion-exchange chromatography and at the same time embarked upon an impressive series of investigations of the chemistry of the acid, leading to a number of critical transformations, the preparation of a range of derivatives and the first synthesis (*76–79*).

The biographical notes on J. F. EYKMAN reveal not only an accomplished experimental scientist but a powerful personality; a man who could agree to differ with JACOBUS VAN'T HOFF but still retain the great man's recognition and appreciation of his ability. The scientific story of

(−)-shikimic acid over the past forty years began with some remarkably elegant observations and experiments by an equally gifted scientist and similarly strong personality, BERNARD DAVIS, Harvard Professor of Medicine and author (1915–1994). In a remarkably frank and eloquent essay 'Science and Politics: Tensions between the Head and the Heart" (*Ann. Rev. Microbiol.*, **46**, 1–33 (1992)), DAVIS portrays his life as a scientist, his strengths and weaknesses, and the controversies in which he unwittingly became embroiled – most particularly the relationships between science and society to which he was increasingly drawn in the 1970s. He also noted that "*although a systematic program, pursuing the shikimate pathway has probably contributed most to my scientific reputation, I have tended not to pursue programs at great length but to skim cream from a variety of problems*".

With the flowering of biochemistry in the twentieth century, there came the realization that for many natural products – fatty acids, lipids, α-amino acids and proteins, nucleotides and nucleic acids – a vital and distinctive role in the life of all organisms could be assigned. The pathways by which these compounds are biosynthesised are similar, if not identical, in all organisms and for this reason these metabolites are frequently referred to as primary metabolites. In contrast an infinitely greater body of natural substances – such as alkaloids, phenols, terpenes, polyenes, polyacetylenes, pigments, antibiotics, mycotoxins, *etc.* – occur sporadically throughout the plant and microbial kingdoms. Moreover many appear to play no explicit role in the life of the producing organism. Because of their apparently secondary function these substances are commonly designated as secondary metabolites – they express the individuality of species in chemical terms. The perceived boundary between these two areas of metabolism is nevertheless often imprecise. Thus informed scientific opinions may differ as to the status of a particular metabolite and as new discoveries are made then the boundary itself may change.

Notwithstanding these caveats the vast majority of natural products still lie, however uncertainly, in this biological "no-man's land". For many the initial euphoria which accompanies their discovery is followed by a period of decline into genteel obscurity as they are consigned to a footnote in a textbook or review. The story of (−)-shikimic acid is well known but bears repetition. Following EYKMAN's discovery and isolation of the acid it fairly promised to find its ultimate resting place, languishing in the guise of "[3*R*-(3α, 4α, 5β)]-3,4,5-trihydroxy-1-cyclohexene-1-carboxylic acid", in *Chemical Abstracts*. Instead it was rescued from scientific oblivion by the beautiful work first of DAVIS (*45, 46*), then of SPRINSON and GIBSON to claim its key position in the pathway of aromatic metabolism which bears its name.

Davis described his discovery of the shikimate pathway:

"In studying biosynthesis, I did undertake one prolonged program: working out many of the steps in a common pathway of aromatic biosynthesis, leading to tyrosine, phenylalanine, tryptophan and p-*aminobenzoate. This path also led to a previously unknown growth factor,* p-*hydroxybenzoate, which others later found to be a precursor of a quinone cofactor. The first intermediate that we identified in the common pathway was an already known but obscure natural plant product, shikimic acid. This intermediate was accumulated by mutants blocked immediately after its production and it supported the growth of those blocked earlier. It is gratifying that the shikimic pathway has given rise to several books and to a review of the past decade with over 500 references.*

I will not dwell on my early contributions to this pathway, but I would like to acknowledge my debt to Roger Stanier. He was studying the breakdown of aromatic compounds by soil bacteria, and he suggested, and supplied, the shikimic acid that turned out to be an intermediate in my pathway (but not in his). The sample was prepared for him by H. O. L. Fischer from the dried fruit of the shikimi tree, obtained from a Chinese pharmacist. The fruit, which contains an alkaloid as well as shikimic acid, is used as a laxative, and the pharmacist originally pretended not to know of it. Stanier fortunately learned why and persisted: The fruit is apparently also used traditionally, in larger doses, to poison one's mother-in-law.

My associates identified many intermediates and enzymes in the aromatic pathway, as well as pathways to several other amino acids .... Our approach could not tell us how the aromatic pathway branched off from the central metabolic pathways (which I named amphibolic, for both catabolic and anabolic). To solve this problem, David Sprinson at Columbia College of Physicians & Sugeons initiated a long, enjoyable collaboration. Using precursors radioactively labeled in specific atoms (some at 5 counts per minute above background!) and then enzymes, he showed that three of the atoms of shikimic acid come from phosphoenolpyruvate and the other four from erythrose-4-phosphate to yield 3-deoxyarabinoheptulosonic-7-P. I admired his patient and thorough approach as an organic biochemist, because I tended to seek problems with intellectual challenges but easy technical solutions."

Since its discovery by Davis the shikimate pathway has given rise to many reviews and to several books, the most recent of which was published in 1993, (*19*, *21*, *27*, *34*, *71*, *85*, *86*, *136*, *155*). This essay is concerned principally with the enzymology of the metabolic pathway, enzyme mechanisms and organisation, and genetic manipulation of enzymes.

2. The Shikimate Metabolic Pathway

2.1. Common Pathway – Enzymes and Intermediates

Intermediary metabolism describes the complex network of chemical reactions which degrade, interconvert and synthesise organic molecules in living systems. The purposes of this intense chemical activity are principally twofold: to release energy and to create new cellular material. A useful distinction which is frequently employed is to separate formally these chemical reactions into two types, namely those which produce energy – catabolic or degradative – and those which lead to the synthesis of new metabolites – anabolic or biosynthetic. The major catabolic sequences are those of glycolysis and the citrate (tricarboxylic, Krebs) cycle; the pentose phosphate pathway serves two purposes, namely the generation of NADPH for reductive biosynthesis and the formation of ribose-5-phosphate for the synthesis of nucleotides. Of the many intermediates involved in these energy releasing sequences a relatively few are then employed as starting points for the synthesis of the quantitatively major components of the cell. These are the four (triose, tetrose, pentose and hexose) sugar phosphates and isomeric metabolites; three α-keto acids – pyruvate, oxaloacetate, and α-ketoglutarate; acetyl coenzyme A, succinyl coenzyme A and phosphoenolpyruvate (PEP). Studies of this latter area of metabolism in which these metabolites are transformed into hundreds of different cell components are referred to as studies of biosynthesis and the reactions themselves as biosynthetic reactions.

Nitrogen is absent from the central energy producing pathways of glycolysis and the tricarboxylic acid cycle and it is presumed that these metabolic sequences were already established in some form before nitrogen was introduced into metabolism. If one adopts a heuristic point of view of evolution then amino acids were probably amongst the first nitrogen-containing organic compounds for which syntheses were developed in living matter. Based on their biosynthetic origins the protein α-amino acids are divisible into just six families; ammonia is the effective form in which nitrogen is incorporated into organic substrates and in a few simple steps eight of the twenty or so α-amino acids may be developed from intermediates in glycolysis or the tricarboxylic acid cycle. Biosynthesis of the remaining α-amino acids occurs *via* pathways of greater complexity.

Estimates vary but under normal conditions of growth approximately one fifth of the carbon fixed by plants is subsequently channeled

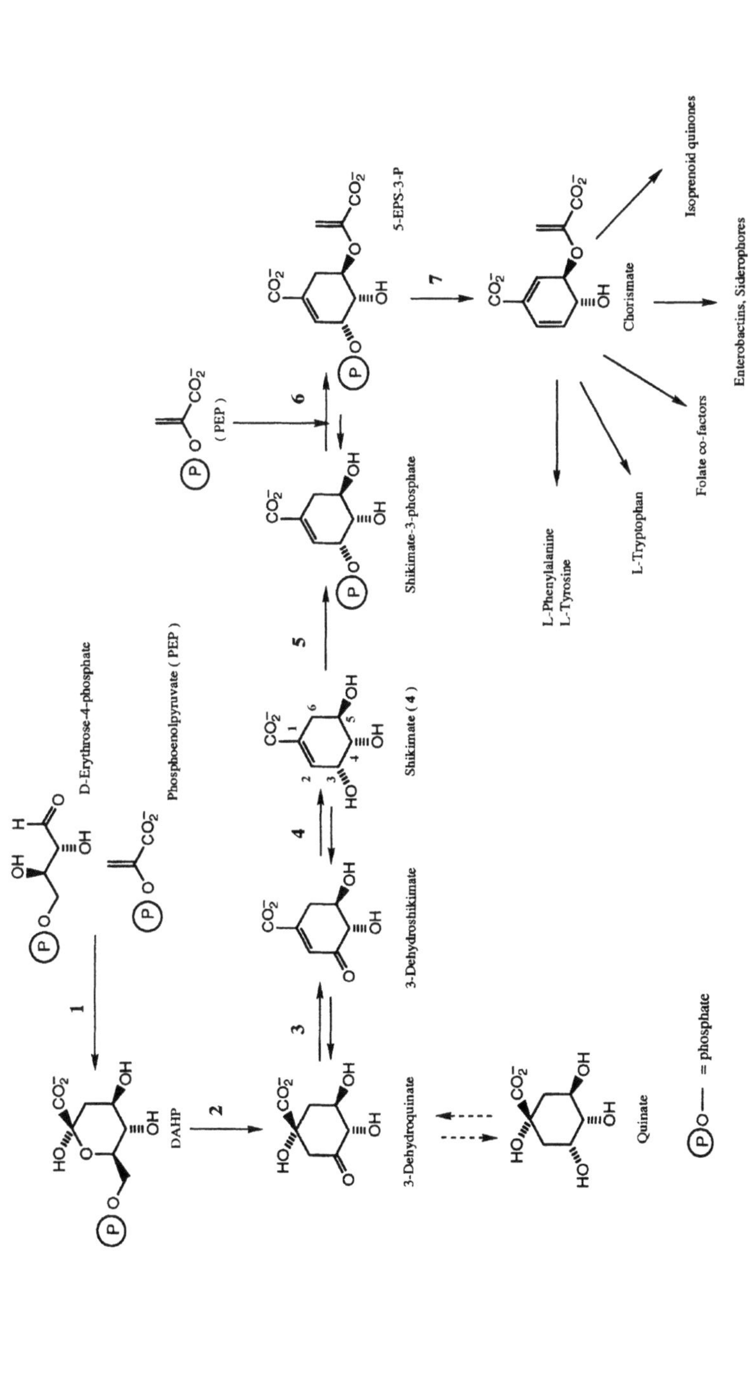

Fig. 1. The shikimate pathway: intermediates of the common pathway; all carboxylic acids are depicted in their anionic forms, *e.g.* shikimate **(4)**

Table 1. *The Shikimate Pathway – Enzymes of the Common Pathway*

Step	Enzyme	Classification	Co-factors
1	3-Deoxy-D-*arabino*-heptulosonic acid-7-phosphate (DAHP) synthase	EC 4.1.2.15	
2	3-Dehydroquinate synthase	EC 4.6.1.3	$^*M^{2+}$ NAD^+
3	3-Dehydroquinase *or* 3-Dehydroquinate dehydratase	EC 4.2.1.10	–
4	Shikimate dehydrogenase	EC 1.1.1.25	NADPH
5	Shikimate kinase	EC 2.7.1.71	ATP
6	5-Enolpyruvylshikimate-3-phosphate (EPSP) synthase *or* 3-Phosphoshikimate 1-carboxyvinyl transferase	EC 2.5.1.19	PEP
7	Chorismate synthase	EC 4.6.1.4	FMN NADPH

* Divalent metal ion (Co^{2+}, Zn^{2+}).

through the *Shikimate Pathway*. In higher plants this metabolic pathway provides a synthetic route to the three protein aromatic α-amino acids – L-phenylalanine (phe, F), L-tyrosine (tyr, Y) and L-tryptophan (trp, W), to the lipid soluble isoprenoid quinones involved in electron transport and, by the oxidative elaboration of L-phenylalanine and to a lesser extent L-tyrosine, to the important structural polymer lignin (*67*). The principal features, enzymes and intermediates of the shikimate pathway are indicated in Fig. 1 and Table 1. Chorismate is usually shown as the branch point of the pathway. However it is now clear that the pathway contains not one but several important branch points and for this reason BENTLEY (*19*) has described the shikimate pathway as "a metabolic tree with many branches". As well as higher plants the pathway is operative in microorganisms. Mammals do not possess the capacity to synthesise aromatic compounds by this route and this fact has led in recent times to a particular interest in the search for synthetic compounds which may selectively block specific enzyme catalysed transformations in this pathway. Researches into the shikimate pathway have, with a little serendipity, brought immense practical dividends in the form of the discovery of the post-emergence herbicide glyphosate (*4, 5, 6*). The observation by AMRHEIN that glyphosate acts by inhibiting the enzyme 5-enolpyruvylshikimate-3-phosphate synthase (EPSP synthase, Fig. 1), has spurred intense efforts to design new or improved compounds of this type.

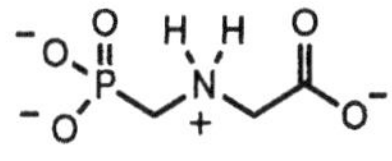

glyphosate

2.2. Common Pathway – Enzymology

Much of the early work which established the details of the chemical intermediates and the enzymes of the common part of the pathway used *Escherichia coli* and in some instances *Klebsiella pneumoniae* (syn. *Aerobacter aerogenes*, *Klebsiella aerogenes*) and the discussion which follows focuses principally on the enzymes from these organisms. The *Escherichia coli* genes and enzymes of the common pathway, their calculated sub-unit M_R values, and organisation within the native enzyme are shown in Table 2 (*136*).

*2.2.1. 3-Deoxy-D-*arabino*-heptulosonate 7-phosphate (DAHP) Synthase*

The first committed step in the biosynthesis of aromatic compounds *via* the shikimate pathway is the condensation of phosphoenolpyruvate (PEP) and D-erythrose 4-phosphate to give 3-deoxy-D-*arabino*-heptulosonate 7-phosphate (DAHP) and inorganic phosphate. The reaction is catalysed by the enzyme 3-deoxy-D-*arabino*-heptulosonate 7-phosphate (DAHP) synthase (E.C.4.2.1.15) which is more conveniently referred to as DAHP synthase.

In *Escherichia coli* there are three DAHP synthase isoenzymes; the phenylalanine sensitive DAHP (phe), the tyrosine sensitive DAHP (tyr) and the tryptophan sensitive DAHP (trp) (Table 3). The respective

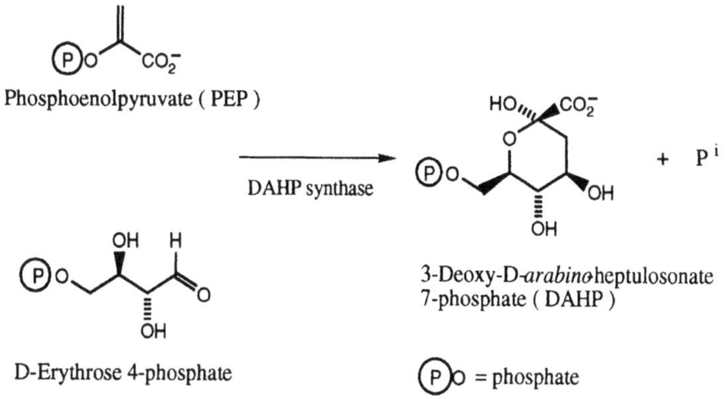

Table 2. *The Common Pathway Enzymes of Escherichia coli* (*136*)

Step	Enzyme	EC number	Gene	M_R(calc.)	Number of amino acids	Quaternary structure
1	DAHP synthase	4.1.2.15				
	DAHP synthase (tyr)		*aro* F	38,804	356	Dimer
	DAHP synthase (phe)		*aro* G	37,997	350	Tetramer
	DAHP synthase (trp)		*aro* H	39,000[a]	347	Dimer
2	3-Dehydroquinate synthase	4.6.1.3	*aro* B	38,880	362	Monomer
3	3-Dehydroquinate dehydratase (3-Dehydroquinase)	4.2.1.10	*aro* D	26,377	240	Dimer
4	Shikimate dehydrogenase	1.1.1.25	*aro* E	29,380	272	Monomer
5	Shikimate kinase	2.7.1.71	*aro* L	18,937	173	Monomer
6	5-Enolpyruvylshikimate-3-phosphate (EPSP) synthase	2.5.1.19	*aro* A	46,112	427	Monomer
7	Chorismate synthase	4.6.1.4	*aro* C	38,183	357	Tetramer

[a] Observed M_R.

Table 3. *The DAHP Isoenzymes of Escherichia coli*

Enzyme	Gene	Number of amino acids	M_R	Native enzyme
DAHP synthase (phe)	*aro* G	350	37,997[a]	Tetramer
DAHP synthase (tyr)	*aro* F	356	38,804[a]	Dimer
DAHP synthase (trp)	*aro* H	347	39,000[b]	Dimer

[a] Calculated value.
[b] Observed value.

ratios of the activities of the three isoenzymes in *Escherichia coli* are approximately 75:25: < 1 (*89, 90*). When fully derepressed the specific activity of DAHP synthase (trp) in cell extracts is only 50 mU/mg of protein whereas the corresponding values for DAHP synthase (tyr) and DAHP synthase (phe) are 560 and 300 mU/mg of protein (*26*). The three isoenzymes from *Escherichia coli* K-12 and *Salmonella typhimurium* have been purified to homogeneity (*89, 90, 95*). DAHP synthase (trp) and DAHP synthase (tyr) exist as dimers with a sub-unit molecular weight of 40,000. On the other hand DAHP synthase (phe) is a tetramer with a sub-unit molecular weight of ~ 35,000. All three DAHP synthases have been shown to contain 1 mole of iron per mole of native enzyme.

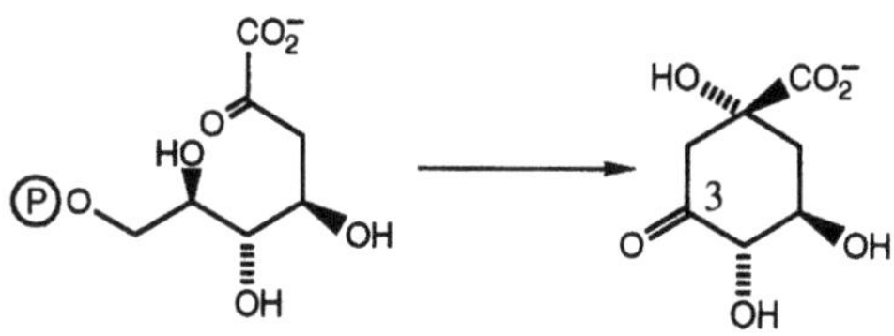

DAHP 3-dehydroquinate

In *Escherichia coli* the three DAHP synthase isoenzymes display a "symmetrical" inhibition pattern in which each isoenzyme is feedback inhibited by, and also repressed by, one of the aromatic amino acids. The synthesis of DAHP synthase (try) is repressed by tyrosine and very high levels of phenylalanine, that of DAHP synthase (phe) is repressed by phenylalanine and trytophan and that of DAHP synthase (trp) is repressed by tryptophan (*24*). Thus the presence of these three isofunctional DAHP synthases, whose activities and rates of synthesis are differentially affected by the individual aromatic amino acids, provides the *Escherichia coli* cell with the ability to modulate synthetic rates in response to the exogenous availability of phenylalanine, tyrosine and tryptophan.

Feedback inhibition of the DAHP isoenzymes is the major quantitative mechanism to control the flow of carbon into the shikimate pathway. Whilst the pattern observed in *Escherichia coli* may be taken as a paradigm for certain microorganisms the enzyme DAHP synthase is now known to possess the largest number of allosteric regulatory patterns described for any one protein (*19*). For example in certain strains of *Bacillus subtilis* (*e.g.*, WB 2802) DAHP synthase and chorismate mutase form a single bifunctional protein and the DAHP synthase is subject to feedback inhibition by chorismic acid and prephenic acid (*82–84*).

The gene DNA sequences (*aro* G, *aro* F, *aro* H) which specify the polypeptide sequences of the three *Escherichia coli* DAHP synthase isoenzymes have been determined. When all three of the sequences are compared they show considerable areas of homology; 41% of the residues are identical, suggesting a common evolutionary origin. It is assumed, since it is absent from all other members of the Gram-negative cluster – containing enteric bacteria, that the DAHP synthase (phe) has evolved most recently. All genera of the contemporary Enterobacteriaceae family possess recently evolved DAHP synthase (phe) in addition to the two other isoenzymes DAHP synthases (tyr and trp) (*19*).

2.2.2. 3-Dehydroquinate Synthase

The enzyme 7-phospho-3-deoxy-D-arabino-heptulosonate phosphate lyase is generally abbreviated as 3-dehydroquinate synthase. It catalyses the ring closure of DAHP to form the first of the alicyclic intermediates in the common pathway (step 2, Fig. 1).

The *Escherichia coli* enzyme was first studied by SPRINSON and his colleagues who observed that the enzyme requires a catalytic amount of NAD^+, (*i.e.*, stoicheiometric with the enzyme), both for catalytic activity and to maintain the structural integrity of the enzyme. No intermediates were detected in the reaction and SPRINSON proposed an interesting and ingenious mechanistic pathway for the enzyme (*vide infra*). The enzyme appears to be synthesised constitutively. Its synthesis is not induced by DAHP nor is it repressed by any of the aromatic amino acids or by chorismate. The yield of enzyme from wild-type *Escherichia coli* cells is however low [in extracts $\sim$ 25 mU/mg of protein, (*136*)] and it has been calculated that this value is about fivefold greater than would be required to meet the needs for aromatic metabolites of cells growing with a doubling time of about 1 hr. Nevertheless in some circumstances 3-dehydroquinate synthase activity may become rate-limiting and cells may accumulate DAHP.

Knowles (*109*) obtained larger quantities of the pure enzyme by sub-cloning the gene for 3-dehydroquinate synthase (*aro* B) from plasmid pLC 29–47 of the Carbon Clarke library of the *Escherichia coli* K-12 genome. This engineered strain overproduces 3-dehydroquinate synthase to the extent that the enzyme constitutes about 5% of the soluble protein of the cell. The molecular weight of the protein was estimated to be 40,000 (by gel-electrophoresis) and 44,000 (using gel-permeation hplc) (*18*). The molecular weight calculated from the 362 amino acid sequence [derived from sequencing the *aro* B gene, (*126*)] is 38,880 and indicates that the enzyme is probably a monomer. The enzyme contains, as isolated, one mole of "tightly bound" Co(II) and rapidly loses activity upon incubation with EDTA giving rise to a stable but inactive apoenzyme. Reconstitution of the apoenzyme with Zn(II) – which is probably the functional metal ion *in vivo* – restores activity to 5% of the level observed with Co(II). 3-Dehydroquinate synthase also binds one mole of NAD^+. Under turnover conditions (pH 7.5, 20°C) with saturating levels of substrate the dissociation rate of NAD^+ is approximately 1 min^{-1}, with a K_m for NAD^+ of 80 nM (*123*).

2.2.3. 3-Dehydroquinate Dehydratase (3-Dehydroquinase)

3-Dehydroquinate dehydratase (3-dehydroquinase) catalyses the third step in the common part of the shikimate pathway (Fig. 1). In bacteria the enzyme is monofunctional whereas in fungi and yeast it is a component of the pentafunctional *arom* enzyme complex (*33*). In plants it is found as a bifunctional enzyme with shikimate dehydrogenase (*33, 50, 131*). The gene coding for the enzyme in *Escherichia coli* (*aro* D) has been cloned and a strain which overproduces the enzyme (× 100) has been constructed; this strain (AB 2848/pK D201) is the most convenient source of the enzyme (*55, 56*). Considerable variations may be noted in the literature in the values reported for the molecular weight of this enzyme; however Coggins and his colleagues (*55, 56*) have indicated that the native enzyme is a dimer with a sub-unit M_R estimated by gel-electrophoresis as 29,000. From the nucleotide sequence of the *aro* D gene the polypeptide chain has 240 amino acids with a calculated sub-unit M_R of 26,377. The Michaelis constant K_M for 3-dehydroquinate in phosphate buffer pH 7.5 is 18 μM. There is additionally good evidence that the *Escherichia coli* enzyme is both structurally and mechanistically related to the 3-dehydroquinate dehydratase (3-dehydroquinase) domain of the *Neurospora crassa arom* multifunctional enzyme.

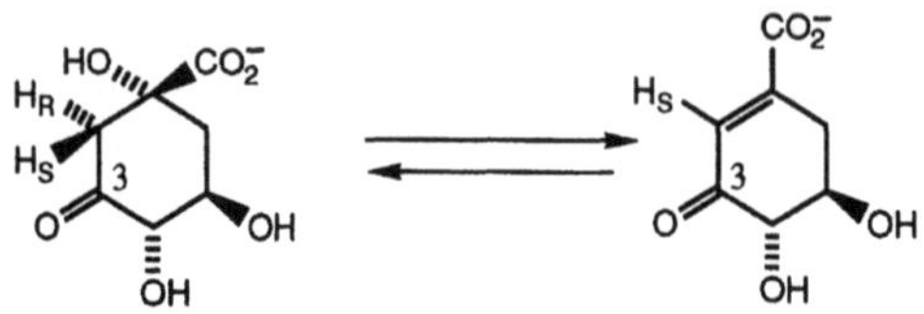

3-Dehydroquinate 3-Dehydroshikimate

Recent work has revealed a second class (class II) of 3-dehydroquinase enzymes which play a biosynthetic role but which do not function (as class I) through an imine intermediate. The type II enzymes (*e.g.*, *Mycobacterium tuberculosis*) occur as highly multimeric proteins with a single type of sub-unit (M_R 12,000–18,000) (*108*). Interestingly the reaction catalysed by the type II 3-dehydroquinase has been shown to proceed exclusively with *anti* stereochemistry of the elimination of water, the opposite stereochemistry of the reaction catalysed by the class I enzyme. It has been established that 3-dehydroquinase occurs in several plant species as a bifunctional polypeptide with shikimate dehydrogenase (*50, 131*). The *Pisum sativum* (pea) 3-dehydroquinase is a borohydride sensitive class I enzyme and its amino acid sequence is similar to that of the corresponding *Escherichia coli* 3-dehydroquinase. In the bifunctional polypeptide chain the 3-dehydroquinase activity resides at the N-terminus.

2.2.4. Shikimate Dehydrogenase (Shikimate Oxido-Reductase)

Shikimate dehydrogenase (shikimate oxido-reductase) catalyses the reversible reduction of 3-dehydroshikimate to shikimate as step four in the common part of the shikimate pathway (Fig. 1). It is an $NADP^+$-specific dehydrogenase enzyme (*156*) and transfers a hydride ion from the "A"-side of the nicotinamide ring in NADPH. In bacteria the enzyme is monofunctional (*30*) whereas in fungi and yeasts it is component of the pentafunctional *arom* enzyme complex (*33*). In plants (corn, pea seedlings, spinach) it occurs as a bifunctional enzyme with 3-dehydroquinase (*50, 131*). Recently the gene encoding the *Escherichia coli* enzyme (*aro* E) has been cloned and sequenced and placed under the control of a powerful promoter to facilitate overexpression, Chaudhuri, Anton and Coggins (*10*) report that the enzyme is a monomeric protein, $M_R = 30{,}000$. From the nucleotide sequence of the *aro* E gene the polypeptide consists of 272 amino acids with a calculated sub-unit M_R of 29,380.

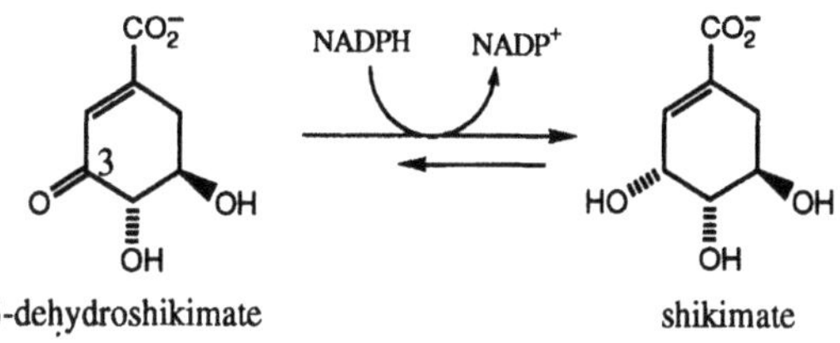

2.2.5. *Shikimate Kinase*

Shikimate kinase catalyses the phosphate transfer from ATP to the C-3 hydroxy group of shikimate. Unusually for an enzyme situated in the middle of a metabolic pathway (step 5 – common pathway, Fig. 1) two shikimate kinases have been described in *Escherichia coli* and *Salmonella typhimurium* (*57*). The one most studied is shikimate kinase II encoded by the *aro* L gene. Strains of *Escherichia coli* lacking shikimate kinase II because of mutations in the *aro* L gene have been isolated and regulation of its synthesis by the aromatic amino acids has been demonstrated (*48*, *49*). The gene *aro* L has been cloned, its nucleotide sequence determined, and shikimate kinase II purified from strains which overexpress this particular isoenzyme (*47*). The complete amino acid sequence for native shikimate kinase II has been deduced (173 amino acids) and an M_R of 18,937 calculated. The amino acid sequence contains regions homologous with other kinases and ATP requiring enzymes. The molecular weight as determined experimentally by gel-filtration was 21,400 and 17,000 by gel-electrophoresis suggesting that the native enzyme is active as a monomer. The apparent K_M for shikimate is 200 μM at 5 mM ATP, and for ATP 160 μM at 1 mM shikimate (*48*). The enzyme activity is dependent on the presence of a divalent metal cation as a cofactor and Mg^{2+} is the most effective ion in this respect.

Shikimate kinase I has also been partially purified from *Escherichia coli* and is likewise thought to have a relative molecular mass in the region of 20,000. The comparable K_M for shikimate is in excess of 5 mM

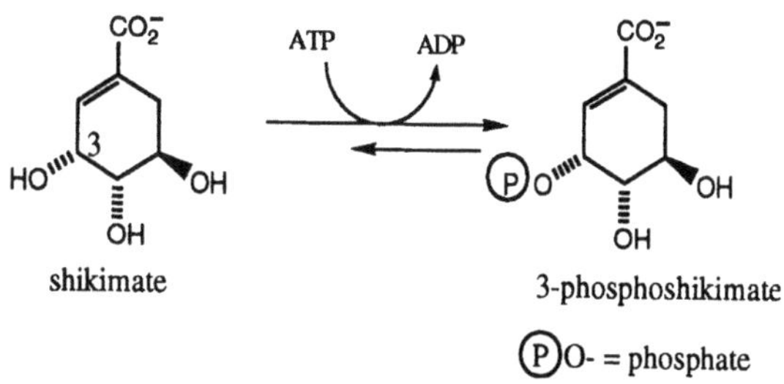

at 5 mM ATP. This difference in affinity for the substrate shikimate suggests that the isoenzyme shikimate kinase II is the one which normally functions in aromatic biosynthesis in the *Escherichia coli* cell and that shikimate kinase I functions only when intracellular levels of shikimate exceed a particular threshold. It has also been suggested that shikimate may in fact be a branch point intermediate for two distinct pathways.

In *Bacillus subtilis* there is a single shikimate kinase which is a component of trifunctional enzyme complex consisting of shikimate kinase and a tightly bound bifunctional complex carrying DAHP synthase and chorismate mutase activities (*96*). The kinase polypeptide is only active in the complex.

2.2.6. 5-Enolpyruvylshikimate-3-phosphate Synthase (5-EPSP Synthase)

The formal name for the enzyme catalysing the reaction in which 5-enolpyruvylshikimate-3-phosphate is formed from phosphoenolpyruvate and shikimate-3-phosphate is *phosphoenolpyruvate: 3-phosphoshikimate 5-O-(1-carboxyvinyl) transferase*; the abbreviated title 5-EPSP synthase is in common use for the enzyme.

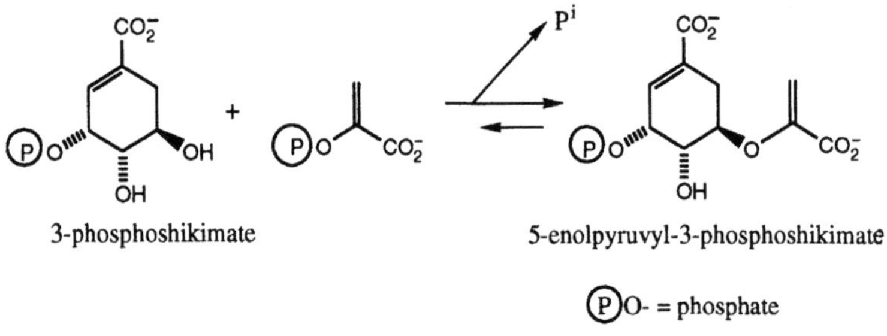

Initial studies on the enzyme were carried out with partially purified preparations from an *Escherichia coli* K 12 mutant by LEVIN and SPRINSON (*116*). The optimal pH for the enzyme was shown to be in the range 5.4–6.2 and LEVIN and SPRINSON also demonstrated, surprisingly perhaps, that the reaction was a reversible one with the equilibrium typically lying 75–80% in favour of the formation of 5-enolpyruvylshikimate-3-phosphate. When a large excess (25x) of inorganic phosphate P^i was added to a mixture of 5-enolpyruvylshikimate-3-phosphate (1 μM) and the enzyme 5-EPSP synthase they showed that phosphoenolpyruvate (PEP) was formed and that the yield could be enhanced

(0.68 μM) by the addition of pyruvate kinase to remove the phosphoenolpyruvate (PEP) as it was produced. An equilibrium constant K $\{[5\text{-EPSP}]\cdot[P^i]/[S\text{-}3\text{-}P]\cdot[PEP]\} = 12$ was estimated for the reaction (pH 6.1, 37 °C). LEVIN and SPRINSON also calculated the free energy of hydrolysis of 5-enolpyruvylshikimate-3-phosphate and showed that it differed little (~ 2 kcal/mole) from that of phosphoenolpyruvate.

5-Enolpyruvylshikimate-3-phosphate synthase occurs at relatively low levels in bacteria. It is synthesised constitutively in *Escherichia coli* and *Salmonella typhimurium* and cell extracts typically possess a specific activity of some 100 mU/mg of protein (*136*). The *Escherichia coli* (*aro* A) gene which encodes 5-enolpyruvylshikimate-3-phosphate synthase has been cloned and sequenced and an overproducing strain constructed which synthesises a 100 fold enhanced level of enzyme (*117*). *Escherichia coli* 5-enolpyruvylshikimate-3-phosphate synthase is a monomeric enzyme (the sub-unit M_R was estimated as 49,000 by polyacrylamide gel electrophoresis and the native molecular weight was given as 55,000 by gel filtration). The complete amino acid sequence has been deduced from the nucleotide sequence of the *aro* gene; the polypeptide chain consists of 427 amino acid residues and the calculated sub-unit M_R is 46,112.

The genes encoding 5-enolpyruvylshikimate-3-phosphate synthase (5-EPSP synthase) enzymes have been isolated from bacteria and fungi and have been sequenced to determine the amino acid sequences of the corresponding enzymes. That of the enzyme from *Escherichia coli* is very similar (only 11% differences) to that of *Salmonella typhimurium* (*117*). In plants 5-enolpyruvylshikimate-3-phosphate synthase is located primarily in the chloroplasts and other plastids. MOUSDALE and COGGINS (*130*) have reported a preparation of the enzyme from peashoot tissue (*Pisum sativum*); as compared with micro-organisms plants are generally a poor source of the enzyme. Like the *Escherichia coli* enzyme the molecular weight of the pea seedling shoot 5-enolpyruvylshikimate-3-phosphate synthase was in the region of 50,000. The extraction and purification procedures may also be applied to the isolation of 5-enolpyruvylshikimate-3-phosphate synthase enzymes from wheat (*Triticum vulgare*), spinach (*Spinacia oleracea*) and lettuce (*Lactuca sativa*) and these enzymes appear to have very similar characteristics to the pea seedling enzyme.

The complete nucleotide sequences of the *c*DNA clones encoding the 5-enolpyruvylshikimate-3-phosphate synthase of petunia (*Petunia hybrida*) and tomato (*Lycopersicon esculentum*) have been determined. From these data it was concluded that the petunia and tomato enzymes are very similar – they differ in the mature peptide in only 30 of the 447 amino acids (93% identity). When comparisons are made between the sequences of the 5-enolpyruvylshikimate-3-phosphate synthase coding regions of fungi, bacteria and plants a consistent pattern begins to emerge.

2.2.7. Chorismate Synthase

The seventh and final step in the common part of the shikimate pathway (Fig. 1), is catalysed by the enzyme chorismate synthase [O^5-(1-carboxyvinyl)-3-phosphoshikimate phosphate lyase]. The enzyme introduces, by an allylic elimination (1, 4) of phosphate, the second of the double bonds into the six membered ring. This enzyme is the least studied of all the enzymes in the pathway, from the mechanistic point of view. It requires a reduced flavin co-factor although the reaction results in no net overall change in redox state. Synthesis of the enzyme is believed to be constitutive and in both *Escherichia coli* and *Salmonella typhimurium* the observation has been made that in cell extracts the specific activity of chorismate synthase is only 10–20% of that of the preceeding enzyme 5-enolpyruvylshikimate-3-phosphate synthase (*26*, *73*).

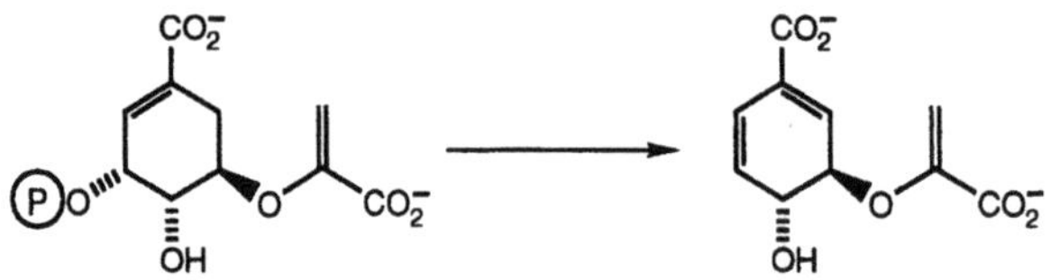

5-enolpyruvylshikimate-3-phosphate chorismate

ⓅO = phosphate

The enzyme has been characterised from three microbial sources, *Escherichia coli* (*128*, *156*), *Neurospora crassa* (*154*), and *Bacillus subtilis* (*82–84*), and from one plant source *Pisum sativum* (*129*). These enzymes display marked differences in their ability to generate the reduced co-factor necessary for catalysis. Both the *Neurospora crassa* and the *Bacillus subtilis* enzymes appear to be associated with a specific flavin reductase (diaphorase) activity that can generate the essential reduced co-factor by the oxidation of nicotinamide nucleotides under aerobic conditions. In contrast the partially purified *Escherichia coli* enzyme and that from *Pisum sativum* could only be assayed under strictly anaerobic conditions in the presence of chemically or enzymically reduced flavin (*70*).

The *Escherichia coli* enzyme, obtained by cloning and overexpressing the *aro* C gene (*128*), has a polypeptide chain consisting of 357 amino acids and a calculated sub-unit M_R of 38,183. The native enzyme is tetrameric. It has no detectable diaphorase activity and will only catalyse the formation of chorismate when it is supplied with exogenous reduced flavin – derived chemically or enzymically. The *Neurospora crassa* enzyme (*156*) is also tetrameric but with a somewhat larger sub-unit M_R of ~ 50,000. The *Neurospora crassa* enzyme has the ability to generate the reduced flavin co-factor, necessary for the reaction, *via* an intrinsic flavin reductase (diaphorase) activity. The evidence presently available suggests that the *Neurospora crassa* enzyme is probably a homotetramer consisting of four identical polypeptide chains. Each chain is bifunctional and carries both chorismate synthase and a diaphorase activity which shows absolute specificity for the physiological substrates [FMN and NADPH]. Coggins and his group have postulated (*156*) on the basis of limited sequence studies and other observations, that the *Neurospora crassa* polypeptide comprises two separate domains: one which is homologous to the *Escherichia coli* enzyme and carries the chorismate synthase activity and the second (only partially characterised) on which the diaphorase activity is located. The smaller sub-unit size of the *Escherichia coli* enzyme is thought to probably reflect a "missing" diaphorase domain.

The *Bacillus subtilis* enzyme displays a bifunctional character and exists as two separable polypeptide chains. These two polypeptides catalyse two distinct chemical reactions: (i) the reduction of either FAD or FMN to reduced flavin by the oxidation of NADPH and (ii) the conversion of 5-enolpyruvylshikimate-3-phosphate into chorismate (*82–84*). The higher plant chorismate synthase has been assayed in intact chloroplasts (*128*) and shows activity only in the presence of exogenously supplied reduced flavin. The evidence to date seems to suggest therefore that chorismate synthase may be found in one of two forms; either a bifunctional form in which the chorismate synthase is associated with diaphorase activity or as a monofunctional enzyme possessing only chorismate synthase activity.

2.3. Pathways Beyond Chorismate – Enzymes and Intermediates

Chorismic acid (**5**) (Fig. 2) was first isolated and identified by Gibson and Gibson (*85, 86*). Its key position in aromatic metabolism was illustrated by the eventual demonstration that in bacteria at least five distinct biochemical pathways utilise it as a precursor to essential metabolites (*136*). These are the synthetic routes to the three aromatic amino acids – L-phenylalanine, L-tyrosine and L-tryptophan; that to the folate

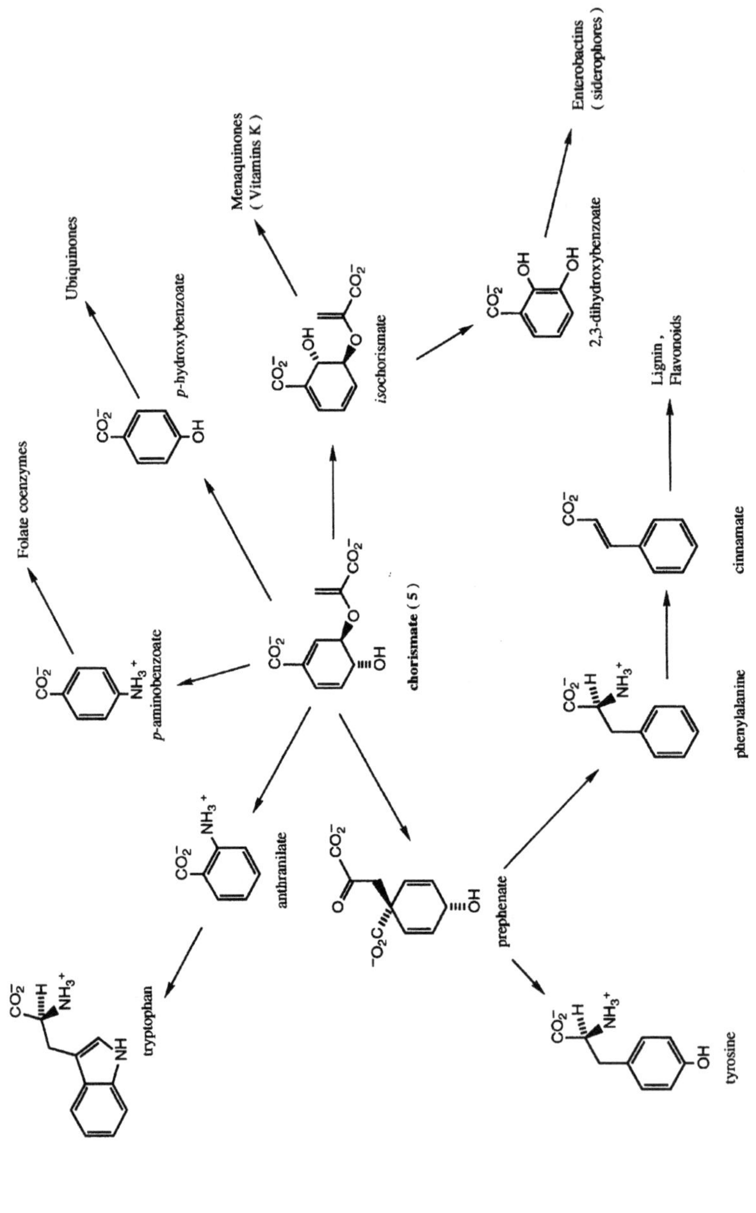

Fig. 2. Chorismate and beyond – primary metabolic pathways

coenzymes *via* *p*-aminobenzoate; the pathways to the ubiquinones *via* *p*-hydroxybenzoate and to the vitamins K (meanquinones and desmethyl-menaquinones) *via* isochorismate; and the pathway to siderophores and the prototype enterobactin (enterochelin) which is derived from 2,3-dihydroxybenzoate.

In higher plants the shikimate pathway plays a significant role not only in the provision of primary metabolites (aromatic amino acids, photosynthetic quinones) but also as the source of the precursors for an extraordinary rich and diverse array of "natural products" including simple phenols, flavonoids, polyphenols (*syn* vegetable tannins), cyanogenic glycosides, alkaloids, glucosinolates, quinones, betalain pigments, lignans and lignins (*86*). In micro-organisms an analogous, though structurally less diverse range of metabolites, is also often produced. Many of these "natural products" appear to have no explicit role in the economy of the organism that produces them. This distinction amongst "natural products" has long been recognised, and substances in this second category, because of their apparently secondary role, are commonly referred to as *secondary metabolites* (*44*). As a source of such metabolites [perceptively designated as "chemist's compounds" by Bu'Lock (*25*)] the shikimate pathway, both its intermediates and its end products, is a particularly rich one, not only in plants – where the three protein α-amino acids (phe, tyr, trp) are the principal precursors – but also in micro-organisms.

2.3.1. Pathways to L-Phenylalanine and L-Tyrosine

The first step in the biosynthesis of the two aromatic amino acids L-phenylalanine (phe) and L-tyrosine (tyr) from chorismate is the transformation to prephenate catalysed by the enzyme chorismate mutase. Prior to 1974 it was assumed that, following prephenate, the *p*-hydroxyphenylpyruvate route to L-tyrosine and the phenylpyruvate route to L-phenylalanine, first indicated by Davis and his co-workers in *Escherichia coli*, were universal. However in 1974 Jensen and his colleagues demonstrated (*103*, *104*) that several species of cyanobacteria transaminated prephenate directly to form the acid labile amino acid L-arogenate (pretyrosine) and that this was an obligatory precursor of L-tyrosine. Subsequent studies have shown that L-arogenate is not only a precursor of L-tyrosine but also of L-phenylalanine in many other organisms. This metabolic dichotomy from prephenate is illustrated in Fig. 3. The bacteria *Escherichia coli* and *Bacillus subtilis* only utilise the biochemical routes *via* phenylpyruvate and *p*-hydroxyphenylpyruvate,

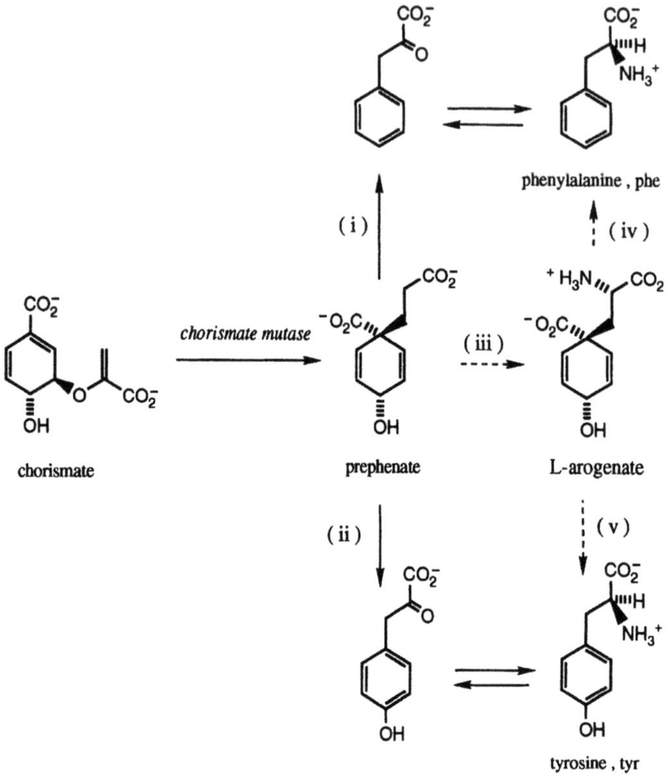

Enzymes : (i) - prephenate dehydratase ; (ii) - prephenate dehydrogenase (iii) - prephenate aminotransferase ; (iv) - L-arogenate dehydratase ; (v) - L-arogenate dehydrogenase .

Fig. 3. Alternative pathways from prephenate to L-phenylalanine and L-tyrosine

whilst the alga *Euglena gracilis* employs L-arogenate as the sole precursor of both L-tyrosine and L-phenylalanine. As indicated by its ubiquitous appearance in the cyanobacteria and the glutamic acid bacteria the combination of the L-arogenate route to L-tyrosine and the phenylpyruvate route to L-phenylalanine is widespread. Surprisingly perhaps in some organisms such as *Pseudomonas aeruginosa* and *Xanthomonas campestris* the coexistence of both sets of pathways is observed.

In higher plants a reliance on the L-arogenate pathways to L-phenylalanine and L-tyrosine appears to be a general characteristic. L-Arogenate dehydrogenase, which converts L-arogenate into L-tyrosine, has been identified in mung bean, tobacco, corn and sorghum

References, pp. 231–240

(*104*). The dehydrogenase activity here is linked to $NADP^+$. JENSEN has similarly reported the presence of prephenate aminotransferase in many plants, (*104*).

This remarkable diversity of post-chorismate pathways to L-phenylalanine and L-tyrosine is further complicated, according to JENSEN's observations, by substrate ambiguities (*104*), which seem to indicate that the existence of dual flow routes to L-phenylalanine and L-tyrosine at the physiological level is not necessarily indicative of the existence of two separate complements of enzymes encoded by two separate sets of genes.

2.3.1.1. Chorismate Mutase – Monofunctional

Chorismate mutase [chorismate pyruvate mutase, EC 5.4.99.5] lies at the branch point in the biosynthetic pathway to the aromatic amino acids (Fig. 3). The enzyme was first characterised by COTTON and GIBSON (*36, 37*) from *Escherichia coli* and *Aerobacter aerogenes* (syn. *Klebsiella aerogenes, Klebsiella pneumoniae*), and is the only characterised enzyme that catalyses a pericyclic process, namely the unimolecular intramolecular rearranement of chorismate to prephenate. In both *Escherichia coli* and *Klebsiella pneumoniae* chorismate mutase is located in two bifunctional enzymes in which its activity is coordinated with that of the prephenate metabolising enzymes – prephenate dehydratase and prephenate dehydrogenase respectively (Fig. 3) (*36, 37, 42, 43, 111, 112*). In other organisms chorismate mutase is found as a single monofunctional enzyme.

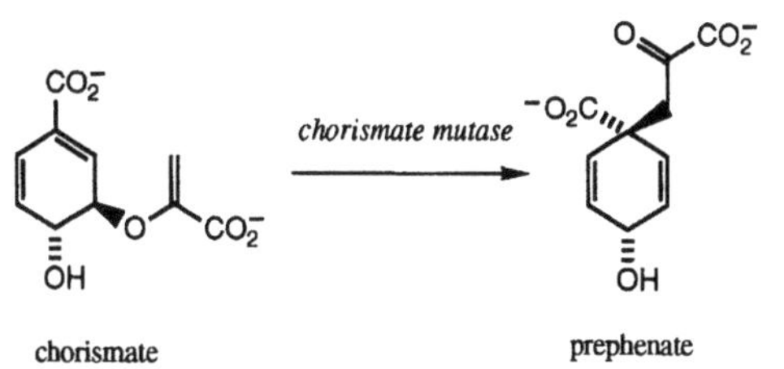

Thus in *Streptomyces* sp. L-phenylalanine is synthesised from prephenate by the action of prephenate dehydratase to give phenylpyruvate and subsequent transamination to give the amino acid, whilst L-tyrosine is formed by transamination of prephenate to give L-arogenate which is then converted into L-tyrosine by arogenate dehydrogenase (Fig. 3) (*103*). Chorismate mutase from *Streptomyces aureofaciens* is a monofunctional protein which is remarkably heat stable. The enzyme shows no evidence of cooperative sub-unit interaction (K_m value for chorismate $= 5.2 \times 10^{-4}$ M) and probably consists of 3 (or 4) similar or possibly identical sub-units. The chorismate mutase of *Brevibacterium flavum* is composed of two inactive components – a tetramer (sub-unit $M_R = 55{,}000$) and a dimer (sub-unit $M_R = 13{,}500$). These are separable by gel filtration but in the presence of the substrate associate to form an active complex.

Studies with *Bacillus subtilis* have revealed an interesting array of aggregated complexes. A bifunctional polypeptide possessing both DAHP synthase (Fig. 1, step 1 and chorismate mutase activities has been obtained from pleiotropic mutant strains derived from the Marburg strain of *Bacillus subtilis.* However chromatographic studies of the same two enzymes derived from the wild-type Marburg strain of *Bacillus subtilis* show that they are readily separable and exist in only one form (chorismate mutase, $M_R = 40{,}000$). X-ray crystallographic analysis of this monofunctional chorismate mutase has been reported very recently, (*31*). The origin of the bifunctional form has been attributed to the artificial generation of bifunctional activity by genetic mutation. The bifunctional enzyme interestingly forms a trifunctional enzyme complex possessing additionally shikimate kinase activity by association with a further small polypeptide.

The picture in relation to chorismate mutase in plants is slowly becoming clearer based upon information derived from studies of at least seven different plant genera (*72*, *137*). Two isoenzymes of chorismate mutase (CM-1 and CM-2) have been purified to homogeneity from mung bean and sorghum and similar isoenzymes have been separated to varying degrees of purity from other plant sources. The reported M_R values of chorismate mutase-1 (CM-1) range between 46,000 for the enzyme in alfalfa to 56,000 for that in sorghum. The corresponding M_R values reported for chorismate mutase-2 (CM-2) vary rather more widely from 36,000 for the mung bean enzyme to 65,000 for the protein in tobacco. In most plants so far examined, although the pattern is not totally consistent (*137*), two isoenzymes of chorismate mutase (CM-1 and CM-2) have been detected with distinctive kinetic characteristics.

At present it appears that higher plants contain a form of chorismate mutase (CM-1) which is strongly regulated by feedback inhibition by L-phenylalanine and L-tyrosine but which is activated by L-tryptophan, with relief of end-product inhibition by the L-tryptophan activation. In most plants a second chorismate mutase isoenzyme (CM-2) has also been observed which is unregulated by the aromatic amino acid end-products. On the basis of these observations the view has been expressed (*137*) that all higher plants contain two pathways for aromatic amino acid metabolism. One organellar pathway is located in the chloroplast and is subject to strong allosteric control reminiscent of that found in prokaryotes (CM-1) and a second pathway is found in the cytosol (CM-2). The plastid pathway is assumed to supply aromatic amino acids for protein synthesis. The cytosolic pathway besides also providing amino acids for the synthesis of proteins is assumed, since it appears to be unregulated, to provide precursors for *secondary metabolism* by a simple overflow mechanism.

2.3.1.2. Chorismate Mutase – Bifunctional

COTTON and GIBSON (*36, 37*) first demonstrated that chromatography of cell extracts of *Aerobacter aerogenes* (syn. *Klebsiella pneumoniae, Klebsiella aerogenes*) and *Escherichia coli* on DEAE cellulose gave two well separated peaks of chorismate mutase activity. Prephenate dehydratase activity was associated with the first peak to be eluted and prephenate dehydrogenase with the second. These enzymes catalyse the immediate post-prephenate stages in the pathways respectively to L-phenylalanine and L-tyrosine (Fig. 3). Subsequent studies have shown that in each case both enzyme activities are associated with a single bifunctional polypeptide (*19, 42, 43, 97, 98, 111, 112, 136*). Chorismate mutase-prephenate dehydratase [chorismate pyruvate mutase prephenate hydrolase (decarboxylating) EC 5.4.99.5/EC 4.2.1.51] has been observed in the bacteria *Escherichia coli, Klebsiella pneumoniae, (Aerobacter aerogenes), Salmonella typhimurium, Xanthomonas campestris, Alcaligenes eutrophus* and many species of *Pseudomonas* (*97*). Phenylpyruvate, synthesised by the enzyme, is then transaminated to give L-phenylalanine. Chorismate mutase - prephenate dehydrogenase [EC 5.4.99.5/EC 1.3.1.12] is analogously a bifunctional enzyme that catalyses two consecutive steps in the biosynthesis of L-tyrosine in the enteric bacteria *Escherichia coli, Klebsiella pneumoniae, (Aerobacter aerogenes),* and *Salmonella typhimurium,* (*97, 98*). Purification of the enzyme has been reported from *Escherichia coli* and *Klebsiella pneumoniae* (*Aerobacter aerogenes*) (*31, 111, 142*). The enzymes from these two sources appear to be very similar. The product, *p*-hydroxyphenylpyruvate, is finally transaminated by a further separate enzyme to give L-tyrosine. In *Escherichia coli* and *Salmonella typhimurium,* chorismate mutase-prephenate dehydrogenase is encoded by the *tyr* A gene. The *Escherichia coli* chorismate mutase-prephenate dehydratase enzyme is encoded by the *phe* A gene.

Kinetic studies with chorismate mutase-prephenate dehydratase suggest that the prephenate dissociates from the mutase active site before recombining at the dehydratase site. This leads to the conclusion that the chorismate mutase and the prephenate dehydratase reactions are catalysed at separate and distinctive sites on the protein which, at most, slightly overlap.

In the case of the chorismate mutase-prephenate dehydrogenase protein, mutants have been isolated that have lost the dehydrogenase activity but retain that of chorismate mutase, (*136*). Kinetic and computer simulation studies support the view that in this bifunctional enzyme the two active sites are spatially distinct but closely juxtaposed, with again the possibility of a slight overlap (*32*). Each reaction is irreversible, however, and the requirement for NAD^+ in the prephenate dehydrogenase reaction allows the two activities to be studied independently.

Both chorismate mutase-prephenate dehydratase and chorismate mutase-prephenate dehydrogenase have very similar amino acid compositions. The enzymes are each homodimers composed of sub-units of similar molecular weights $\sim$ 40,000. Both enzymes are subject to end-product inhibition; chorismate mutase-prephenate dehydratase by L-phenylalanine and chorismate mutase-prephenate dehydrogenase by L-tyrosine in the presence of NAD^+. In both cases the second activity involving prephenate is the more sensitive to inhibition ($\sim$ 90%) whilst the chorismate mutase activity is only partially inhibited ($\sim$ 50%) (*97*). The two enzymes are partially related. DNA sequencing of the genes that encode the enzymes has revealed that they are homologous at their N-terminus only, with 22 of the first 56 residues identical. This homologous region reflects a common function implying that these two protein segments have a common evolutionary origin and are responsible for the chorismate mutase activity.

In the synthesis of the three aromatic amino acids chorismate mutase-prephenate dehydratase and chorismate mutase-prephenate dehydrogenase compete with anthranilate synthase for the same substrate-chorismate (Fig. 2). It is therefore of interest to compare the affinities of each of these enzymes for this substrate. Chorismate mutase-prephenate dehydratase from *Escherichia coli* has a K_m for chorismate of 45 μM, whereas chorismate mutase-prephenate dehydrogenase from the same bacterial organism has K_m values of 92 μM for chorismate and 50 μM for prephenate. By comparison the K_m of anthranilate synthase for chorismate mutase is 1.2 μM (*13*). This marked difference in affinities suggests that under conditions of substrate (chorismate) limitation or depletion chorismate would be preferentially diverted down the L-tryptophan pathway rather than towards L-phenylalanine or L-tyrosine and it may also go some way towards explaining why aromatic auxotrophs with incomplete blocks in any of the common pathway reactions seem able to meet their requirement for L-tryptophan but not that for L-phenylalanine and L-tyrosine.

2.3.1.3. Monofunctional Prephenate Dehydratase and Prephenate Dehydrogenase

In *Escherichia coli* the prephenate utilising enzymes occur as bifunctional polypeptides with chorismate mutase. In other organisms the enzyme activities often exist as separate entities and are isolable as such. Thus the formation of phenylpyruvate from prephenate is catalysed by prephenate hydrolase (decarboxylase) generally referred to as prephenate dehydratase. A monofunctional prephenate dehydratase has been isolated from *Bacillus subtilis* and purified to homogeneity (*65*). This enzyme displays an interesting characteristic of allosteric activation and interconversion between inactive dimers ($M_R = 55{,}000$) and active octamers ($M_R = 210{,}000$) under the influence of small effector molecules. In the presence of positive effectors (activators such as L-methionine, L-leucine and prephenate) the maximally activated octamer is formed [this regulation by metabolites (L-methionine, L-leucine) of seemingly unconnected biosynthetic pathways is an example of "metabolic interlock"]. The enzyme is also converted reversibly into the variably activated dimeric form by inhibitors such as L-phenylalanine and L-tyrosine. Purification of monofunctional prephenate dehydrogenase enzymes has not been extensive. These enzymes may employ NAD^+ or $NADP^+$ or both as co-factors (*64*). The NAD^+ dependent enzymes, such as is found in *Bacillus subtilis*, appear to be more sensitive to L-tyrosine feedback control and inhibition. The *Bacillus subtilis* enzyme is also inhibited by L-tryptophan and L-phenylalanine (remote effectors), *p*-hydroxphenylpyruvate (product), and chorismate (an apparent substrate analogue) (*29*).

2.3.1.4. Aminotransferases

The final reactions in the biosynthesis of L-phenylalanine and L-tyrosine from the corresponding phenylpyruvates involve transamination (Fig. 3), with glutamate as the amino donor (*19*, *136*). Particular attention has been drawn to the roles played by three enzymes – the branched-chain amino acid aminotransferase (specified by *ilv* E), the aromatic amino acid aminotransferase (specified by *tyr* B), and the aspartate aminotransferase (specified by *asp* C). The aminotransferase reaction is freely reversible and it seems that under normal physiological conditions L-phenylalanine and L-tyrosine biosyntheses are primarily under the guidance of the aromatic amino acid aminotransferase. Only when the pool sizes of the phenylpyruvate and *p*-hydroxyphenylpyruvate become high does the aspartate aminotransferase begin to contribute to the biosynthesis of the two aromatic amino acids. The branched-chain amino acid aminotransferase is probably only involved in aromatic amino acid biosynthesis in *tyr* B and *asp* C mutants of *Escherichia coli.*

2.3.1.5. L-Arogenate ("Pretyrosine") Metabolism

The observation in 1974 by Jensen and his colleagues (*148*), that species of cyanobacteria were able to transaminate prephenate to form L-arogenate has led to the discovery of the existence of a remarkable diversity of routes in the late stages of biosynthesis of L-phenylalanine and L-tyrosine. These alternatives are depicted in Fig. 3. Virtually every combination of possible enzyme catalysed routes to L-phenylalanine and L-tyrosine has been observed. The situation is made more complicated by the existence of substrate and co-factor ambiguity. Thus the dehydrogenase reactions are commonly satisfied by either NAD^+ or $NADP^+$ (*135*). Likewise, in a number of cases, a single dehydrogenase or dehydratase enzyme is able to act upon both prephenate and arogenate (*104*). As further information is gained the full implications of this duality of pathways, post-prephenate to L-phenylalanine and L-tyrosine, will doubtless be realised.

Extensive studies have been carried out with the prephenate aminotransferase from *Nicotiana silvestris* (*22, 23*). Microbial prephenate aminotransferases are frequently able to transaminate phenylpyruvate and *p*-hydroxyphenylpyruvate just as well, if not better, than prephenate. The *Nicotiana* enzyme shows highest activity at pH 8.0 to 8.6 and will utilise – although less effectively – L-aspartate as amino donor. The relative molecular mass of the prephenate aminotransferase, determined after gel filtration, was 88,000.

Two distinct types of L-arogenate dehydratase have been described (*63*). L-Arogenate dehydratase has been partially purified from *Euglena gracilis*, *Pseudomonas aeruginosa*, *Pseudomonas diminuta*, and *Xanthomonas campestris*. The enzyme from *Pseudomonas diminuta* ATCC 13184 (*63*) has a broad pH optimum between pH 7.0 and 8.0, with an apparent K_m for L-arogenate of 0.63 mM. The enzyme is strongly inhibited by L-phenylalanine. At 0.136 mM L-arogenate, 100% inhibition of the enzyme is obtained with 0.25 mM L-phenylalanine. Analogues of L-phenylalanine are also potent inhibitors of L-arogenate dehydratase.

L-Arogenate dehydrogenase has been partially purified from three species of coryneformbacteria and purified to homogeneity from *Streptomyces phaeochromogenes* and *Phenylobacterium immobile.* Both enzymes are dimers [$M_R = 66{,}300$ and 69,000, respectively] and NAD^+ dependent. BONNER and JENSEN (*23*) have summarised data compiled with a range of L-arogenate dehydrogenases which have been partially purified. L-Arogenate dehydrogenase may utilise NAD^+, $NADP^+$ or both as co-factors. The NAD^+ dependence generally predominates in higher plants, coryneformbacteria and cyanobacteria, but is variable in pseudomonas. Feedback inhibition has been demonstrated by L-tyrosine in both plants and bacteria.

2.3.2. Pathway to L-Tryptophan

The biosynthetic pathway leading from chorismate to L-tryptophan consists of five enzymically controlled steps which, in those organisms which have been studied to date, are invariant (see Fig. 4, Table 4).

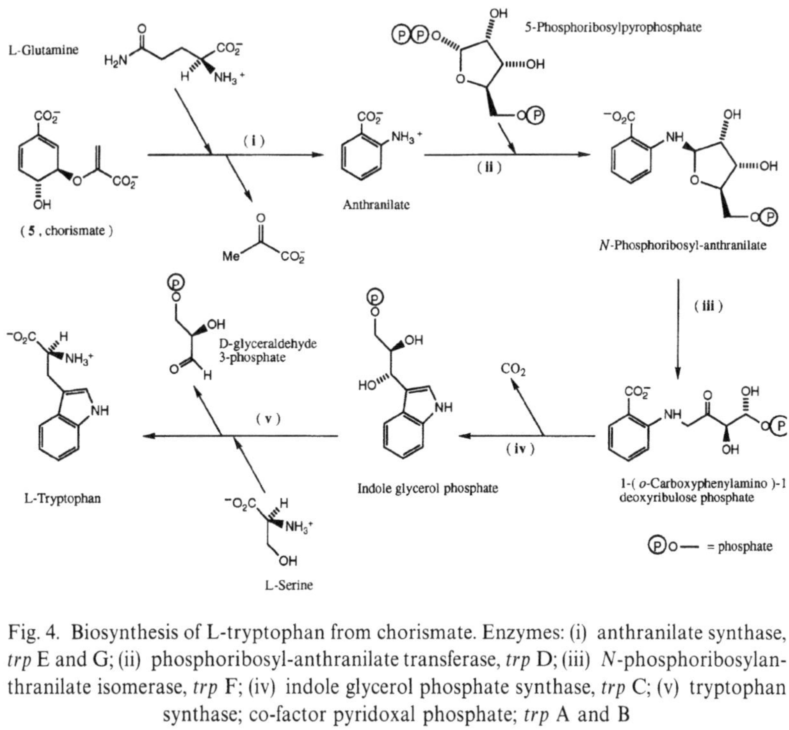

Fig. 4. Biosynthesis of L-tryptophan from chorismate. Enzymes: (i) anthranilate synthase, *trp* E and G; (ii) phosphoribosyl-anthranilate transferase, *trp* D; (iii) *N*-phosphoribosylanthranilate isomerase, *trp* F; (iv) indole glycerol phosphate synthase, *trp* C; (v) tryptophan synthase; co-factor pyridoxal phosphate; *trp* A and B

Table 4. *Enzymes of the L-Tryptophan Pathway*

Enzyme	EC number	Gene	Substrates	Products	Co-factors
(i) Anthranilate synthase	4.1.3.27	*trp* E *trp* G	Chorismate, Glutamine	Anthranilate, Glutamate, Pyruvate	Mg^{2+}
(ii) Anthranilate-5′-phosphoribose-1-pyrophosphate phosphoribosyl transferase	2.4.1.18	*trp* D	Anthranilate, 5-Phosphoribose-1-pyrophosphate	*N*-(5′-Phosphoribosyl)-anthranilate, Pyrophosphate	Mg^{2+}
(iii) *N*-(5′-Phosphoribosyl)-anthranilate isomerase		*trp* F	*N*-(5′-Phosphoribosyl)-anthranilate	1-(*o*-Carboxyphenylamino)-1-deoxyribulose-5-phosphate	
(iv) Indole-3-glycerol phosphate synthase	4.1.1.48	*trp* C	1-(*o*-Carboxyphenylamino)-1-deoxyribulose-5-phosphate	Indole-3-glycerol phosphate, Carbon dioxide	
(v) L-Tryptophan synthase	4.2.1.20	*trp* A *trp* B	Indole-3-glycerol phosphate, L-Serine, Pyridoxal phosphate	L-Tryptophan, D-Glyceraldehyde-3-phosphate	Pyridoxal phosphate

Different organisms are, however, far from invariant in the organisation and regulation of the genes and enzymes used to catalyse the reactions in the pathway in bacteria and fungi (*40*, *99*). Genetic information relating to L-tryptophan biosynthesis in plants, algae and other eukaryotes is still limited and fragmentary, although eukaryotes appear to manifest a trend toward fewer separate genes and toward multifunctional proteins.

Diverse forms of nomenclature for the *trp* structural genes may be encountered in the literature. In the picture which has emerged a set of seven activity domains make up the L-tryptophan pathway in all organisms. These activity domains are designated according to the individual activities in bacteria and following the conventions adopted by CRAWFORD (*40*):

(i) E domain: catalyses the anthranilate synthase reaction with ammonia as the amino donor.
(ii) G domain: interacts with the E domain and provides the glutamine amido-transferase activity for the glutamine dependent anthranilate synthase reaction.
(iii) D domain: catalyses the phosphoribosylanthranilate transferase (PR transferase) reaction.
(iv) F domain: catalyses the *N*-phosphoribosylanthranilate isomerase (PRA isomerase) reaction.
(v) C domain: catalyses the indole glycerol phosphate synthase (IGP synthase) reaction.
(vi) A domain: catalyses the conversion of indole glycerol phosphate into indole.
(vii) B domain: catalyses the conversion of indole into L-tryptophan.

During the course of evolution of eukaryotic organisms, fusion of genes has produced multifunctional proteins. Different patterns of fusion have produced multifunctional complexes with different combinations of these activity domains. The enzymes of the L-tryptophan pathway appear to be more highly organised in eukaryotic than in prokaryotic organisms. In the eukaryotic micro-organisms which have been studied, the activity domains are located upon fewer polypeptide chains than in most prokaryotes. A higher order of active site organisation also appears to have emerged in eukaryotic organisms.

Although sequence information for enzymes of the L-tryptophan pathway is restricted the available evidence seems to suggest substantial conservation of domain structure throughout the eukaryotic and prokaryotic world. It appears that the basic structural plan of the molecules has been conserved through evolution, despite a considerable divergence of sequence at most positions and the occurrence of several different gene fusions.

2.3.2.1. Anthranilate Synthase

Anthranilate synthase catalyses the first reaction in the L-tryptophan biosynthetic pathway – from chorismate to anthranilate. The enzyme contains two functional domains, designated component I (α) and component II (β). Component I (α) can synthesise anthranilate directly from chorismate and high concentrations of ammonia. In bacteria it also contains the L-tryptophan binding site utilised for feedback inhibition.

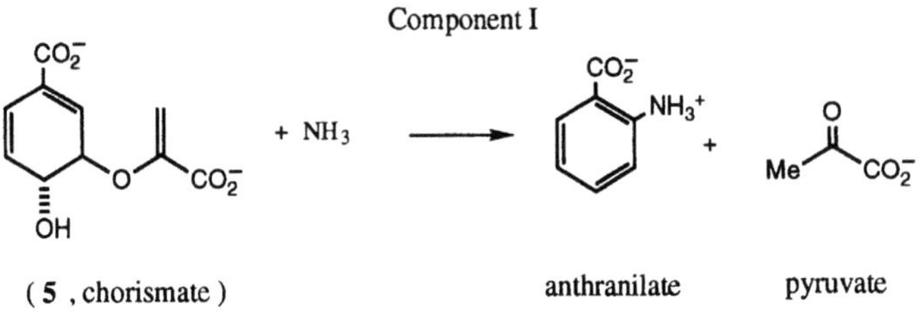

(**5**, chorismate)

Component II (β) has glutaminase activity and appears to activate glutamine by forming a bound γ-glutamyl thioester and "nascent ammonia". The fully constituted enzyme (I plus II) then catalyses the reaction of chorismate with glutamine to give anthranilate.

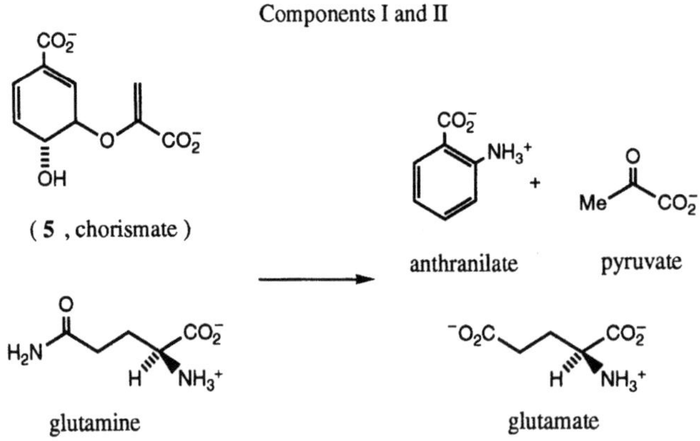

In many bacterial species a very similar enzyme *p*-aminobenzoate synthase catalyses the synthesis of *p*-aminobenzoate from chorismate and glutamine. Although less well studied this enzyme also possesses two sub-units which have considerable similarities in function to their anthranilate synthase counterparts.

Component II (β) of anthranilate synthase exhibits the distinctive and interesting property of existing in differently fused arrangements with other enzymes in the L-tryptophan pathway (i.e., multifunctional proteins). These fusion patterns distinguish anthranilate synthase enzymes from different organisms. Some of the arrangements for component II (β) which have been clearly defined are as follows:

(i) *Serrattia marcescens* – unfused.
(ii) *Escherichia coli, Salmonella typhimurium, Klebsiella pneumoniae* – fused to anthranilate-5′-phosphoribose-1-pyrophosphate phosphoribosyl transferase (step ii, Fig. 4).
(iii) *Neurospora crassa* – fused to *N*-(5′-phosphoribosyl)-anthranilate isomerase and indole-3-glycerol phosphate synthase, (steps iii and iv, Fig. 4).
(iv) *Euglena* – fused to anthranilate synthase component I (α).

Anthranilate synthase [unfused] has been isolated from *Serrattia marcescens* strain Hy 150, a L-tryptophan auxotroph lacking indole-3-glycerol phosphate synthase (*162*). *Serratia marcescens* anthranilate synthase is a tetramer containing two component I(α) sub-units (M_R = ~ 60,000) and two component II (β) sub-units (M_R = 20,956). The component II (β) sub-unit has a polypeptide chain of 192 residues and cysteine-83 is the active site residue at which the covalently bound glutaminyl intermediate is formed. Its formation requires prior binding of chorismate to the component I (α) sub-unit of the enzyme complex. K_m values of 0.5 mM (glutamine) and 4 μM (chorismate) have been reported for the enzyme which is subject to allosteric inhibition by L-tryptophan.

Anthranilate synthase of *Euglena gracilis* is distinctive. The enzyme activity is contained on a single polypeptide chain (*19*). Its relative molecular mass (gel-filtration) was determined as 80,000 ± 2,500 and it possesses both ammonia and glutamine dependent activity. It has been suggested that the original physiological reaction in the formation of anthranilate was that between chorismate and ammonia and that the glutamine dependent enzyme arose from the ammonia utilising ancestral enzyme by aggregation with a molecule that had evolved to catalyse the cleavage of the amide bond of glutamine. The existence of catalysis of the ammonia dependent reaction in all known anthranilate synthases, the identity (in some organisms) of the glutamine binding sub-unit of anthranilate synthase with that of a second glutamine utilising enzyme which aminates chorismate (*p*-aminobenzoate synthase) and the presence of glutaminase activity are all consistent with this idea. Component II (β) activity has been noted in a variety of associative patterns with other enzymes of this pathway (*vide supra*). Where enzyme fusion has

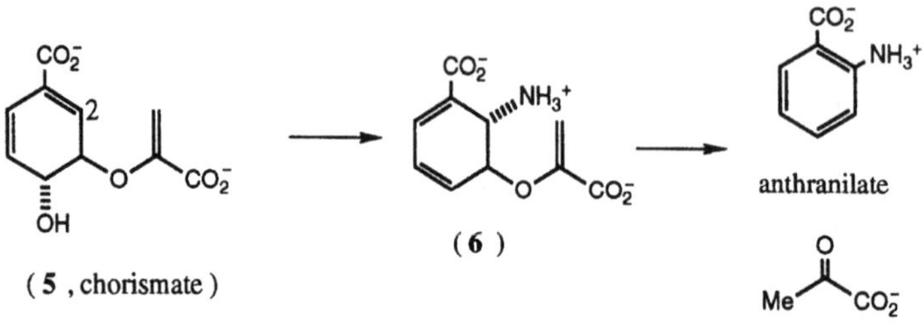

(**5**, chorismate) (**6**) anthranilate pyruvate

occurred it has never been observed with component I (α). Whilst the observed covalent linkage of components I (α) and II (β) in *Euglena* may be simply a chance event in an unusual organism in an evolutionary *cul de sac*, the *Euglena* enzyme may nevertheless represent an evolutionary trend towards the loss of the ability to use ammonia in the anthranilate synthase reaction. *In vitro* the *Euglena* enzyme is much more efficient in its utilisation of glutamine than of ammonia, much more so than other enzymes from other sources. It has been speculated that although the ability to incorporate ammonia is still of value to many mico-organisms it is far less so for *Euglena*, thus permitting the development of a more compact, efficient enzyme.

Although the role of glutamine as the amino group donor is relatively clear the further details of anthranilate formation are still not entirely firm. It has been established that the nitrogen atom becomes attached to C-2 of chorismate (**5**), and that the C-2 hydrogen atom is *not* incorporated into the molecule of pyruvate formed simultaneously in the reaction; the third methyl hydrogen atom in the pyruvate is in fact derived by *re*-face addition to the enolpyruvyl group of (**5**) by a proton from the solvent (*11*). LEVIN and SPRINSON (*116*) suggested the intermediacy of the amino-enolpyruvate (**6**) in this reaction and two syntheses of the 2,3-trans isomer have been reported (*151*). Although (**6**) was shown to be a chemically and kinetically competent intermediate in the biosynthesis of anthranilate it has not been detected as an accumulated intermediate.

2.3.2.2. Anthranilate-5′-phosphoribose-1-pyrophosphate Phosphoribosyl Transferase, *N*-(5′-Phosphoribosyl)-anthranilate Isomerase and Indole-3-glycerolphosphate Synthase

The attention given to the study of the first and the last enzymes in the L-tryptophan pathway from chorismate has perhaps inadvertently diverted interest and attention away from the three enzymes which catalyse the central chemical transformations in the pathway (Steps ii, iii and iv, Fig. 4). CREIGHTON and YANOFSKY (*41*) have reviewed the properties of these enzymes.

The enzyme anthranilate-5′-phosphoribose-1-pyrophosphate phosphoribosyl transferase (EC 2.4.2.18) catalyses the second step in L-tryptophan biosynthesis (Fig. 4, Table 4). The enzyme displays a surprising multiplicity of forms within a single bacterial family, the Enterobacteriaceae. In *Escherichia coli* and *Salmonella typhimurium* it is part of the

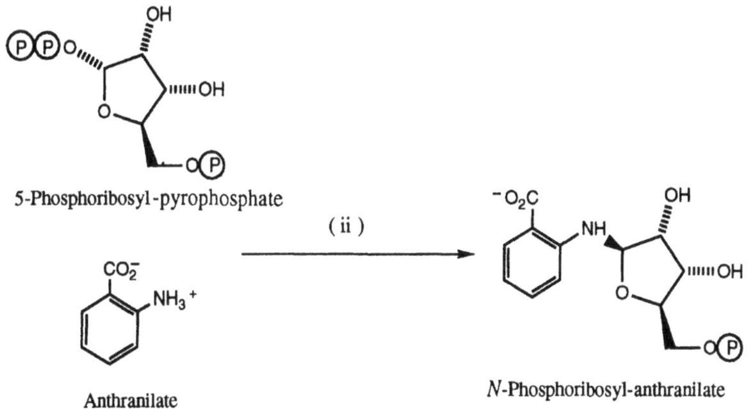

Enzyme (ii) - anthranilate-5'-phosphoribose-1-pyrophosphate phosphoribosyl transferase .

multifunctional enzyme with anthranilate synthase (*17*) (*vide supra*). The transferase enzyme is however separate in *Erwinia carotovora*, *Hafnia alvei* and *Serrattia marcescens*. Amino acid analysis shows that the *Erwinia* and *Hafnia* sequences include a thirteen residue leader sequence missing from the *Serrattia* enzyme. Beyond this point there is a strong homology. Anthranilate-5′-phosphoribose-1-pyrophosphate phosphoribosyl transferase from *Hafnia alvei* ATCC 13337 has been crystallised and preliminary X-ray data presented. The enzyme has also been prepared in a dimeric form (M_R = 83,000) from an overproducing strain of *Saccharomyces cerevisiae* (*94*).

N-(5′-Phosphoribosyl)-anthranilate isomerase catalyses the third step in the L-tryptophan biosynthetic pathway from chorismate (Fig. 4). The reaction is in a formal chemical sense an example of the Amadori rearrangement and is essentially irreversible. Formation of the indole ring occurs by cyclisation to C-2 of the original ribose and expulsion of the carboxyl group of the anthranilate. The enzyme occurs as a separate monofunctional protein in some organisms such as *Pseudomonas putida*,

Bacillus subtilis and *Saccharomyces cerevisiae*, but in other organisms it is frequently found associated with other enzymes of the pathway. Thus in *Brevibacterium flavum*, along with indole-3-glycerol phosphate synthase it forms a readily dissociable multienzyme complex whilst in the fungi *Neurospora crassa* and *Aspergillus nidulans* it occurs as part of a "post-chorismate multienzyme complex" which also carries anthranilate synthase and indole-3-glycerol phosphate synthase activities (*122*). In the enteric bacteria *Escherichia coli, Salmonella typhimurium, Serrattia marcescens* *N*-(5′-phosphoribosyl)-anthranilate isomerase and indole-3-glycerol phosphate synthase occur as a monomeric bifunctional protein (M_R = 49,5000), (*122, 107*). The three dimensional structure of the *Escherichia coli* enzyme has been determined at 2.8 A resolution by X-ray crystallography (*138*): the first native bifunctional enzyme for which this has been achieved. The two catalytic activities reside on distinct functional domains of similar folding, that of an eightfold parallel β-barrel with α-helices on the outside connected by β-strands. The active sites do not face each other, making "channeling" of the substrate between them virtually impossible. The amino acid sequence consists of 452 amino acids and the indole-3-glycerol phosphate synthase activity is located in the N-terminal domain.

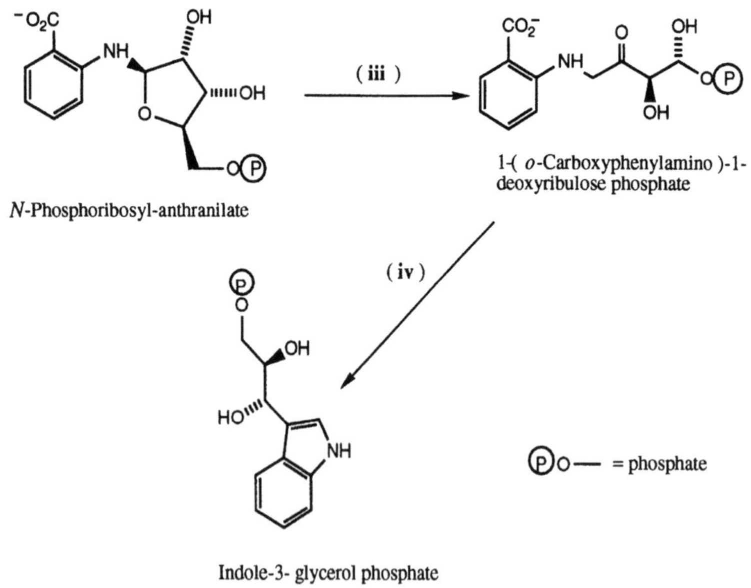

Enzymes: (iii) *N*-(5′-phosphoribosyl)-anthranilate isomerase and (iv) indole-3-glycerolphosphate synthase

Gene sequences of both the *N*-(5′-phosphoribosyl)-anthranilate isomerase and the indole-3-glycerol phosphate synthase enzymes have been determined for several organisms in addition to *Escherichia coli*. Assignment of these sequences for ten different enzymes in the light of the tertiary structure of the *Escherichia coli* enzyme revealed several features of interest (*138*). Regions of high sequence homology are associated with the catalytically active sites [66.7% for the indole-3-glycerol phosphate synthase and 52.2% for the *N*-(5′-phosphoribosyl)-anthranilate isomerase sites were invariant].

2.3.2.3. Tryptophan Synthase

Tryptophan synthase (EC 4.2.1.20) catalyses the final reaction sequence in the biosynthesis of L-tryptophan (Step v, Fig. 4). The enzyme has an $\alpha_2\beta_2$ composition in all prokaryotes examined and a fused $[\alpha\beta]_2$ structure in ascomycetes, and performs *in vivo* two reactions (a) and (b) in a concerted fashion, thereby converting indole-3-glycerol phosphate into L-tryptophan.

In the physiological reaction these two sequences are coupled in such a way that indole, produced in the α reaction, does not freely exchange with indole in solution. The two partial reactions (a) and (b) are also catalysed by the separate α ($M_R = 29{,}000$) and β_2 ($M_R = 2 \times 43{,}000$) sub-units, respectively. Each sub-unit's activity is stimulated, by one or two orders of magnitude, upon combination with the other sub-unit. The β sub-unit usually exists as a dimer (β_2) and contains two moles of the co-factor pyridoxal phosphate per dimer.

The $\alpha_2\beta_2$ complex, which is more stable than the separate α and β_2 sub-units, has been purified from many bacterial strains (*125*) as have the separate α sub-unit and the β_2 sub-unit from mutant strains of *Escherichia coli* (*41*). The $\alpha_2\beta_2$ complex from *Salmonella typhimurium* has proved to be of particular interest because it crystallises in a form suitable for X-ray analysis (*100*).

In *Saccharomyces cerevisiae* tryptophan synthase is a homodimer of two sub-units of $M_R = 76{,}000$, a protein in which the A and B domains are fused and connected by a region of some 28 amino acid residues. There is considerable homology between the yeast enzyme and the α and β chains of the enzyme from enteric bacteria. One unusual feature however is that the genes are fused in the order A.B, whilst in all other prokaryotes studied the genes are in the same operon but the order is reversed.

(a , α reaction)

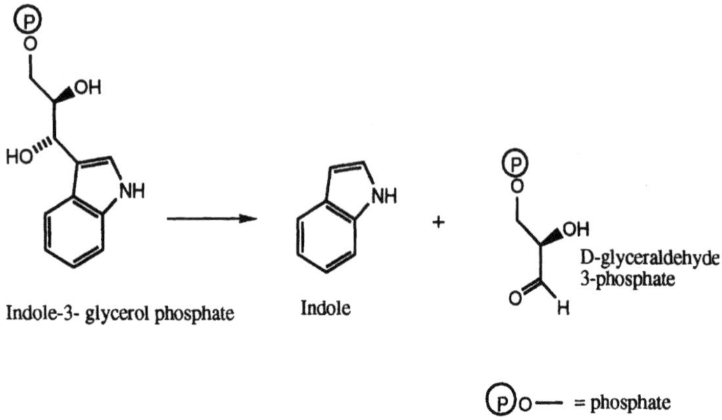

(b , β reaction)

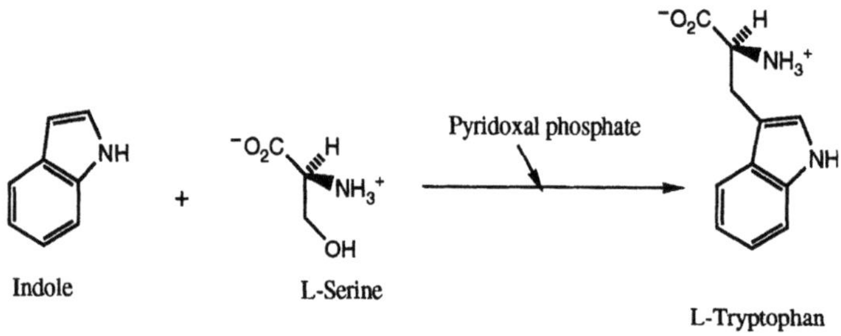

The three-dimensional structure of the $\alpha_2\beta_2$ complex of tryptophan synthase from *Salmonella typhimurium* has very recently been determined by X-ray crystallography at 2.5 A resolution (*100*). The four polypeptide chains are arranged almost linearly in an αββα order to form a complex ~ 150 A in length. The overall folding of the polypeptide chain of the smaller α sub-unit, which catalyses the cleavage of the indole-3-glycerol phosphate, is that of 8 fold [α/β] barrel. This particular structural motif was first observed in triosephosphate isomerase and has since been observed in several other enzymes, including that of both domains of the bifunctional enzyme – *N*-(5′-phosphoribosyl)-anthranilate isomerase: indole-3-glycerol phosphate synthase – from *Escherichia coli* (*vide supra*) (*138*). There is a high level of structural homology between triosephosphate isomerase and the α sub-unit of tryptophan synthase, and since both enzymes participate in reactions involving a common substrate,

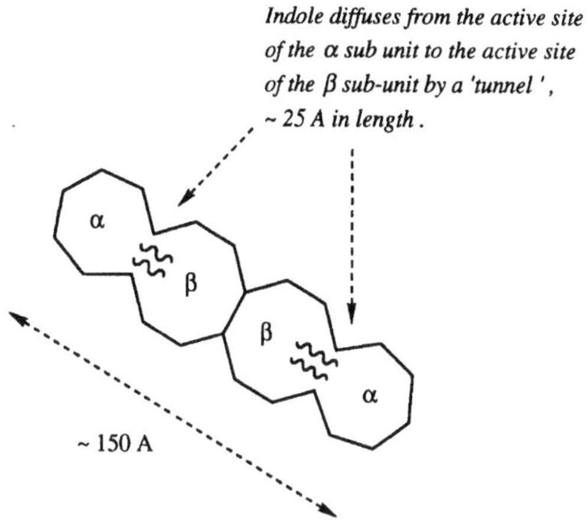

Pictorial representation of the $\alpha_2\beta_2$ complex of tryptophan synthase from *Salmonella typhimurium* (*100*).

D-glyceraldehyde-3-phosphate, it is tempting to speculate that this structural motif may have been preserved for functional reasons.

The β sub-unit contains two domains of nearly equal size separated by a plane approximately parallel to the dyad axis that relates the two β sub-units in the complex. Involved in the catalytic action of this sub-unit is the co-enzyme pyridoxal phosphate and this is located at the interface between the two structural domains near the centre of each β sub-unit. The phosphate group of the pyridoxal phosphate is linked *via* hydrogen bonding to a glycine rich region of the polypeptide backbone. The aldehyde group of the co-enzyme forms a Schiff's base with a "cisoid" configuration with the ε-amino group of lysine (*87*) on the β sub-unit. Open space found adjacent to the deeply buried co-enzyme and lined with highly conserved residues is presumed to be the most probable binding site for the L-serine molecule.

2.3.3. *Folate Coenzymes, Isoprenoid Quinones and Enterochelin*

The major branch-point intermediate, chorismate (**5**), may be directed into a variety of metabolic pathways (Figs. 1 and 2) and it has been suggested (*105*) that the group of enzymes (or at least some of them) responsible for these transformations may be encoded by a family of evolutionary related genes which have since diverged to produce the observed range of end-products of reaction.

2.3.3.1. *p*-Aminobenzoate Synthase

p-Aminobenzoate synthase and anthranilate synthase are enzymes that perform different, but related, functions and possibly share a common evolutionary origin. Both enzymes use glutamine and chorismate as substrates and produce glutamate, pyruvate and either *p*-aminobenzoate or anthranilate. Anthranilate synthase catalyses the first step in the synthesis of L-tryptophan (Fig. 4), whilst *p*-aminobenzoate synthase catalyses a reaction in the biosynthesis of the folate co-enzymes (Fig. 2). Various studies have suggested that these enzymes are structurally and functionally related and probably arose *via* the duplication and subsequent divergence of pre-existing genetic information. Both enzymes are composed of two non-identical sub-units. The larger sub-unit (component I, α, M_R approximately 50,000) catalyses the synthesis of the aminobenzoate from chorismate and ammonia, whilst the smaller sub-unit is a glutamine amidotransferase (component II, β, M_R approximately 20,000) which confers on the enzyme complex the ability to utilise the amide nitrogen of glutamine as the nitrogen source in the chorismate amination. Detailed characterisation of the enzyme *p*-aminobenzoate synthase has not yet been accomplished due to the extremely low levels of the enzyme in wild type cells. However characterisation of the genes encoding *p*-aminobenzoate synthase in *Escherichia coli* has advanced more rapidly. Nucleotide sequence determination has indicated (*105*) that p*abB* encodes the large sub-unit ($M_R = 53{,}400$) and that p*abA* specifies the small sub-unit ($M_R = 21{,}700$). The analogy between *p*-aminobenzoate synthase and anthranilate synthase has received additional support at the physiological level by the observation, in some micro-organisms, that the two enzymes share a common glutamine amidotransferase sub-unit (*105*).

Very recently, using the organism *Escherichia coli*, plasmids were constructed that overproduce (250 to 500 fold) the *p*-aminobenzoate synthase sub-units (*132*). Partial purification showed that they formed a diffusible intermediate, possibly (7), which is then converted by a second enzyme (X, $M_R = 49{,}000$) into *p*-aminobenzoate. This enzyme may have some similarities to chorismate lyase which is encoded by the *ubi*C

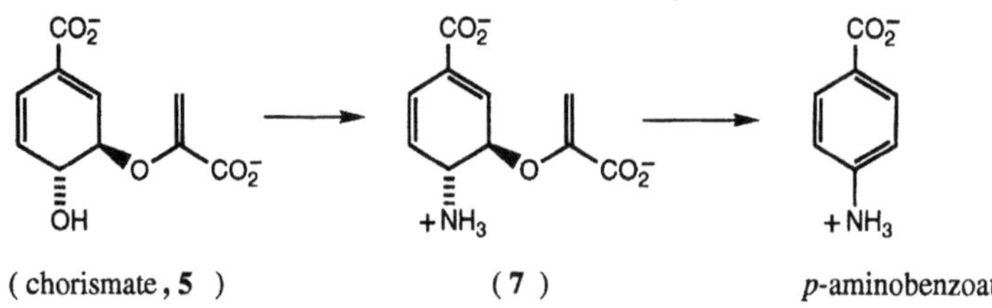

(chorismate, **5**) (**7**) *p*-aminobenzoate

gene and catalyses the transformation of chorismate to *p*-hydroxybenzoate (*115*).

2.3.3.2. Isochorismate Synthase and Enterochelin (Enterobactin)

The role of isochorismate synthase in the biosynthesis of 2,3-dihydroxybenzoate and the siderophore enterochelin (enterobactin) was first discovered by GIBSON and his colleagues (*159–161*). The enzyme was finally purified to homogeneity following sub-cloning of the appropriate *entC* gene and overproduction of the enzyme (*118, 119*). Isochorismate synthase is active as a monomer ($M_R = 43{,}000$) and catalyses the reversible interconversion of chorismate and isochorismate. Continuous coupled spectrophotometric assays of both the forward and backward reactions gave K_m values of 14 μM for chorismate and 5 μM for isochorismate and an equilibrium constant of 0.56 favouring chorismate and corresponding to a free energy difference of 0.36 kcal/mole between the isomers. A similar value was obtained by "watching" the interconversion in an NMR experiment. The origin of the hydroxyl group, introduced at C-2, in the isochorismate synthase reaction was determined by WALSH and his colleagues to be from solvent by conducting the reaction in $H_2{}^{18}O$ (*119, 152*). Similar results were obtained by GOULD and EISENBERG (*74*), using the enzyme obtained from *Enterobacter aerogenes* 62-1 (syn. *Aerobacter aerogenes*).

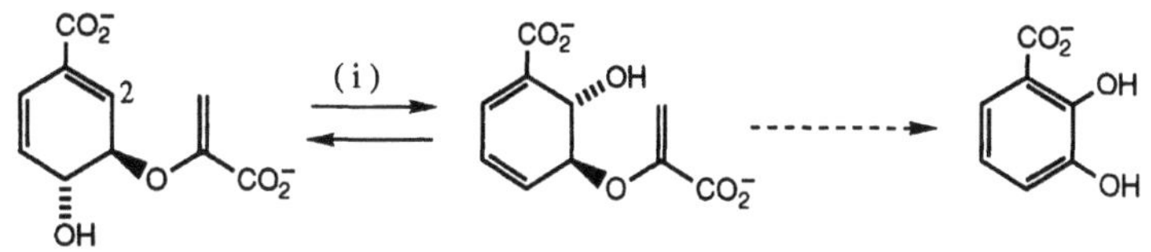

chorismate (**5**) isochorismate

(i) - *isochorismate synthase*, Mg^{2+}

The regions of greatest similarity between isochorismate synthase and the chorismate binding proteins of anthranilate synthase and *p*-aminobenzoate synthase are found towards the carboxyl termini where the residue similarities approach or exceed 40% (*132, 134*). The amino termini are characterised by relatively few similarities. These observations provide considerable support for the hypothesis that these enzymes are encoded by a family of genes that have a common evolutionary origin and their structural similarities reflect their ability to bind the substrate chorismate (**5**) in a (possibly very similar) reactive conformation. However it may be also noted that two other enzymes which utilise chorismate as their substrate, chorismate mutase and chorismate lyase, do not exhibit these same sequence similarities.

Iron is the most abundant transition metal found in the biosphere; it is required for almost all living organisms from bacteria to hominids. In humans iron is of medical concern because of its effects on infectious organisms and in diseases of iron overload and iron deficiency. Because iron in its most common form is not readily available as a nutrient (the solubility product of ferric hydroxide is 10^{-38}) strategies have been evolved by many living organisms to acquire sufficient quantities of this vital metal. Many micro-organisms thus produce siderophores – low molecular weight chelating agents that bind and solubilise ferric iron. Of all the natural siderophores so far investigated, the one that binds iron most strongly under physiological conditions is a siderophore biosynthesised by *Escherichia coli* named enterobactin or enterochelin, a cyclic trimer of *N*-2,3-dihydroxybenzoyl-L-serine (**9**). The siderophore employs the three 2,3-dihydroxybenzoyl nucleii to give a charged octahedral triscatecholate complex with a Δ-*cis* configuration. The (proton independent) formation constant with ferric iron for the macrocycle of 10^{49} M^{-1} is the highest reported for a siderophore (*120*).

Enterobactin is released and exported from the cell to scavenge and capture Fe^{3+}. The enterobactin-iron complex is transported back across the outer and inner membranes of the bacterial cell by permease and transport systems. Release of the iron is then thought to be achieved by two possible mechanisms. The first of these involves a ferric-enterobactin esterase which functions to disassemble the macrocycle by hydrolysis into the individual 2,3-dihydroxybenzoyl L-serine residues. A second mechanism for iron release which has been proposed is that a reductase reduces the iron to the ferrous (Fe^{2+}) state, which is known to bind less strongly to the enterobactin molecule. The *ent* genes (enterobactin biosynthesis), *fep* genes (iron permeation) and the *fes* genes (ferric enterobactin esterase) which regulate these various activities are clustered in several regulons in the *Escherichia coli* chromosome and are transcriptionally regulated by intracellular levels of iron.

In *Escherichia coli* enterobactin biosynthesis requires the expression of a number of genes – *ent C*, *ent B* and *ent A* genes code for the enzymes involved in the biosynthesis of 2,3-dihydroxybenzoate from chorismate. The subsequent condensation of 2,3-dihydroxybenzoate with L-serine followed by cyclisation to give enterobactin requires four polypeptides, (probably functioning as a complex-enterobactin synthase), specified by *ent D*-*ent G* genes.

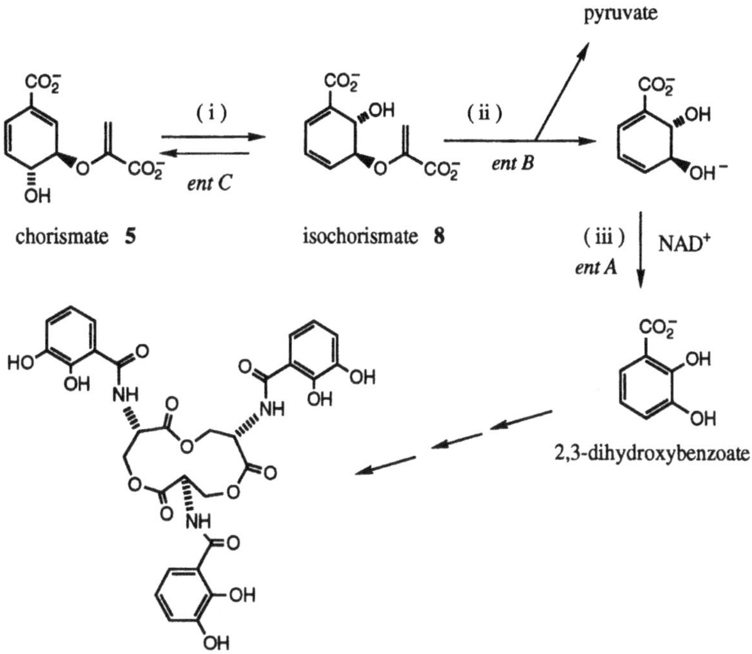

Enterobactin (Enterochelin) **9**

Enzymes : (i) - isochorismate synthase (EC 5.4.99.6) , *ent C* ; (ii) - isochorismatase (EC 3.3.2.1) , *ent B* ; (iii) - 2,3-dihydro-2,3-dihydroxybenzoate dehydrogenase (EC 1.3.1.28) , *ent A* .

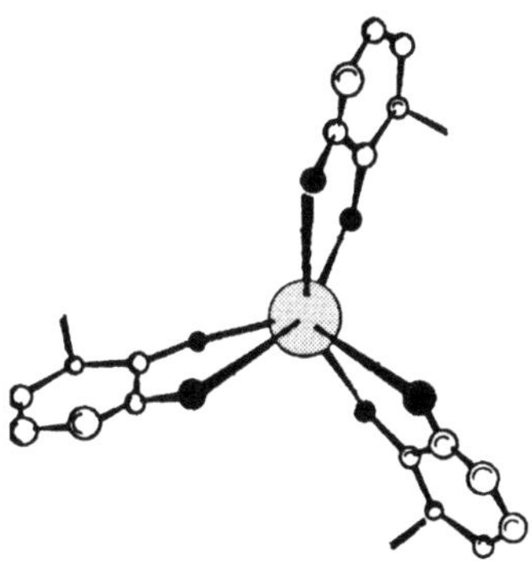

Enterobactin (Enterochelin) iron (III) complex

Studies with *Escherichia coli* mutants have shown that 2,3-dihydroxybenzoate is derived from chorismate by the action of three enzymes. The first of these is isochorismate synthase. Isochorismatase, the second enzyme in the sequence, catalyses the hydrolytic cleavage of the enolpyruvyl side-chain of isochorismate to give 2,3-dihydro-2,3-dihydroxybenzoate and pyruvate (*141*). The enzyme 2,3-dihydro-2,3-dihydroxybenzoate dehydrogenase catalyses the final step in the biosynthesis of 2,3-dihydroxybenzoate in an NAD^+-dependent, alcohol dehydrogenation reaction (*118*).

2.3.4. Phenylpropanoid Metabolism in Plants

2.3.4.1. L-Phenylalanine Ammonia Lyase (PAL)

The discovery of the enzyme L-phenylalanine ammonia lyase (PAL; EC 4.3.1.5 in barley seedlings by KOUKOL and CONN in 1961 (*113*) pointed to the gateway and the route by which the aromatic amino acids L-phenylalanine and L-tyrosine were diverted from protein synthesis into a whole range of aromatic metabolites, including lignin, in plants. The enzymic deamination of L-phenylalanine catalysed by L-phenylalanine ammonia lyase (PAL) is the first committed step in the biosynthesis of the phenylpropanoid skeleton in higher plants and takes place stereospecifically with the loss of the *pro*-3*S* hydrogen from the L-amino acid to give *trans*-cinnamate (*87*, *157*). The enzyme derived from many sources, notably grasses and certain fungi, also acts upon L-tyrosine with the same stereospecificity to give *trans*-*p*-coumarate. Interestingly a distinctive enzyme possessing L-tryptophan ammonia lyase activity has not, to date, been described from any plant source.

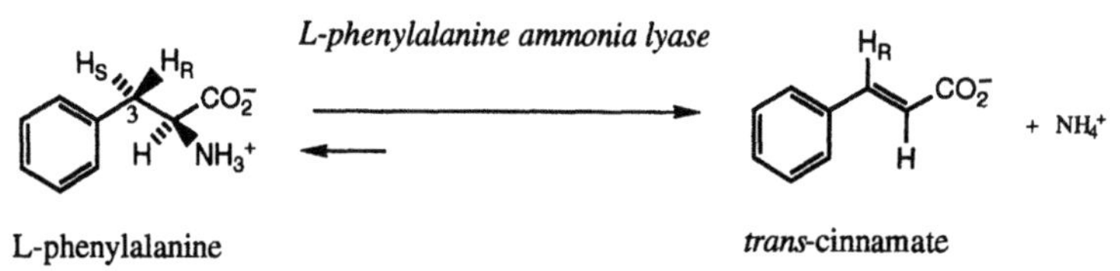

L-Phenylalanine ammonia lyase (PAL) is not an enzyme which is easily and rigorously purified. This may account for the variations in M_R and certain kinetic parameters which have, from time to time, been reported for the enzyme. It seems likely that the normal sub-units are either very similar or identical in size. The relative molecular mass (M_R) data reported for the enzyme from parsley appears to be typical of that recorded for the enzyme from a range of other plants (*163*). The parsley enzyme is tetrameric with a sub-unit M_R of 83,000 and an overall M_R of 330,000; $K_m^H = 240\ \mu M$, $K_m^L = 32\ \mu M$; Hill coefficient = 0.60. The pH optimum for L-phenylalanine ammonia lyase (PAL) at saturating substrate concentrations is approximately 8.7 with a bell-shaped pH activity curve. At pH 8.5 and 30°C the equilibrium constant for the elimination of ammonia is 4.1 M. The turnover number for L-phenylalanine ammonia lyase (PAL) is $\sim$ 3/sec., suggesting that there is an intrinsically slow step in the mechanism. Labelling studies indicate that there are two active sites per tetramer. Chemical inhibition of L-phenylalanine ammonia lyase (PAL) activity may be achieved by the use of typical carbonyl group reagents such as sodium borohydride and potassium cyanide.

2.4. Metabolic "Costs" of Aromatic Amino Acid Biosynthesis

Stoichiometric coupling by ATP underlies all metabolic functionality. By analogy with a money based economy ATKINSON (*12*), has estimated the metabolic costs of the production of a whole series of metabolites in terms of a metabolic unit of exchange – the ATP equivalent. Metabolic relationships are discussed in terms of ATP coupling coefficients – defined as the number of moles of ATP (or ATP metabolic equivalents) produced per mole of substrate converted in the reaction or sequence. Thus in the shikimate pathway the metabolic cost (in ATP equivalents) of the synthesis of chorismate from 2 moles of phosphoenolpyruvate (PEP, 2×16 ATP equivalents) and one of D-erythrose-4-phosphate (26 ATP equivalents) has been estimated as **60** ATP equivalents, and that of the synthesis of the three aromatic α-amino acids as L-phenylalanine (**65** ATP equivalents), L-tyrosine (**62** ATP equivalents) and L-tryptophan (**78** ATP equivalents). These metabolic costs are high, particularly when compared with those of all other protein α-amino acids, *e.g.* alanine (**20** ATP equivalents), serine (**18** ATP equivalents), cysteine (**19** ATP equivalents), aspartate (**21** ATP equivalents) and glutamate (**30** ATP equivalents), and this may give some hint at one of the underlying reasons for the presumed loss of the enzymic apparatus for this pathway by mammals during the course of evolution. When an organism obtains its energy and carbon by the consumption of other organisms or the products of those organisms, some of the fundamental pathways of intermediary metabolism become superfluous to its needs. It is therefore very striking that vertebrates retain only the ability to produce α-amino acids that may be synthesised from intermediates in glycolysis or the citrate (TCA) cycle in just one or two steps. They have lost all the synthetic pathways, such as the Shikimate pathway, which embody long reaction sequences, require many enzymes, and are costly in metabolic terms (ATP equivalents).

3. Enzyme Mechanisms

In terms of the chemical intermediates involved, the common pathway (Fig. 1) is the same in all species which have been examined thus far. This part of the shikimate pathway has nevertheless proved to be a veritable gold mine of ingenious and unusual enzymology which has intrigued chemists and enzymologists alike. Many of the mechanistic questions posed have attracted considerable interest and attention; some of the most significant of these are noted below. It is of particular interest to note the number of cases where reaction pathways appear to be dictated by the nature of chemical intermediates and their inherent chemistry. As Sir ROBERT ROBINSON once pointed out: "*Even enzymes are unlikely to disregard stereochemistry or the mode of electronic displacements in molecules.*"

3.1. 3-Deoxy-D-*arabino*-heptulosonate-7-phosphate (DAHP) Synthase

The question of the precise mode of action of the DAHP synthase reaction, (Fig. 1, Step 1), is one which remains to be fully answered. Initial kinetic studies pointed to a "ping-pong" reaction in which one of the substrates reacts with the enzyme with the release of a product prior to the addition of the second substrate. DELEO and SPRINSON (*51*) interpreted the reaction as involving the intermediate formation of an enolpyruvyl enzyme (represented as a carboxylate ester) but the details of this scheme, in which the methylene group of PEP is first transformed to a methyl group before regeneration as a methylene group, are seemingly vitiated by the stereochemical studies of ONDERKA, CARROLL and FLOSS (*66*). More recently GANEM has suggested (*71*), a novel mechanism for the mode of action of the enzyme DAHP synthase which involves the transfer of an enzyme-bound sulphur containing enolpyruvyl intermediate (Fig. 5). This proposal is in concert with observations and suggestions made contemporaneously concerning the enzyme phosphoenolpyruvate-UDP-*N*-acetyl-D-glucosamine-3-enolpyruvyl transferase. This enzyme catalyses the first step in the biosynthesis of the *N*-acetyl muramyl pentapeptide in cell-wall synthesis, in which the enolpyruvyl group is transferred from PEP to the hydroxyl group at C-3 of *N*-acetyl-2-amino-2-deoxy-D-glucose.

In the GANEM mechanism (Fig. 5) an enzymic thiol group reversibly conjugates PEP to the enzyme surface (Step i). After binding of the second substrate (Step ii), the key step of sulphur migration and concomitant 1,2 phosphate shift (Step iii) leads to a further complex from which

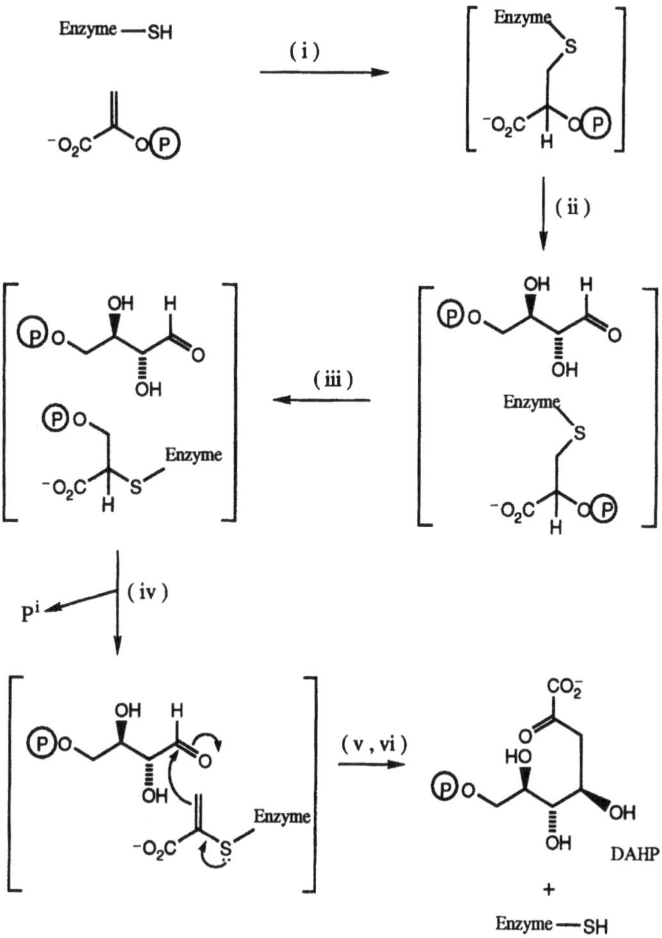

Fig. 5. Suggested mechanism of action of DAHP synthase (*71*)

phosphate is now released by β-elimination (Step iv). The delayed release of inorganic phosphate is consistent with the kinetics and the required C–O cleavage noted earlier (*51*). Condensation and hydrolytic removal of the product would then occur (Steps v and vi).

3.2. 3-Dehydroquinate Synthase (7-Phospho-3-deoxy-D-*arabino*-heptulosonate Phosphate Lyase)

3-Dehydroquinate synthase has attracted particular attention because of the apparent need of only one enzyme to accomplish the series of chemical stages in the cyclisation processs which transforms DAHP to 3-dehydroquinate (Fig. 1, Step 2). It is a sequence in which "inherent chemistry" is seen to play a very large part.

The *Escherichia coli* enzyme was first studied by SPRINSON and his colleagues (*121, 146*) who observed that the enzyme requires a catalytic amount of NAD^+, (*i.e.*, stoichiometric with the enzymc), both for catalytic activity and to maintain the structural integrity of the enzyme. No intermediates were detected in the reaction and SPRINSON proposed an interesting and ingenious mechanistic pathway for the enzyme, (Fig. 6). In this sequence loss of orthophosphate from DAHP was facilitated by oxidation at C-5 (NAD^+ to NADH). The second step of the sequence involves the β-elimination of inorganic phosphate and in the third step the enzyme bound NADH reduces the ketone (regenerating the same configuration at C-5 as in DAHP). Ring opening of the enolpyranose and then reclosure by attack of the enolate carbon (C-7) on the carbonyl

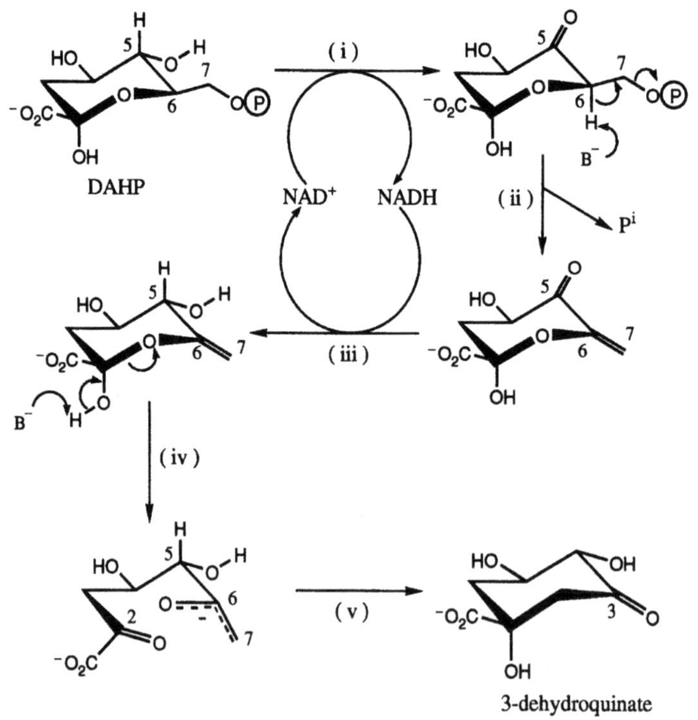

Fig. 6. Sprinson's proposed mechanistic pathway for the enzyme 3-dehydroquinate synthase (*109*)

group at C-2, in an aldol like reaction, produces 3-dehydroquinate and completes the reaction.

SPRINSON's view in fact was a very perceptive one of the way this enzyme functioned; as KNOWLES also remarked later (*109*), echoing the words of ROBINSON, "*Chemically and logically this pathway is very attractive, and has many features that would surely have been included had an organic chemist been responsible for the design of dehydroquinate synthase.*"

KNOWLES (*69*) sought to solve these questions surrounding the mode of action of 3-dehydroquinate synthase in a wide ranging study; the basis of his approach was to prepare a series of substrate analogues that, by virtue of minor structural alterations, were unable to complete the hypothetical reaction scheme put forward in Fig. 6. BARTLETT and SATAKE (*16*) similarly looked at the final stages of ring opening of the enol-pyranose intermediate and the subsequent internal aldol condensation (Steps iv and v, Fig. 6). They concluded that there was no good reason in fact to invoke enzymic catalysis for these final biosynthetic transformations. SPRINSON's proposed mechanism was thus vindicated; KNOWLES' final judgement on this enzyme was therefore an interesting one (*109*):

"*It seems likely that dehydroquinate synthase is not, after all, an enzyme of unprecedented catalytic versatility and prowess. Perhaps it is merely a dehydrogenase, for the catalytic activity of which a divalent metal cation and enzyme bound NAD^+ are both necessary and sufficient. The enzyme-catalysed oxidation of C-5 of the substrate DAHP may be followed by the (now facile) passive loss of P_i in an ElcB process to produce the enone. Reduction to the enolpyranose could complete the enzyme's catalytic involvement, for the loss of the enol-pyranose from the active site would allow the rapid and stereoselective rearrangement of this species to the final product, dehydroquinate. ... in the overall transformation that is mediated by dehydroquinate synthase, nature has neatly and ingeniously exploited several kinetically feasible and thermodynamically favourable processes. The superficially impressive enzyme that mediates the concatenation of catalytic steps outlined may be no more, in reality, than a relatively banal dehydrogenase.*"

3.3. 5-Enolpyruvylshikimate-3-phosphate (5-EPS-3-P) Synthase

The mechanism of action of the enzyme 5-enolpyruvylshikimate-3-phosphate synthase (Fig. 1, step 6) has been the object of considerable attention over the past decade because of its agronomic importance as

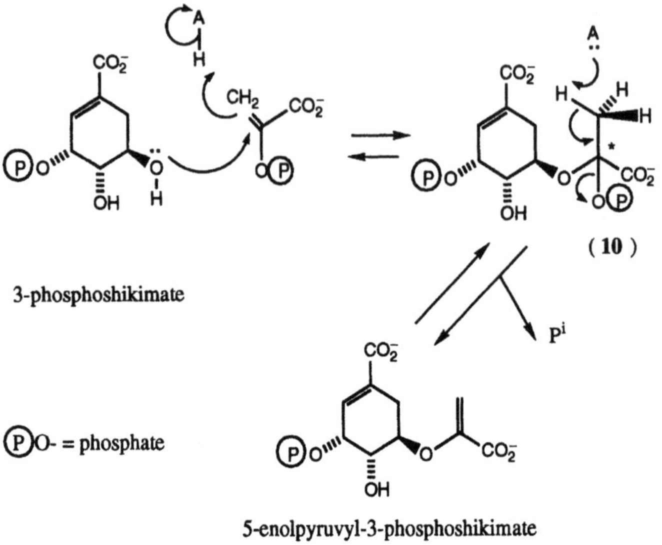

Fig. 7. A postulated mechanism for the mode of action of the enzyme 5-enolpyruvylshikimate-3-phosphate synthase (5-EPSP synthase) (*116*)

the target of the non-selective post-emergence herbicide glyphosate (*N*-phosphonomethylglycine). Glyphosate is believed to act by competing with the substrate phosphoenolpyruvate to bind at the active site of the enzyme. A mechanism for the mode of action of the enzyme was originally put forward by LEVIN and SPRINSON (*116*) (Fig. 7). Although other mechanisms have been suggested a picture resembling most nearly the SPRINSON mechanism has emerged from kinetic studies and the isolation and characterisation of the putative intermediate (**10**) (*7–9*).

Initial experiments (*7–9*) demonstrated the presence of an acid labile intermediate (decomposing to give pyruvate and shikimate-3-phosphate). If the enzyme was denatured by quenching under basic conditions (neat triethylamine) the intermediate was stable and could be isolated by ion-exchange hplc (*7–9*). Its structure (**10**) was confirmed by ^{1}H, ^{31}P and ^{13}C NMR after the isolation of 300 μg of the intermediate from enzymic synthesis. The Pennsylvania State-Monsanto group concluded that the tetrahedral adduct, as first proposed by SPRINSON (Fig. 7) was a true intermediate on the pathway, and that the reaction proceeds by an addition-elimination mechanism involving the nucleophilic attack of the C-5 hydroxyl group of shikimate-3-phosphate at C-2 of phosphoenolpyruvate, thus laying to rest the suggestion of a covalent enzyme-enolpyruvyl type of intermediate. The absolute stereochemistry of the new tetrahedral centre (*) created in the intermediate (**10**) is not yet known.

3.3.1. Glyphosate

Glyphosate (**11a**, *N*-phosphonomethylglycine) is the herbicidal component of "Roundup" which is widely used as a broad spectrum, non-selective, post-emergence weed killer. Its herbicidal properties are combined with the properties of good translocation in plants, rapid inactivation by soil micro-organisms and low toxicity to organisms other than plants.

$pK_1 = < 2$
$pK_2 = 5.86$

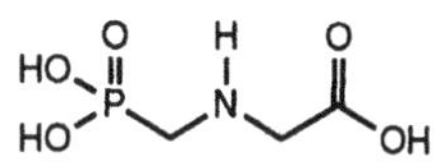

$pK = 2.32$

(**11a** , glyphosate)

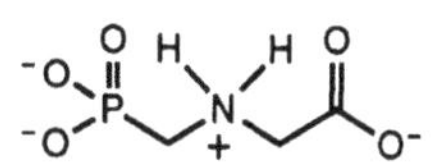

(**11b**)

The enzyme 5-enolpyruvylshikimate-3-phosphate synthase was later shown by STEINRUCKEN and AMRHEIN (*3–6*) to be the primary target of glyphosate and the means whereby it exerted its herbicidal action. Glyphosate has been shown, with all the 5-enolpyruvylshikimate-3-phosphate synthases from different organisms studied to date, to be a competitive inhibitor of the substrate phosphoenolpyruvate (PEP). K_i values for glyphosate encompassing the range 0.1 to 12.0 μM have been reported with the enzyme from various organisms (*3–6*). Resistance to glyphosate in micro-organisms and in plant tissue culture has generally been found to be correlated with either an increase in the production of glyphosate *sensitive* 5-enolpyruvylshikimate-3-phosphate synthase or alternatively the formation of a glyphosate *insensitive* 5-enolpyruvylshikimate-3-phosphate synthase (*3–6*, *19*). Tissues and cultured cells often accumulate shikimic acid or its 3-phosphate in the presence of glyphosate (*19*).

The discovery of glyphosate owes little to rational design. Despite its importance as a herbicide and the attention directed to this comparatively simple molecule (*18*) its mode of action remains something of an enigma. Numerous analogues of glyphosate (**11a**) have been prepared, but only three derivatives show even modest inhibition of 5-enolpyruvylshikimate-3-phosphate synthase. The pH of the medium strongly influences the sensitivity of 5-enolpyruvylshikimate-3-phosphate synthase to glyphosate; between pH 6 and 8 there is a striking increase in the inhibition. On the basis of this evidence and the pK values of the various ionisable groups in glyphosate, AMRHEIN has proposed (*3–6*) that the kinetically active form of the inhibitor is the ionised form (**11b**). Independently both AMRHEIN (*3–6*) and ABELES (*1*) speculated that glyphosate may act as a transition state analogue of phosphoenolpyruvate in which the positive charge on the imino group of the herbicide mimics the developing carbocation characteristics of the normal substrate following protonation at the methylene carbon atom (Fig. 8); location of

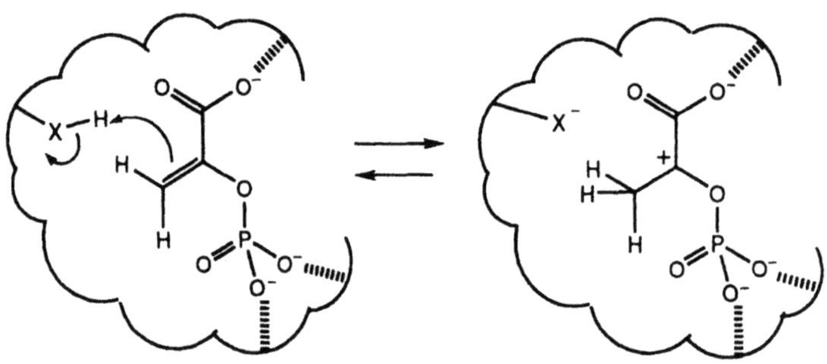

(a) - *Suggested mode of binding of phosphoenolpyruvate to the active site of 5-enolpyruvylshikimate-3-phosphate synthase . Formation of the enzyme bound protonated intermediate , which is held in position prior to reaction with shikimate-3-phosphate , by ionic interactions.*

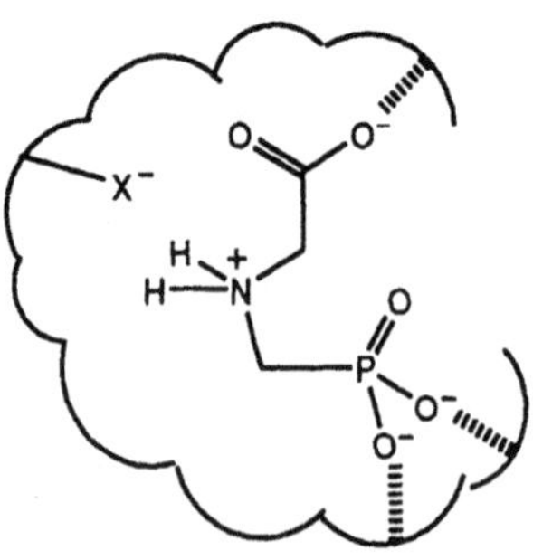

(b) - *Glyphosate , it is proposed , fits into the active site of the enzyme in a spatial orientation which models that of the protonated natural substrate .*

Fig. 8. Proposed models for the binding of phosphoenolpyruvate and the herbicide glyphosate to the active site of 5-enolpyruvylshikimate-3-phosphate synthase (*1*, *3–6*)

glyphosate at the active site of the 5-enolpyruvylshikimate-3-phosphate synthase is, it is argued, engineered by the correct geometric disposition of the carboxylate and phosphonate groups on the inhibitor, binding to the carboxylate and phosphate sites normally utilised by the substrate phosphoenolpyruvate on the enzyme.

ANDERSON and her colleagues have nevertheless indicated that one must also consider, when formulating hypotheses for the mechanism of 5-enolpyruvyl-shikimate-3-phosphate synthase inhibition by glyphosate, the potential for the interaction between glyphosate and shikimate-3-phosphate. In the view of these workers (*7*) the most likely explanation of glyphosate inhibition is that the herbicide and shikimate-3-phosphate combine at the active site of 5-enolpyruvylshikimate-3-phosphate synthase to give a complex whose overall geometry closely resembles that of the tetrahedral intermediate (**10**). Glyphosate, it was supposed, interacts with the same active-site residues which catalyse the formation and breakdown of this intermediate.

3.4. Chorismate Synthase

The final step (7) of the common part of the shikimate pathway is mediated by the enzyme chorismate synthase and is the conversion of 5-enolpyruvylshikimate-3-phosphate (5-EPS-3-P) into chorismate (Fig. 1). This transformation proceeds by an overall *trans* 1,4-elimination of phosphate and with the abstraction of the C-6 *pro*-R hydrogen atom. The precise mechanism of this reaction, just as that of the first enzyme in the sequence (DAHP synthase), continues to baffle and intrigue cogniscenti. Incubation of each of the stereospecifically fluoro-substituted derivatives (**12a**, **12b**) of the substrate 5-enolpyruvylshikimate-3-phosphate (5-EPS-3-P) with chorismate synthase derived from *Neurospora crassa* showed that both acted as competitive inhibitors of the enzyme, with (**12a**) having an affinity an order of magnitude greater than (**12b**) (*14*).

Pre-steady state kinetic experiments indicate that loss of phosphate is not a fast step prior to the rate determining step (*88*). Combined with the data from the observed kinetic isotope effects for the hydrogen release

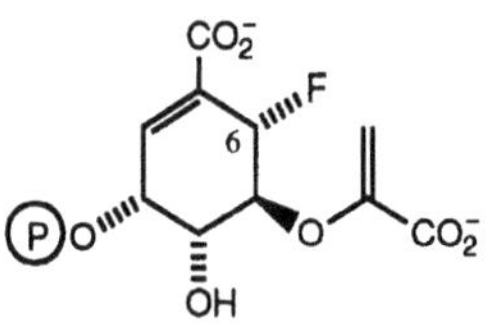

(**12a**)

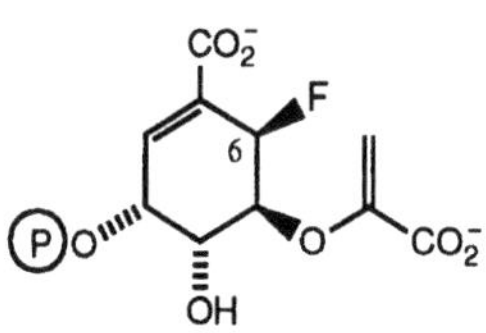

(**12b**)

from C-6, ABELL and his collaborators suggested (*2*) that these features are best accommodated by a concerted mechanism. Uncertainty however still surrounds the precise role of the flavin co-factor. Evidence for the presence of a flavin monoradical intermediate has been presented and this, it has been speculated, may not accumulate when rapid removal of a hydrogen radical from C-6 of the substrate 5-enolpyruvylshikimate-3-phosphate (5-EPS-3-P) is possible (*139*, *140*).

3.5. Chorismate Mutase

Although not an enzyme of the *common pathway* of aromatic amino acid metabolism chorismate mutase, because of its uniqueness, has attracted considerable attention. The intramolecular rearrangement of chorismate to prephenate catalysed by the enzyme chorismate mutase appears to be the only example in primary/intermediary metabolism of a 3,3-sigmatropic rearrangement of the Claisen type. Stereochemical studies by KNOWLES and BERCHTOLD (*36*, *144*), have demonstrated that both the enzymic and the thermally induced non-enzymic reactions proceed through a transition state of chair like geometry resembling (**14**). The non-enzymic rearrangement shows a secondary isotope effect at C-3, the site of bond breaking, but none at C-9, the site of new bond formation, which suggests either a stepwise reaction or a very unsymmetrical transition state. Various studies have also examined the way in which the different participating structural elements in the chorismate-prephenate thermal rearrangement affect the ease of the transformation. The general consensus appears to be that the non-enzymic reaction is dissociative and involves a transition state which is dipolar in character.

On the other hand although various proposals have been made from time to time regarding the mechanism of action of the enzyme chorismate mutase (*80*, *86*, *101*) neither mechanistic details of the enzyme catalysed rearrangement nor the origins of the 2×10^6-fold rate enhancement over the uncatalysed process are known precisely; however an educated guess can be made! Very recently LIPSCOMB and his colleagues (*31*) have reported the crystal structure of the chorismate mutase from the Marburg strain of the organism *Bacillus subtilis* (see Sect. 2.3.1.1). This enzyme is monofunctional, non-allosteric, unaffected by end-product amino acids and effectors and exhibits Michaelis-Menten kinetics (*75*, *143*). With just 127 amino acid residues per sub-unt the *Bacillus subtilis* chorismate mutase is the smallest enzyme of this class which is known. The chorismate mutase monomer consists of a single domain in which the polypeptide chain is folded into a five-stranded mixed β-sheet packed against an 18 residue α-helix and a two turn 3_{10} helix. The C-terminal tail consists of one 3_{10} helix which is directed away from the β-sheet on the face opposite the helices into the solvent region. The electron density map clearly shows that the enzyme exists as a homotrimer with the three monomers packing to form a pseudo-αβ-barrel; the sub-units of the trimer are related by pseudo-3-fold symmetry. The core of the homotrimer consists of large portions of the β-sheet from each monomer, together forming a closed barrel of β-strands. Amino acid side-chains protruding into the core of the barrel are almost entirely hydrophobic in character, (Ile-3, Val-43, Leu-46, Pro-72, Met-76 and Met-92). Near the surface of the trimer the interfaces between adjacent sub-units form three equivalent clefts, the walls of which are formed by different regions of the two sub-units (Fig. 9).

The active-site of the *Bacillus subtilis* chorismate mutase was located and identified using data from the complex formed by the enzyme with an *endo*-oxabicyclic transition state analogue (**13**, $K_i = 3$ μM), reported earlier by Bartlett and Johnson (*15*). The inhibitor thus binds, with little resultant change in conformation of the enzyme, to the surface cleft(s) described above. Various interactions were observed between the enzyme and the inhibitor; hydrophobic contacts (between the inhibitor

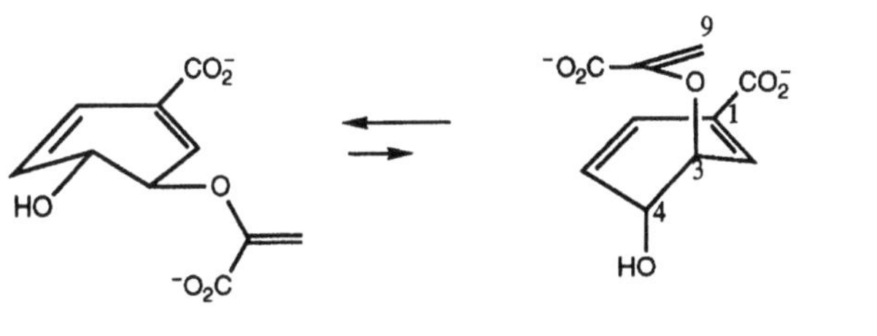

chorismate - *pseudo-diequatorial* chorismate - *pseudo-diaxial* , **14**

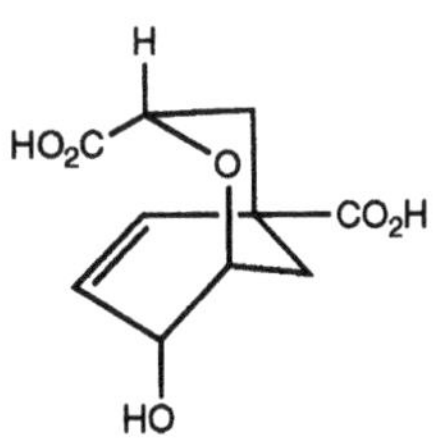

(**13**)

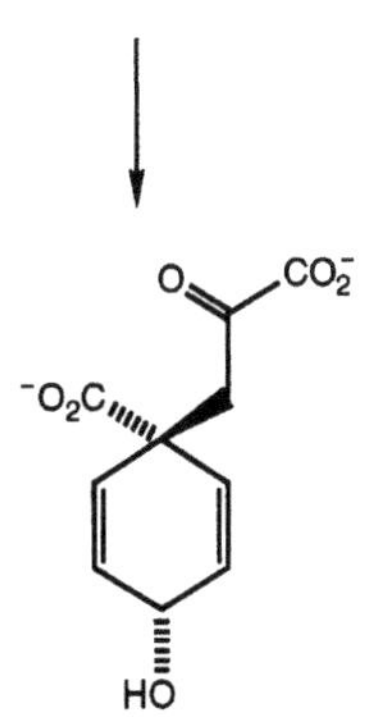

prephenate

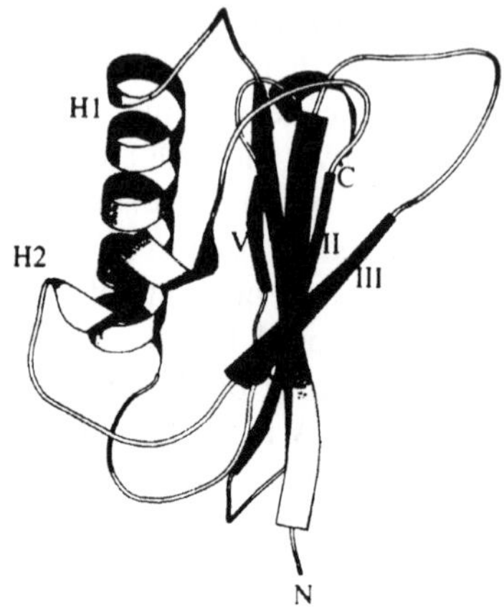

(a) - Ribbon drawing of chorismate mutase monomer

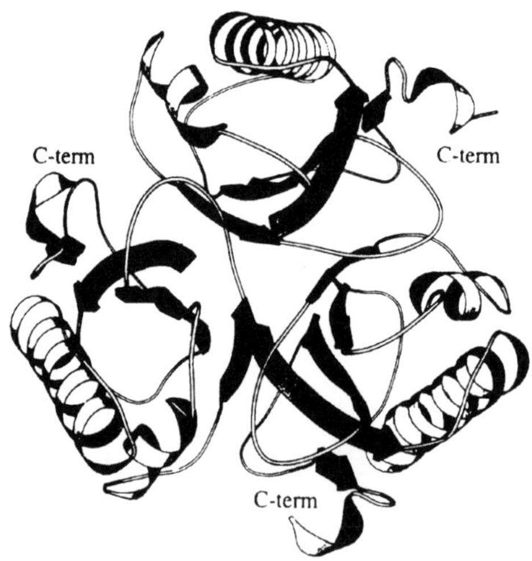

(b) - Ribbon drawing of chorismate mutase trimer in a view looking down the barrel / pseudo-threefold axis

Fig. 9. Ribbon drawings of the crystal structure of the chorismate mutase from *Bacillus subtilis* (*31*): (a) monomer; (b) trimer in a view looking down the pseudo-threefold axis of the barrel

and side-chains Phe-57 and Ala-59 from one monomer and Leu-115 from the adjacent monomer), ionic contacts between both carboxy groups and arginine side-chains (Arg-7 and Arg-90 for one carboxy and Arg-116 for the other) of the protein and polar contacts involving the hydroxy group (side chains of Glu-78 and Cys-75) and the ether oxygen (side chain of Arg-90) of the inhibitor. However no functional groups, capable of *proton transfer* to the ether oxygen of the inhibitor, were observed strongly suggesting that the enzyme catalysed rearrangement of chorismate to prephenate is a pericyclic process analogous to the uncatalysed reaction. The enzyme chorismate mutase, in this view, serves as a template to stabilise the transition state of the rearrangement (resembling **14**), by specific contacts and interactions as noted above and thereby lowers the activation energy for the intramolecular reaction.

Copley and Knowles (*35*) had earlier shown that up to 20% of chorismate molecules are present in solution in the pseudo-diaxial form (**14**) and that conformational equilibration between this and the pseudo-diequatorial conformer is rapid on the NMR time scale. Although a detailed analysis of the catalytic mechanism is now awaited this evidence suggests that the enzyme selects the appropriate pseudo-diaxial conformer (**14**) from the equilibrium mixture of conformers to initiate the catalytic transformation. A pictorial representation of the location of the active-sites in chorismate mutase and the binding interactions which are thought to stabilise the putative transition state of the chorismate-prephenate rearrangement is shown in Fig. 10.

It is worth noting that in fact very similar models for the active-site of chorismate mutase had been suggested earlier on the basis of "indirect" methods of active-site analysis. A model was proposed in the author's laboratory for example [*J. Chem. Soc. Perkin Trans. I* 2765 (1987)] in which emphasis was placed on the complementarity of the active-site to the pseudo-bisaxial conformer of chorismate (**14**). Studies of the interaction of chorismate analogues and of various enzyme inhibitors suggested that both carboxy groups were bound by ionic interactions and the C-4 hydroxyl group by hydrogen-bonding to a suitable partner. This latter group was proposed, on the basis of inhibition studies with iodoacetamide, to be the thiol group of a cysteine residue. It was additionally suggested that hydrophobic and/or π-electron interactions between the diene system of chorismate and an aromatic amino acid residue on the enzyme would also assist in stabilisation of the transition state complex.

Recent work has utilised the immune system as a prolific source of specific receptor molecules for the construction of "enzyme-like" catalysts – the so-called catalytic antibodies. One approach has been to

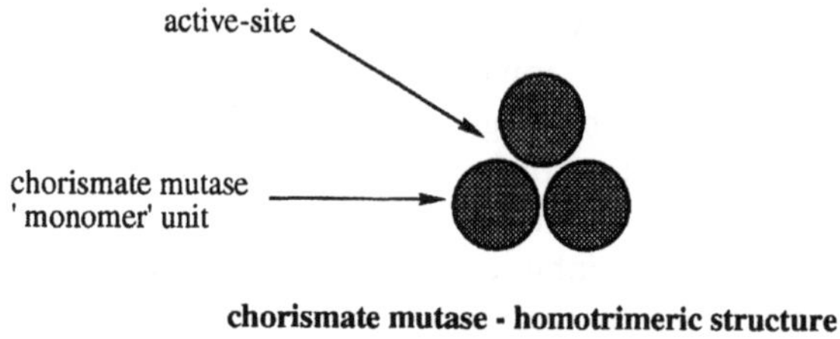

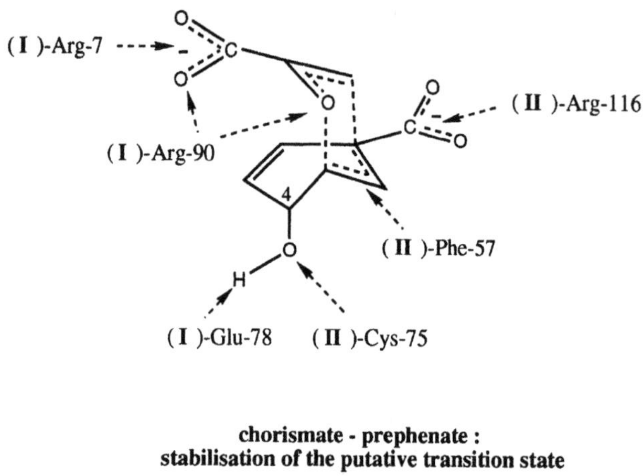

Fig. 10. Pictorial representation of (i) homotrimeric structure and active-sites of chorismate mutase and (ii) the binding interactions which are believed to stabilise the putative transition state of the chorismate-prephenate rearrangement; "monomer" units are designated as (**I** and **II**)

generate catalytic antibodies which are complementary to a rate determining transition state in a reaction. Monoclonal antibodies have been elicited (*102*) to the *endo*-bicyclic transition state analogue (**13**) using the derivative (**15**). One of eight antibodies was found to catalyse the Claisen

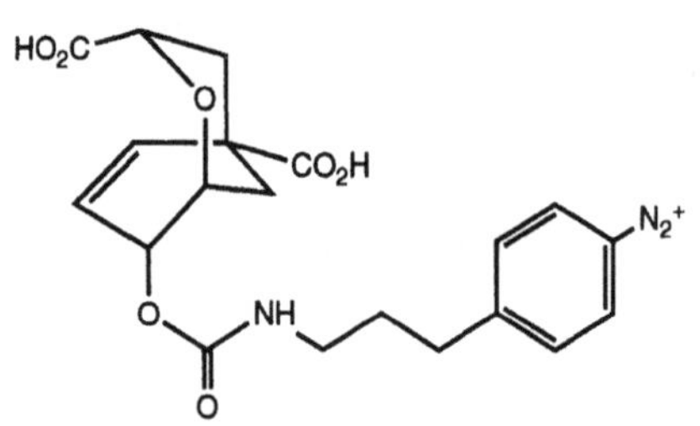

(**15**)

rearrangement of chorismate to prephenate, (K_m for chorismate at 10 °C = 260 μM). The antibody catalysed rearrangement when observed in D_2O showed no detectable isotope effect and the racemic compound (**13**) competitively inhibited the reaction, (K_i at 10 °C = 9 μM). A direct comparison with the rate of the uncatalysed thermal rearrangement gave a value of $k_{cat}/k_{uncat} = 1 \times 10^4$ at 10 °C, compared to a value of 3×10^6 for

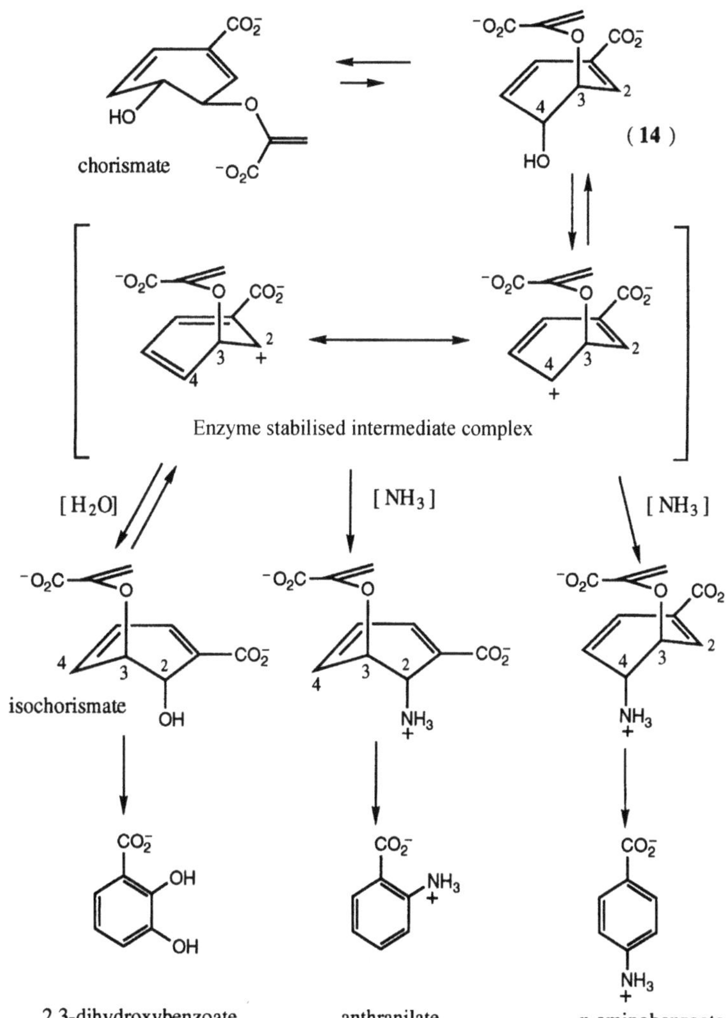

Fig. 11. Suggested enzyme catalysed pathways to *o*- and *p*-aminobenzoates and 2,3-dihydroxybenzoate having a "common" carbocation intermediate

the rearrangement induced, under the same conditions, by the chorismate mutase from *Escherichia coli*. Very similar results were presented simultaneously by HILVERT and his colleagues using the same transition state analogue but a different linkage technique (*28, 91*). Values of $k_{cat} = 1.2 \times 10^{-3}\ s^{-1}$ and $K_m = 5.1 \times 10^{-5}$ M were recorded for a catalytic antibody for which the substrate (**13**) acted as a competitive inhibitor ($K_i = 0.6\ \mu M$). It seems very probable therefore that the catalytic antibody catalyses the molecular rearrangement of chorismate by providing an environment complementary to the conformationally restricted transition state related to the pseudo-diaxial conformer of chorismate (**14**).

Attention was drawn earlier to the suggestion that the group of enzymes which are responsible for directing chorismate (**5**) into a range of metabolic pathways shown in Fig. 2, may be encoded by a family of evolutionary related genes. Evidence in support of that thesis is available for anthranilate synthase, isochorismate synthase and *p*-aminobenzoate synthase (*105*). It is tempting to speculate that these enzymes may therefore utilise particular transition states of reaction which are analogous, at least in the chemical sense, to one another and *perhaps* to that for chorismate mutase (*vide supra*). Such a speculative scheme is proposed in Fig. 11 and involves the initial generation of a common carbocation intermediate (or such species appropriately stabilised by the enzyme) by loss of the C-4 hydroxyl group from the pseudo-bisaxial conformer of chorismate (**14**). Stereospecific capture of the carbocation intermediate by either water (at C-2) or "ammonia" (at C-2 or C-4) would lead, after loss of the enolpyruvate side chain, to 2,3-dihydroxybenzoate or to the aminobenzoates. In this scheme loss of the C-4 hydroxyl group and capture of the carbocation by the incoming nucleophile would occur from the same face of the molecule. Nucleophile and leaving group could, in principle, be coordinated in the various steps in this process to a (the same) metal ion.

4. Multifunctional Enzymes

The shikimate pathway has proved to be a rich source of enzymes which form parts of multifunctional enzyme complexes or multifunctional proteins, (multifunctional enzymes). Numerous examples have now been recorded where two or more enzyme activities are located within one polypeptide chain. Most notable of these are the different forms of aggregation of enzymes involved in the biosynthesis of

L-tryptophan (Fig. 4) and in the common part of the shikimate pathway itself (Fig. 1). In all cases examined the observed patterns of molecular assembly are very much dependent on the type of organism under examination. The following major operational advantages are usually put forward for the occurrence of multifunctional enzymes: (i) enhancement of catalytic activity; (ii) substrate channelling; (iii) co-ordinate regulation of enzyme activity; (iv) protection of unstable intermediates; (v) co-ordinate expression of the genes coding for the individual enzyme activities; (vi) essential spatial organisation of the various enzymatic functions (*69*). Despite their undoubted attractiveness as theoretical concepts detailed kinetic evidence is, in many cases, still required to support the ideas of catalytic facilitation (i) and substrate channelling (ii).

In *Neurospora crassa* the five genes coding for the central five enzyme catalysed steps (2–6) of the common pathway (Fig. 1) are tightly linked in an "operon-like" cluster. This situation however contrasts strongly with that observed in *Escherichia coli* where the five corresponding genes have been demonstrated to be widely scattered about the genome. The *Neurospora crassa* enzymes were found to co-purify and were initially believed to constitute a classic example of a multi-enzyme complex, composed of different non-covalently associated sub-units capable of catalysing each of the distinct reactions. In *Escherichia coli*, on the other hand, the five enzyme activities could be resolved and showed no tendency to aggregate. A definitive purification of the *arom* complex from *Neurospora crassa*, has now been described (*33*). All five enzyme activities co-purify in constant ratio and the resulting enzyme is homogeneous with a sub-unit M_R of 165,000. Remarkably perhaps, the turnover numbers for each of the five enzyme activities lie within a very narrow range of 18 to 51 sec^{-1}. It is now therefore quite firmly established that the *arom* multienzyme complex consists of two identical pentafunctional polypeptide chains and that the *arom* gene cluster, since it encodes a single species of polypeptide chain, is really a single gene.

Genetic and biochemical studies have now revealed the presence of an *arom* gene cluster in *Neurospora crassa*, *Aspergillus nidulans*, *Saccharomyces cerevisiae*, other fungal and yeast species, and in *Euglena gracilis*. The nucleotide sequence of the *Saccharomyces cerevisiae* ARO1 gene, which encodes the *arom* multifunctional enzyme, has been determined by COGGINS and his group (*56*) and similar work has also been reported for the *arom* polypeptide from *Aspergillus nidulans*. The amino acid sequence deduced for the pentafunctional *arom* polypeptide from yeast is 1588 amino acids in length and has a calculated M_R of 174,555, corresponding closely to the M_R of 159,698 for the combined *Escherichia*

Table 5. *The Five Escherichia coli Enzymes Corresponding to the Saccharomyces cerevisiae arom Complex Activities*

Step	Enzyme	*E. coli* gene	M_R (Calc.)	Length*
2	3-Dehydroquinate synthase	*aro* B	38,880	362
3	3-Dehydroquinase	*aro* D	26,377	240
4	Shikimate dehydrogenase	*aro* E	29,380	272
5	Shikimate kinase	*aro* L	18,937	173
6	5-Enolpyruvylshikimate-3-phosphate (EPSP) synthase	*aro* A	46,112	427

* Number of amino acids. Total M_R (Calc.) = 159,689; total (amino acids) = 1475. Taken from COGGINS *et al.* (*33, 56*).

coli enzymes (Table 5). Functional regions within the polypeptide chain were identified by comparison with the sequences of the five monofunctional *Escherichia coli* enzyme. The order of functional regions is the same for the *Saccharomyces cerevisiae* and *Aspergillus nidulans* pentafunctional polypeptides, *viz.* 3-dehydroquinate synthase [Step 2]; 5-enolpyruvylshikimate-3-phosphate (EPSP) synthase [Step 6]; shikimate kinase [Step 5]; 3-dehydroquinase [Step 3]; and shikimate dehydrogenase [Step 4].

The yeast *arom* polypeptide chain contains 1588 amino acid residues, some 113 more than the total found in the five separate corresponding *Escherichia coli* enzymes. Many of these extra amino acids occur in the regions linking the various domains and these connector regions are probably essential for the retention of the structural integrity of the multifunctional protein. The amino acid sequences of the *Saccharomyces cerevisiae* and *Aspergillus nidulans* multifunctional proteins have mosaic structures with defined and recognisable regions that are closely related to the monofunctional *Escherichia coli* counterparts. The most likely explanation for the origin of the pentafunctional fungal *arom* polypeptides is that they have arisen by fusion of ancestral *Escherichia coli*-like genes. The question of whether any specific and special kinetic and catalytic properties are associated with the *arom* polypeptide remains unresolved.

The yeast *arom* polypeptide chain-organisation of the enzyme activities: the numbers **2**, **3**, **4**, **5**, **6** refer to the enzyme activities, Fig. 1

The shikimate common pathway enzymes are less well characterised in higher plants than in micro-organisms but it is clear from much of the data now available that five of the enzymes occur separately whilst two [3-dehydroquinate dehydratase (3-dehydroquinase) and shikimate dehydrogenase, Steps 3 and 4], occur on a bifunctional polypeptide chain (*131*). Parenthetically it is interesting to note that limited proteolysis of the *Neurospora crassa arom* complex by trypsin or subtilisin (*145*) yields a stable fragment (M_R 68,000) which contains the dehydroquinase and shikimate dehydrogenase active sites. The two enzyme activities occur next to each other on the *Neurospora crassa arom* polypeptide (*vide supra*) and apparently do not require the intact polypeptide chain for activity. The molecular assembly and organisation of the enzymes in the pre-chorismate part of the shikimate pathway thus differ markedly according to the type of organism under consideration.

5. Genetic Engineering

The traditional approach of isolating proteins from natural sources often provides only limited quantities for possible structural (crystallisation, X-ray analysis), kinetic and mechanistic studies; more recent work using "engineered" bacteria have helped to minimise if not entirely overcome many of these problems. The enteric bacterium *Escherichia coli* is used to carry (host) and amplify "foreign" DNA sequences during recombinant DNA manipulations. "Foreign" DNA (present in a plasmid vector) can be readily introduced into the bacterium and amplified *in vivo* by simply allowing the organism to multiply. This change in experimental practice from classical isolation to engineered overexpression shifts the emphasis on expertise from that of protein biochemistry to that of molecular biology.

The technique of protein overproduction has been widely employed to obtain enhanced yields of many shikimate pathway enzymes. Typical examples are: 3-dehydroquinate synthase (*123*), 3-dehydroquinate dehydratase (3-dehydroquinase) (*55*) shikimate dehydrogenase (shikimate oxido-reductase) (*10*), shikimate kinase II (*47*), 5-enolpyruvyl-shikimate-3-phosphate (5-EPS-3-P) synthase (*117*), chorismate synthase (*156*), chorismate mutase (monofunctional) (*149*), chorismate mutase-prephenate dehydratase (P-protein) and chorismate mutase-prephenate dehydrogenase (T-protein) (*20, 150*), *p*-aminobenzoate synthase (*132*) and isochorismate synthase (*119*). Recent work has also demonstrated how the techniques of genetic engineering may be employed with whole organisms. This work is exemplified here by ongoing studies of the genetic manipulation of microbial aromatic amino acid biosynthesis and anthocyanin pigmentation in plants.

5.1. Biocatalytic Syntheses of Aromatics from D-Glucose

In the organism *Escherichia coli* the first enzyme in the shikimate pathway, DAHP synthase (Fig. 1) occurs as three isoenzymes, encoded by three unlinked genes, *aro*F (sensitive to tyrosine), *aro*G (sensitive to phenylalanine) and *aro*H (sensitive to tryptophan), each of which has been cloned. Carbon flow through the pathway is controlled by modulation of DAHP synthase; although all three DAHP synthases are transcriptionally regulated, feedback inhibition is quantitatively the major control mechanism *in vivo* (*133*). Genes *insensitive* to tyrosine and phenylalanine, respectively, have also been cloned. Thus for example an *aro*F allele encoding a tyrosine insensitive DAHP synthase has been cloned from an *Escherichia coli* mutant (*154*). The derived DAHP synthase was *not* inhibited by tyrosine, even at concentrations that reduced the wild-type enzyme activity by more than 90%. This change, it was demonstrated, was the result of the change in identity of one single amino acid residue at position 148 from proline to leucine. Residues 147–149 are thought to form part of an aromatic amino acid binding pocket, common to at least two of the three *Escherichia coli* DAHP isoenzymes.

The percentage of D-glucose that organisms are able to convert into aromatic metabolites is an important factor that will help to determine the commercial and industrial viability of whole organism biocatalytic syntheses. Frost and his colleagues have focused efforts (*52–54*) in this area on the amplification of the catalytic activity of the first enzyme in the pathway with an *aro*G gene encoding an isoenzyme of DAHP synthase which was insensitive to feedback inhibition, and by enhancing the bioavailability of the precursor D-erythrose-4-phosphate. Consideration of which enzymes influence the *in vivo* concentration of this metabolite led to the prediction that transketolase (*tkt*) is the pivotal enzyme which determines the concentration of D-erythrose-4-phosphate. These observations led to the use of an *Escherichia coli* mutant D 2704 with amplified DAHP synthase and transketolase activity. Increasing the flow of D-glucose equivalents into the common pathway of aromatic amino acids by such a device however creates a situation in which individual common pathway enzymes become rate limiting, *viz*, 3-dehydroquinate synthase, shikimate kinase 5-enolpyruvyl-shikimate-3-phosphate (5-EPS-3-P) synthase and chorismate synthase become rate limiting but 3-dehydroquinate

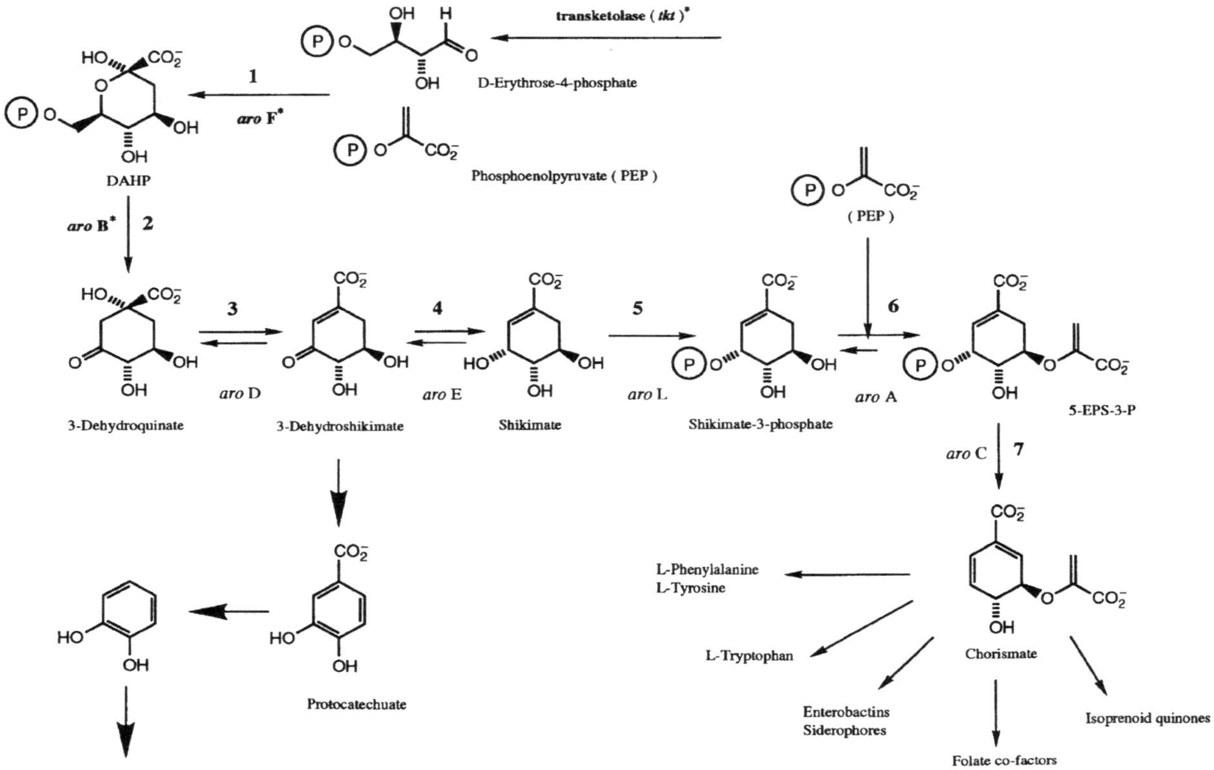

Fig. 12. The "not-so-common" pathway of aromatic biosynthesis in *Escherichia coli* (*52–54*)

dehydratase (3-dehydroquinase) and shikimate dehydrogenase (shikimate oxido-reductase) do not.

In analogous experiments Frost and Draths (*68*) increased the percentage of D-glucose directed into aromatic biosynthesis in *Escherichia coli* by transformation with the plasmid pKD 136. In addition to transketolase (*tkt*) and an isoenzyme of DAHP (*aro*F), pKD 136 carries an *aro*B locus which prevents the accumulation of 3-deoxy-D-*arabino*-heptulosonic acid (DAH). Expression of pKD 136 by various *Escherichia coli* mutants indicated however that approximately 90% of the D-glucose equivalents directed into aromatic biosynthesis were "lost" after the formation of 3-dehydroshikimate. Further experiments suggested that an additional pathway of metabolism (the "Not-So-Common Pathway", Fig. 12) had been induced in which protocatechuate and catechol are formed (*via* 3-dehydroshikimate dehydratase and protocatechuate decarboxylase).

5.2. Anthocyanin Biosynthesis – Genetic Manipulation of Flower Colour

Classical methods of flower breeding, despite their many achievements, also have their limitations (*127*). Chief amongst these is the limited gene pool of any species. Thus for example no one plant species possesses the genetic capacity to produce flowering varieties which span the full spectrum of colours; one does not find blue carnations, roses, tulips, and chrysanthemums nor orange petunias. Another limitation is that the plant breeders are unable to alter traits in a directed manner. Thus the new combinations of genetic complements, consequent upon the crossing of two plants, may well alter flower colour but also change the delicate balance of the other factors which determine plant growth and shape. In the process good commercial characteristics such as uniform growth and synchronous flowering may therefore be deleteriously affected. Recent advances in molecular biology, especially in gene isolation, manipulation and transfer between species are making possible the alteration of plant properties, with commercial and aesthetic value, in a highly directed fashion. Several enterprises have begun flower breeding using recombinant DNA technology; many of these are intimately associated with processes designed to change anthocyanin pigmentation.

The anthocyanin biosynthetic pathway (Figs. 13 and 14) is now well established, in most of its stages from the two precursors – L-phenylalanine and malonyl coenzyme-A – although the precise number of enzymes and their sequence of operation from the dihydroflavonol

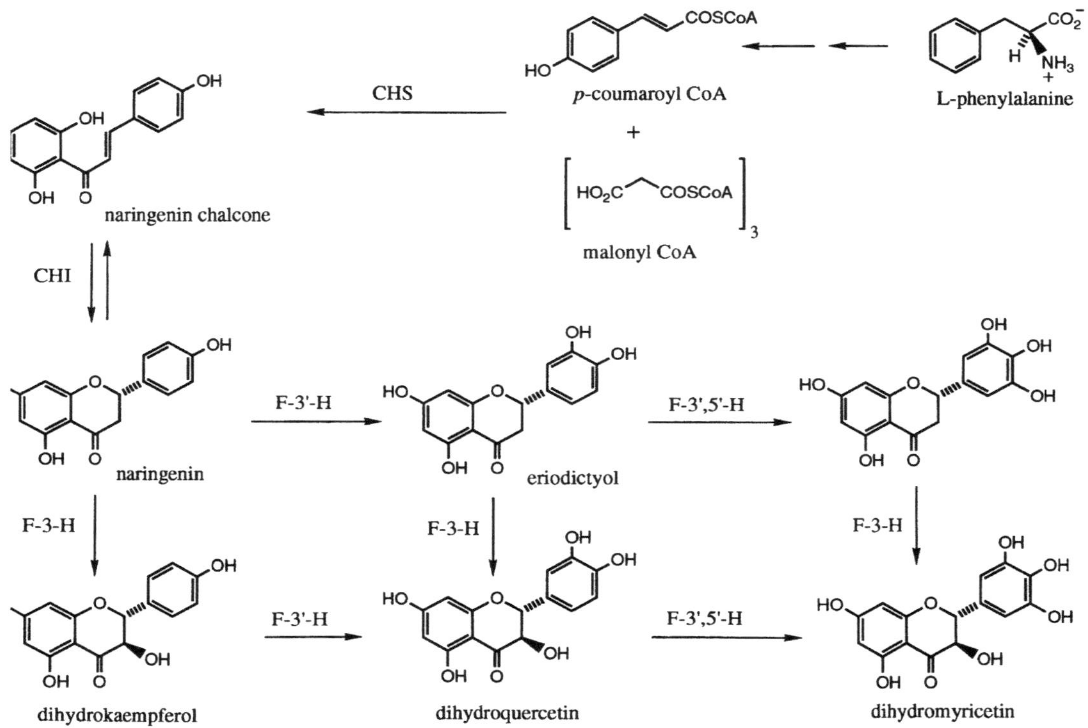

Fig. 13. Biosynthetic pathways to the dihydroflavonols – precursors of the anthocyanin pigments in plants. Enzymes: *chalcone synthase* (CHS); *chalcone isomerase* (CHI); *flavanone-3-hydroxylase* (F-3-H); *flavonoid-3'-hydroxylase* (F-3'-H) and *flavonoid-3',5'-hydroxylase* (F-3',5'-H)

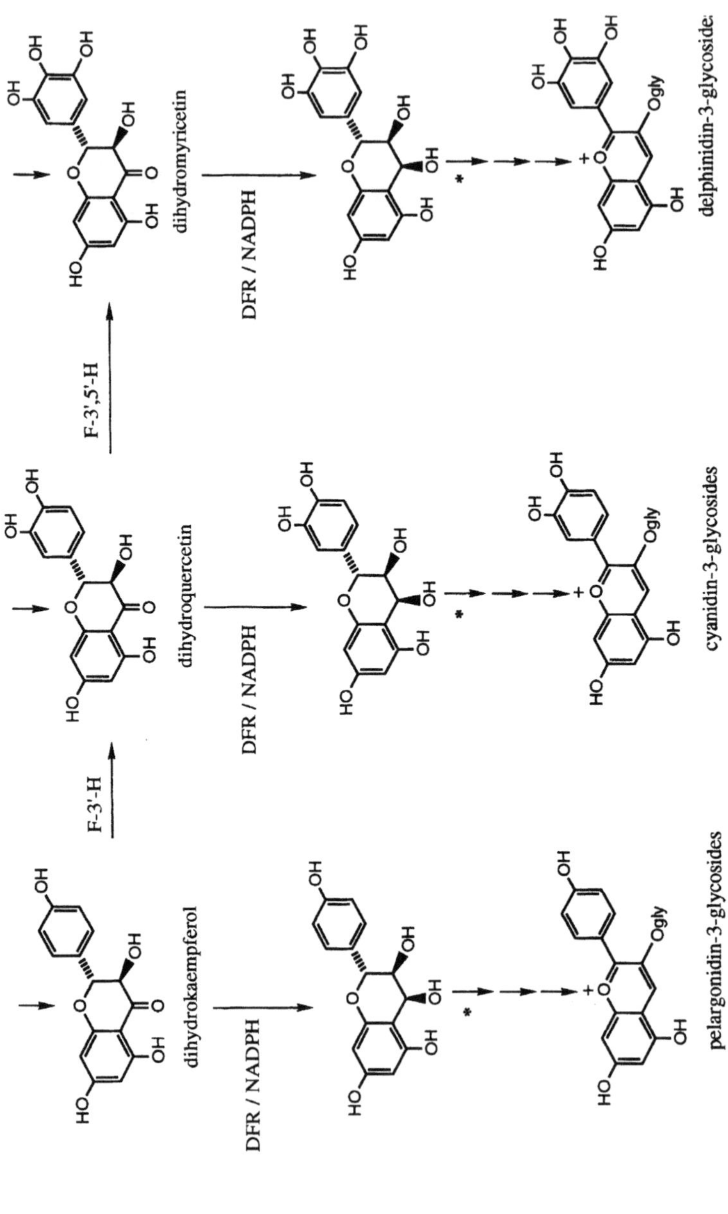

Fig. 14. Biosynthetic conversion of the dihydroflavonols into anthocyanins. Enzyme: *dihydroflavonol-4-reductase* (DFR). *Asterisk indicates at least 3 chemical steps – loss of water, oxidation and glycosylation – occur between the flavan-3,4-diol and the anthocyanidin-3-glycosides. Enzyme details are not yet clear

stage to the anthocyanin (Fig. 14) is still not entirely clear (*86*). Since chalcone is the central C_{15} intermediate for the synthesis of *all* flavonoids the enzyme *chalcone synthase* (CHS) can be regarded as the key enzyme of flavonoid biosynthesis. The enzyme catalyses the stepwise condensation of three molecules of malonyl coenzyme-A ("acetate" units) and one of *p*-coumaroyl coenzyme-A to yield tetrahydroxychalcone (naringenin chalcone). A sequence of steps catalysed by the enzymes *chalcone isomerase* (CHI) and *flavanone-3-hydroxylase* (F-3-H) leads to the dihydroflavonol, dihydrokaempferol, which is the ultimate precursor of anthocyanins of the pelargonidin class; the first stage in this pathway (Fig. 14) is undoubtedly reduction to the flavan-3,4-diol catalysed by the enzyme *dihydroflavonol-4-reductase* (DFR).

Differences in the hydroxylation patterns in ring "B" of the anthocyanidins, which determine the degree of "blueness" of the final pigment are controlled by the microsomal cytochrome P450 enzymes – *flavonoid-3′-hydroxylase* (F-3′-H) and *flavonoid-3′,5′-hydroxylase* (F-3′,5′-H). Aryl hydroxylations may occur in a number of sequences and there is, in principle, a network of pathways from the flavanone naringenin to the other dihydroflavonols (dihydroquercetin and dihydromyricetin, Figs. 13, 14) which are the precursors of respectively the cyanidin and delphinidin classes of anthocyanin pigments. One such network is shown in Fig. 13.

There have been several reports in recent years of targeted gene expression in the anthocyanin biosynthetic pathway to bring about changes in floral pigmentation (*39*, *92*, *93*, *114*, *127*). The key enzyme in the biosynthetic pathway which initiates the diversion of metabolic intermediates into pigment biosynthesis is *chalcone synthase* (CHS) which catalyses the synthesis of naringenin, from which all other flavonoids are derived. To avoid the accumulation of pigments – to produce, in other words, white-flowering varieties – the first committed step in flavonoid biosynthesis, and hence the enzyme *chalcone synthase* (CHS), must be suppressed. Two techniques – "sense" and "anti-sense" suppression – have been described (*114*, *127*), in petunia and chrysanthemum, respectively. Reducing gene expression using RNA that contains the complementary sequence to a given RNA (and hence is termed "anti-sense" RNA), is a strategy developed to mimic mutations. Constitutive expression of an "anti-sense" chalcone synthase gene in transgenic petunia plants results, with high frequency, in an altered flower pigmentation due to a reduction of the messenger RNA for the enzyme and ultimately of the enzyme itself. Whilst the normal petunia hybrids had uniformly coloured corollas, flowers of the petunia transgenic plants displayed a number of pigmentation patterns ranging from entirely white flowers to those in which the corolla had alternating coloured and colourless (white) sectors (*114*).

Floral pigmentation can alternatively be changed by the introduction into the plant of an enzyme which was previously absent. The most notable example of the application of this technique is again from petunia (*124*). In petunia cyanidin and delphinidin derivatives are synthesised as pigments (Fig. 14) but *no* pelargonidin derivatives are formed. This is due to the substrate specificity of the petunia *dihydroflavonol-4-reductase* (DFR) which is able to reduce dihydroquercetin and dihydromyricetin but is *unable* to reduce dihydrokaempferol (Fig. 14). The same enzyme in *Zea mays* (maize) has a different substrate specificity and is able to reduce the dihydrokaempferol substrate. A gene encoding the *dihydroflavonol-4-reductase* (DFR) from maize was introduced into a petunia mutant RL01 which normally has a white flower and, because of various mutations, accumulates dihydrokaempferol. A number of transgenic plants were produced and many of these displayed flowers with a strong brick-red colouration. Analysis showed these plants to contain pelargonidin glycosides as major components of the pigments present. These observations convincingly demonstrate that the *dihydroflavonol-4-reductase* (DFR) of maize does not show the same narrow substrate specificity as the analogous enzyme from petunia. Its presence in the transgenic petunia plants enables them to convert the substrate dihydrokaempferol to the corresponding flavan-3,4-diol which is then processed further to pelargonidin glycosides and gives rise to the brick-red colour of the flowers.

A similar strategy is in the course of development by those anxious to see a blue rose. Although the complex supramolecular structures of commelinin and protocyanin which lend the blue colour to flowers of *Commelina communis* and cornflower respectively are both based upon a "B" ring 3′,4′-dihydroxylation pattern in the anthocyanin part of these structures, there is a general view that the structural requirements necessary for the formation of a blue flower colour are:

(i) the accumulation of delphinidin derived anthocyanins,
(ii) the accumulation of flavonoid copigments,
(iii) a vacuolar pH > 5.0.

Further chemical embroidery (such as acylation, methylation and glycosylation) may also be important in the formation of the final pigment structure. Anthocyanins from roses have been analysed extensively and a recent survey of 670 cultivars found only glucosides of the anthocyanidins – pelargonidin, paeonidin and cyanidin. To date none of the anthocyanins detected in roses have the typical 3′,4′,5′-tri-hydroxylation pattern in

References, pp. 231–240

ring "B" of delphinidin and its derivatives. Recently the first major steps were taken towards the modification of the types of anthocyanin pigments in roses using recombinant DNA methods (*93*). Differences in the "B" ring hydroxylation patterns of the various anthocyanins are controlled by the cytochrome P-450 enzymes – *flavonoid-3'-hydroxylase* (F-3'-H) and *flavonoid-3',5'-hydroxylase* (F-3',5'-H), Figs. 13, 14. An Australian-Japanese-French consortium has very recently reported the isolation of complementary DNA clones of two different *flavonoid-3',5'-hydroxylase* (F-3',5'-H) genes that are expressed in petunia flowers (*93*). The implanting of the *flavonoid-3',5'-hydoxylase* (F-3',5'-H) genes into cyanidin or pelargonidin producing rose cultivars should, it is thought, divert the flux of flavonoid biosynthesis towards the formation of delphinidin glucosides and hence change the flower colour towards the blue pigmentation which is desired.

The doctrine of the language of flowers is redolent of a bygone age; writing in 1856 from Boston in a book "Floras Interpreter and Fortuna Flora", SARAH JANE HALE suggested that a burgundy rose meant "you have simplicity and beauty", a white rose meant that "I am sad" and a yellow rose said "let us forget". Whilst it is no doubt interesting and amusing to speculate what message a blue rose might convey midst the hustle and bustle of the late 20th century, there is no doubt that, consonant with these times, there is a yearning in the Far East for roses that grow bright blue straight from the earth – a yearning which it is said will put a price tag of at least £50 on each rose bloom!

Notes Added in Proof

Chorismate Mutase

LIPSCOMB and his colleagues [CHOOK, Y. M., J. V. GRAY, H. KE, and W. N. LIPSCOMB: The Monofunctional Chorismate Mutase from *Bacillus subtilis*. Structure Determination of Chorismate Mutase and Its Complexes with a Transition State Analog and Prephenate, and Implications for the Mechanism of the Enzymatic Reaction, J. Mol. Biol., **240**, 476–500 (1994)] have now published full and detailed information on the X-ray structure determination of the monofunctional chorismate mutase from *Bacillus subtilis* and its complexes with the *endo*-oxabicyclic transition state analogue (**13**) and the rearrangement product prephenate (see Sect. 3.5: *Chorismate Mutase*). In the crystal the chorismate mutase trimers are tightly packed. Each trimer forms hydrogen bonds to approximately 120 water molecules, most of which are located at the surface of the enzyme. However in each trimer there are three water molecules buried in the hydrophobic β-sheet core. The structure of the active-sites of the enzyme and its complexes with the *endo*-oxabicyclic transition state analogue (**13**) and the rearrangement product prephenate provide a structural basis for an interpretation of the biochemical studies that indicate a pericyclic mechanism for the conversion of chorismate to prephenate. Aside from the ordering of the C-terminal regions of the proteins there is little change in the active-site structure upon binding the transition state analogue (**13**) and prephenate. The authors conclude that it is unlikely that the active-site that captures chorismate is significantly different from the three structures reported and thus that chorismate is bound to the enzyme in a pseudo-chair-like conformation with its enolpyruvyl side-chain positioned over the cyclohexadiene ring, (*cf.* conformer **14** and Fig. 10). The absence of active-site groups capable of protonating the ether oxygen of chorismate, and the absence of a nucleophilic group which might be capable of generating a covalent intermediate with the enzyme, almost rule out the possibilities of electrophilic and nucleophilic catalysis of the enzymic rearrangement. The structure of the active-site of chorismate mutase thus strongly implies that the enzyme-catalysed rearrangement of chorismate to prephenate is a pericyclic process, very similar to the "uncatalysed" *in vitro* reaction in which the chemistry derives from an inherent property of the chorismate molecule.

The active-site of chorismate mutase has a disposition of charged groups which complements and is electrostatically appropriate to stabilise the polar transition state of this pericyclic reaction – the guanidinium group of Arg-90 could thus stabilise the developing partial negative charge on the ether oxygen of the activated complex and the π-electrons of Phe-57 and the carboxylate of Glu-78 could help to stabilise a partial positive charge at C-5 of the substrate. Stabilisation of the transition state by the enzyme thus contributes to the observed rate acceleration of the enzymatic reaction over the "uncatalysed" one. If the enzymatic reaction is concerted, it is very asynchronous and this same conclusion was also reached by GRAY and KNOWLES from FTIR studies of the mechanism of action of the enzyme [GRAY, J. V., and J. R. KNOWLES: Monofunctional Chorismate Mutase from *Bacillus subtilis*: FTIR Studies and the Mechanism of Action of the Enzyme. Biochemistry, **33**, 9953–9959 (1994)].

The rearrangement of chorismate to prephenate is also catalysed by antibodies raised against the *endo*-oxabicyclic transition state analogue (**13**). A crystal structure of the antibody obtained by HILVERT and his colleagues, which has a k_{cat}/k_{uncat} of ~ 250, has now been obtained. The structure of its ligand binding site is apparently very different from the active-site of the *Bacillus subtilis* chorismate mutase [HAYNES, M. R., E. A. SURA, D. HILVERT, and I. A. WILSON: Routes to Catalysis: The Structure of a Catalytic Antibody and Comparison with its Natural Counterpart. Science, **263**, 646–652 (1994)]. There appear to be many fewer strong polar contacts between ligand and protein than in the enzyme and the antibody binding site seems to be far less specific for the pseudo-chair-like conformation (**14**) of the chorismate substrate. The antibody is a less efficient catalyst for the rearrangement, (there is an approximate rate acceleration difference of 10^4 fold between the two proteins), probably due to its chemically and structurally less specific active-site.

Shikimate Pathway in Plants

SCHMID and AMRHEIN have published a comprehensive and detailed review of the molecular organisation of the shikimate pathway in higher plants which highlights the question of whether there are one, two or three sites for the synthesis of the aromatic amino acids in higher plants; the authors focus principally on the results which have been obtained using the techniques of molecular biology and which bear on this problem [SCHMID, J., and N. AMRHEIN: Molecular Organisation of the Shikimate Pathway in Higher Plants. Phytochemistry, **39**, 737–749 (1995)]. In plants proteins are synthesised in three different compartments: in the cytoplasm, in the plastids and in the mitochondria. Therefore the aromatic amino acids must either be synthesised *in situ* in the respective protein synthesising compartment or they are synthesised outside this compartment and are imported. In the case of mitochondria there is no evidence that aromatic amino acids are synthesised in these organelles. On the other hand it is well documented that each of the three aromatic amino acids can be synthesised in plastids but it still remains a matter of debate whether there is, in addition, a complete pathway in the cytoplasm of the cell. The shikimate pathway appears to be the only pathway of amino acid biosynthesis for which questions regarding its subcellular location remain to be fully answered. The presence of a cytosolic pathway is disputed by some groups; conversely others have speculated that those aromatic acids synthesised in the plastids are primarily used in protein synthesis whereas those synthesised in the cytosol are utilised in the biosynthesis of the very wide range of secondary plant products which are derived from the shikimate pathway. SCHMID and AMRHEIN present a wide ranging review of the present state of knowledge of the individual enzymes of the pathway in plants, nevertheless they conclude that the recent isolation and characterisation of cDNAs and genes coding for enzymes of the shikimate pathway in higher plants have confirmed that plastids are the major, if not the only, site of aromatic acid biosynthesis in plants.

Biocatalytic Syntheses

Growing environmental concerns are causing chemical manufacturers to rethink existing procedures employed to prepare chemicals utilised in our everyday lives. In this context FROST and his group in Michigan continue to explore the possibility of using enzymes of the shikimate pathway, in genetically engineered organisms, to generate commercially viable routes to chemicals of importance to the chemical industry, [see Sect. 5.1 and Fig. 12; DRATHS, K. M., and J. W. FROST: Environmentally Compatible Synthesis of Catechol from D-Glucose. J. Amer. Chem. Soc., **117**, 2395–2400 (1995); DRATHS, K. M., and J. W. FROST: Sweetening Chemical Manufacture. Chemistry in Britain, **31**,

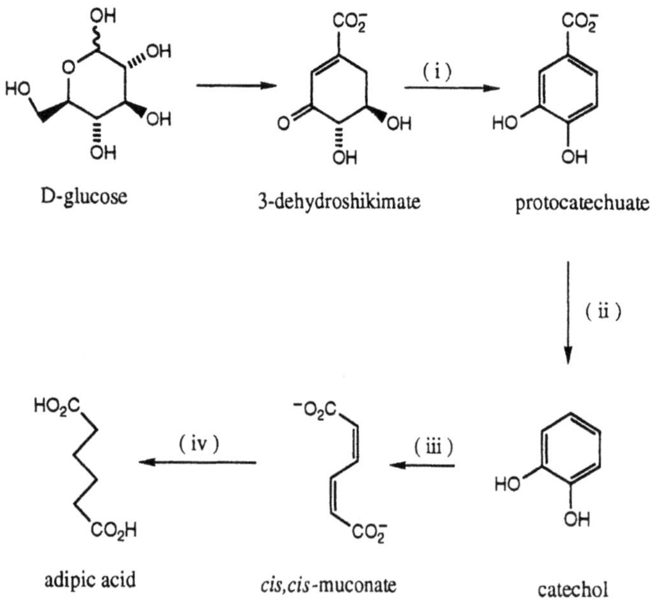

Biocatalytic synthesis of catechol and adipic acid: (i) *3-dehydroshikimate dehydratase*; (ii) *protocatechuate decarboxylase*; (iii) *catechol-1,2-dioxygenase*, NADPH, O_2; (iv) H_2/Pt on C

206–210 (1995)]. A biocatalytic alternative to the currently employed industrial synthesis of catechol from benzene *via* cumene and phenol has been fully described. Genes from *Klebsiella pneumoniae* encoding 3-dehydroshikimate dehydratase (*aroZ*) and protocatechuic decarboxylase (*aroY*) were introduced into an *Escherichia coli* construct that synthesises elevated levels of 3-dehydroshikimic acid. One of the resultant genetically modified organisms produces on a one litre scale, from 56 mM D-glucose, catechol at a concentration of $\sim$ 18.5 mM. This scheme has been further developed to produce adipic acid (a precursor of nylon-6,6 and currently manufactured from benzene at the rate of $\sim 8.8 \times 10^9$ kg per annum world wide) by the introduction of the gene for a further enzyme – catechol-1,2-dioxygenase, from *Acinetobacter calcoaceticus* – into the genetically modified *Escherichia coli* organism. The catechol-1,2-dioxygenase cleaves the catechol ring to give *cis,cis*-muconic acid, which accumulates extracellularly and which is sufficiently pure to be hydrogenated directly to give adipic acid [Draths, K. M., and J. W. Frost: Environmentally Compatible Synthesis of Adipic Acid from D-Glucose. J. Amer. Chem. Soc., **116**, 399–400 (1994)].

References

1. Abeles, R.H., D.L. Anton, L. Hedstrom, and S.M. Fish: Mechanism of Enolpyruvylshikimate-3-phosphate Synthase Exchange of Phosphoenolpyruvate with Solvent Protons. Biochemistry, **22**, 5903 (1983).

2. Abell, C., S. Balasubramanian, and J.R. Coggins: Observation of an Isotope Effect in the Chorismate Synthase Reaction. J. Amer. Chem. Soc., **112**, 8581 (1990).

3. Amrhein, N.: Specific Inhibitors as Probes into the Biosynthesis of Aromatic Amino Acids. In: The Shikimic Acid Pathway (Recent Advances in Phytochemistry, Vol. 20) (E.E. Conn, ed.), pp. 83–117. New York: Plenum Press. 1986.

4. Amrhein, N., and H.C. Steinrucken: Enolpyruvylshikimate-3-phosphate Synthase of *Klebsiella pneumoniae*. Purification and Properties. Eur. J. Biochem., **143**, 341 (1984).

5. Amrhein, N., and H.C. Steinrucken: Enolpyruvylshikimate-3-phosphate Synthase of *Klebsiella pneumoniae*, II: Inhibition by Glyphosate [N-(Phosphonomethyl)-glycine]. Eur. J. Biochem., **143**, 351 (1984).

6. Amrhein, N., H.C. Steinrucken, B. Deus, and P. Gherke: The Site of Inhibition of the Shikimate Pathway by Glyphosate. Plant Physiol., **66**, 830 (1980).

7. Anderson, K.S., and K.A. Johnson: Kinetic and Structural Analysis of Enzyme Intermediates: Lessons from EPSP Synthase. Chem. Rev., **90**, 1131 (1990).

8. Anderson, K.S., J.A. Sikorski, and K.A. Johnson: Evaluation of 5-Enolpyruvylshikimate-3-phosphate Synthase Substrate and Inhibitor Binding by Stopped-Flow and Equilibrium Fluorescence Methods. Biochemistry, **27**, 1604 (1988); A Tetrahedral Intermediate in the EPSP Synthase Reaction Observed by Rapid Quench Kinetics. Biochemistry, **27**, 7395 (1988).

9. ANDERSON, K.S., J.A. SIKORSKI, A.J. BENESI, and K.A. JOHNSON: Isolation and Structural Elucidation of the Intermediate in the EPSP Synthase Enzymatic Pathway. J. Amer. Chem. Soc., **110**, 6577 (1988).
10. ANTON, I.A., S. CHAUDHURI, and J.R. COGGINS: Shikimate Dehydrogenase from *Escherichia coli*. Methods in Enzymology, **142**, 315 (1987).
11. ASANO, Y., J.J. LEE, T.L. SHIEL, F. SPREAFICO, C. KOWL, and H.G. FLOSS: Steric Course of the Reactions Catalysed by 5-Enolpyruvylshikimate-3-phosphate Synthase, Chorismate Mutase and Anthranilate Synthase. J. Amer. Chem. Soc., **107**, 4314 (1995).
12. ATKINSON, D.E.: Cellular Energy Metabolism and Its Regulation. New York: Academic Press. 1977.
13. BAKER, I.T., and I.P. CRAWFORD: Anthranilate Synthase. Partial Purification and Some Kinetic Studies on the Enzyme from *Escherichia coli*. J. Biol. Chem., **241**, 5577 (1966).
14. BALASUBRAMANIAN, S., G.M. DAVIES, J.R. COGGINS, and C. ABELL: Inhibition of Chorismate Synthase by (6*R*)- and (6*S*)-6-Fluoro-5-enolpyruvylshikimate-3-phosphate. J. Amer. Chem. Soc., **113**, 8945 (1991).
15. BARTLETT, P.A., and C.R. JOHNSON: An Inhibitor of Chorismate Mutase Resembling the Transition-State Conformation. J. Amer. Chem. Soc., **107**, 7792 (1985).
16. BARTLETT, P.A., and K. SATAKE: Does Dehydroquinate Synthase Synthesise Dehydroquinate? J. Amer. Chem. Soc., **110**, 1628 (1988).
17. BAUERLE, R., J. HESS, and S. FRENCH: Anthranilate Synthase-Anthranilate Phosphoribosyltransferase Complex and Sub-Units of *Salmonella typhimurium*. Methods in Enzymology, **142**, 366 (1987).
18. BENDER, S.L. J.W. FROST, J.T. KADONGA, and J.R. KNOWLES: Dehydroquinate Synthase from *Escherichia coli*: Purification, Cloning and Construction of Overproducers of the Enzyme. Biochemistry, **23**, 4470 (1984).
19. BENTLEY, R.: The Shikimate Pathway – A Metabolic Tree with Many Branches. Crit. Rev. Biochem. Mol. Biol., **25**, 307 (1990).
20. BHOSEDALE, B.S., J.I. ROOD, M.K. SNEDDON, and J.F. MORRISON: Production of Chorismate Mutase-Prephenate Dehydrogenase by a Strain of *Escherichia coli* Carrying a Multicopy *tyr*A Plasmid. Isolation and Properties of the Enzyme. Biochim. Biophys. Acta, **717**, 6 (1982).
21. BOHM, B.A.: Shikimic Acid (3,4,5-Trihydroxy-1-cyclohexene-1-carboxylic Acid). Chem. Rev., **65**, 435 (1965).
22. BONNER, C., and R.A. JENSEN: Prephenate Aminotransferase. Methods in Enzymology, **142**, 479 (1987).
23. BONNER, C., and R.A. JENSEN: Arogenate Dehydrogenase. Methods in Enzymology, **142**, 488 (1987).
24. BROWN, K.D., and R.L. SOMERVILLE: Repression of Aromatic Amino Acid Biosynthesis in *Escherichia coli* K-12. J. Bacteriol., **108**, 386 (1971).
25. BU'LOCK, J.D.: The Biosynthesis of Natural Products – An Introduction to Secondary Metabolism. Maidenhead: McGraw-Hill. 1965.
26. CAMARKIS, H., D.E. TRIBE, and J.A. PITTARD: Constitutive and Repressible Enzymes of the Common Pathway of Aromatic Biosynthesis in *Escherichia coli* K-12: Regulation of Enzyme Synthesis at Different Growth Rates. J. Bacteriol., **127**, 1085 (1979).
27. CAMPBELL, M.M., M. SAINSBURY, and P. A. SEARLE: The Biosynthesis and Synthesis of Shikimic Acid, Chorismic Acid and Related Compounds. Synthesis, 179 (1993).
28. CAMPBELL, A.P., T.M. TARASOW, W. MASSEFSKI, P.E. WRIGHT, and D. HILVERT: Binding of a High Energy Substrate Conformer in Antibody Catalysis. Proc. Natl. Acad. Sci. (U.S.A.), **90**, 8663 (1993).

29. CHAMPNEY, W.S., and R.A. JENSEN: The Enzymology of Prephenate Dehydrogenase in *Bacillus subtilis*. J. Biol. Chem., **245**, 3763 (1970).
30. CHAUDHURI, S., and J.R. COGGINS: The Purification of Shikimate Dehydrogenase from *Escherichia coli*. Biochem. J., **226**, 217 (1985).
31. CHOOK, Y.M., H. KE, and W.N. LIPSCOMB: Crystal Structures of the Monofunctional Chorismate Mutase from *Bacillus subtilis* and Its Complex with a Transition-State Analog. Proc. Natl. Acad. Sci. (U.S.A.), **90**, 8600 (1993).
32. CHRISTOPHERSON, R.I., E. HEYDE, and J.F. MORRISON: Chorismate Mutase–Prephenate Dehydrogenase from *Escherichia coli*: Spatial Relationship of the Mutase and Dehydrogenase Sites. Biochemistry, **22**, 1650 (1983).
33. COGGINS, J.R., M.R. BOOCOCK, S. CHAUDHURI, J.M. LAMBERT, J. LUMSDEN, G.A. NIMMO, and D.S.S. SMITH: The *arom* Multifunctional Enzyme from *Neurospora crassa*. Methods in Enzymology, **142**, 325 (1987).
34. CONN, E.E. (ed.): The Shikimic Acid Pathway (Recent Advances in Phytochemistry, Vol. 20). New York: Plenum Press. 1986.
35. COPLEY, S.D., and J.R. KNOWLES: The Conformational Equilibrium of Chorismate in Solution: Implications for the Mechanism of the Non-enzymic and the Enzyme Catalysed Rearrangement of Chorismate to Prephenate. J. Amer. Chem. Soc., **107**, 5008 (1987).
36. COPLEY, S.D., and J.R. KNOWLES: The Uncatalysed Claisen Rearrangement of Chorismate to Prephenate Prefers a Transition-State of Chair-like Geometry. J. Amer. Chem. Soc., **107**, 5306 (1987).
37. COTTON, R.G.H., and F. GIBSON: The Biosynthesis of Phenylalanine and Tyrosine: Enzymes Converting Chorismic Acid into Prephenic Acid and Their Relationships to Prephenate Dehydratase and Prephenate Dehydrogenase. Biochim. Biophys. Acta, **100**, 76 (1965).
38. COTTON, R.G.H., and F. GIBSON: The Biosynthesis of Phenylalanine and Tyrosine in the Pea (*Pisum sativum*): Chorismate Mutase. Biochim. Biophys. Acta, **100**, 76 (1965).
39. COURTNEY-GUTTERSON, N., C. NAPOLI, C. LEMIEUX, A. MORGAN, E. FIROOZABADY, and K.E.P. ROBINSON: Modification of Flower Colour in Florist's Chrysanthemum: Production of a White-flowering Variety Through Molecular Genetics. Biotechnology, **12**, 268 (1994).
40. CRAWFORD, I.P.: Synthesis of Tryptophan from Chorismate: Comparative Aspects. Methods in Enzymology, **142**, 293 (1987).
41. CREIGHTON, T.E., and C. YANOFSKY: Chorismate to Tryptophan (*Escherichia coli*): Anthranilate Synthetase, PR Transferase, PRA Isomerase, InGP Synthetase and Tryptophan Synthetase. Methods in Enzymology, **17A**, 365 (1970).
42. DAVIDSON, B.E.: Chorismate Mutase–Prephenate Dehydratase from *Escherichia coli*. Methods in Enzymology, **142**, 432 (1987).
43. DAVIDSON, B.E., E.H. BLACKBURN, and T.A.A. DOPHEIDE: Chorismate Mutase–Prephenate Dehydratase from *Escherichia coli* K-12. J. Biol. Chem., **247**, 4441 (1972).
44. DAVIES, J.: Secondary Metabolites: Their Function and Evolution (Ciba Foundation Symposium, No. 171). Chichester: Wiley. 1992.
45. DAVIS, B.D.: Aromatic Biosynthesis, I: The Role of Shikimic Acid. J. Biol. Chem., **191**, 315 (1951).
46. DAVIS, B.D.: Biochemical Explorations with Biochemical Mutants. Harvey Lectures, **50**, 230 (1954/1955).
47. DE FEYTER, R.: Shikimate Kinases from *Escherichia coli* K-12. Methods in Enzymology, **142**, 355 (1987).

48. DE FEYTER, R., and J. PITTARD: Purification and Properties of Shikimate Kinase II from *Escherichia coli* K-12. J. Bacteriol., **165**, 336 (1986).
49. DE FEYTER, R., B. DAVIDSON, and J. PITTARD: Nucleotide Sequence of the Transcription Unit Containing the *aro*L and *aro*M Genes from *Escherichia coli* K-12. J. Bacteriol., **165**, 233 (1986).
50. DEKA, R.K., I.A. ANTON, B. DUNBAR, and J.R. COGGINS: The Characterisation of the Shikimate Pathway Enzyme Dehydroquinase from *Pisum sativum*. FEBS Lett., **349**, 397 (1994).
51. DE LEO, A.B., and D.B. SPRINSON: Mechanism of 3-Deoxy-D-*arabino*-heptulosonate-7-phosphate (DAHP) Synthetase. Biochem. Biophys. Res. Comm., **32**, 873 (1968).
52. DELL, K.A., and J.W. FROST: Identification and Removal of Impediments to Biocatalytic Synthesis of Aromatics from D-Glucose. Rate-Limiting Enzymes in the Common Pathway of Aromatic Amino Acid Biosynthesis. J. Amer. Chem. Soc., **115**, 11581 (1993).
53. DRATHS, K.M., and J.W. FROST: Synthesis Using Plasmid-Based Catalysis: Plasmid Assembly and 3-Deoxy-D-*arabino*-heptulosonate Production. J. Amer. Chem. Soc., **112**, 1657 (1990).
54. DRATHS, K.M., D.L. POMPLIANO, D.L. CONLEY, J.W. FROST, A. BERRY, G.L. DISBROW, R.J. STAVERSKY, and J.C. LIEVENE: Biocatalytic Synthesis of Aromatics from D-Glucose: The Role of Transketolase. J. Amer. Chem. Soc., **114**, 3956 (1992).
55. DUNCAN, K., S. CHAUDHURI, and J.R. COGGINS: 3-Dehydroquinate Dehydratase from *Escherichia coli*. Methods in Enzymology, **142**, 320 (1987).
56. DUNCAN, K., R.M. EDWARDS, and J.R. COGGINS: The Pentafunctional *arom* Enzyme of *Saccharomyces cerevisiae* Is a Mosaic of Monofunctional Domains. Biochem. J., **246**, 375 (1987).
57. ELY, R., and J. PITTARD: Aromatic Amino Acid Biosynthesis: Regulation of Shikimate Kinase in *Escherichia coli* K-12. J. Bacteriol., **138**, 933 (1979).
58. FISCHER, H.O.L., and G.L. DANGSCHAT: Abbau der Chinasäure zur Zitronensäure. Helv. Chim. Acta, **17**, 1196 (1934).
59. FISCHER, H.O.L., and G.L. DANGSCHAT: Konstitution der Shikimisäure. Helv. Chim. Acta, **17**, 1200 (1934).
60. FISCHER, H.O.L., and G.L. DANGSCHAT: Abbau der Shikimisäure zur Aconitsäure. Helv. Chim. Acta, **18**, 1204 (1935).
61. FISCHER, H.O.L., and G.L. DANGSCHAT: Zur Konfiguration der Shikimisäure. Helv. Chim. Acta, **18**, 1206 (1935).
62. FISCHER, H.O.L., and G.L. DANGSCHAT: Über die Konfiguration der Shikimisäure und ihren Abbau zur Glucodesonsäure. Helv. Chim. Acta, **20**, 705 (1937).
63. FISCHER, R., and R.A. JENSEN: Arogenate Dehydratase. Methods in Enzymology, **142**, 495 (1987).
64. FISCHER, R., and R.A. JENSEN: Prephenate Dehydrogenase (Monofunctional). Methods in Enzymology, **142**, 503 (1987).
65. FISCHER, R., and R.A. JENSEN: Prephenate Dehydratase (Monofunctional). Methods in Enzymology, **142**, 507 (1987).
66. FLOSS, H.G., D.K. ONDERKA, and M. CARROLL: Stereochemistry of the 3-Deoxy-D-*arabino*-heptulosonate-7-phosphate Reaction and the Chorismate Synthase Reaction. J. Biol. Chem., **247**, 736 (1972).
67. FREUDENBERG, K.: Biogenesis and Constitution of Lignin. J. Pure Appl. Chem., **5**, 9 (1962).

68. FROST, J.W., and K.M. DRATHS: Conversion of D-Glucose into Catechol: The Not-So-Common Pathway of Aromatic Biosynthesis. J. Amer. Chem. Soc., **113**, 9361 (1991).
69. GAERTNER, F.H.: Unique Catalytic Functions of Enzyme Clusters. Trends Biochem. Sci., **3**, 63 (1978).
70. GAERTNER, F.H.: Chorismate Synthase: A Bifunctional Enzyme from *Neurospora crassa*. Methods in Enzymology, **142**, 362 (1987).
71. GANEM, B.: From Glucose to Aromatics: Recent Developments in Natural Products of the Shikimate Pathway. Tetrahedron, **34**, 3353 (1978).
72. GILCHRIST, D.G., and J.A. CONNELLY: Chorismate Mutase from Mung Bean and Sorghum. Methods in Enzymology, **142**, 450 (1987).
73. GOLLUB, E.G., H. ZALKIN, and D.B. SPRINSON: Correlation of Genes and Enzymes and Studies of the Regulation of the Aromatic Pathway in *Salmonella*. J. Biol. Chem., **242**, 5323 (1967).
74. GOULD, S.J., and R.L. EISENBERG: The Origin of the C-2 Hydroxyl in the Isochorismate Synthase Reaction. Tetrahedron, **47**, 5979 (1991).
75. GRAY, J.V., B. GOLINELLI-PIMPANEAU, and J.R. KNOWLES: Monofunctional Chorismate Mutase from *Bacillus subtilis*. Purification of the Protein, Molecular Cloning of the Gene and Overexpression of the Gene Product in *Escherichia coli*. Biochemistry, **29**, 376 (1990).
76. GREWE, R., and W. LORENZEN: Die Überführung der Shikimisaüre in Chinasäure. Chem. Ber., **86**, 928 (1953).
77. GREWE, R., and A. BOKRANZ: Shikimisäure und Diazomethan. Chem. Ber., **88**, 49 (1955).
78. GREWE, R., H. JENSEN, and M. SCHNOOR: Darstellung und Eigenschaften des Shikimialkohols. Chem. Ber., **89**, 898 (1956).
79. GREWE, R., and H. BUTTNER: Darstellung und Eigenschaften des Shikimialdehydes. Chem. Ber., **91**, 2452 (1958).
80. GUILFORD, W.J., S.D. COPLEY, and J.R. KNOWLES: On the Mechanism of the Chorismate Mutase Reaction. J. Amer. Chem. Soc., **109**, 5013 (1987).
81. HARRIS, J., C. KLEANTHOUS, J.R. COGGINS, A.R. HAWKINS, and C. ABELL: Different Mechanistic and Stereochemical Courses for the Reactions Catalysed by Type I and Type II Dehydroquinases. J. Chem. Soc. Chem. Commun., 1080 (1993).
82. HASAN, N., and E.W. NESTER: Purification of Chorismate Synthase from *Bacillus subtilis*. J. Biol. Chem., **253**, 4993 (1978).
83. HASAN, N., and E.W. NESTER: Purification and Characterisation of NADPH-Dependent Flavin Reductase, an Enzyme Required for the Activation of Chorismate Synthase in *Bacillus subtilis*. J. Biol. Chem., **253**, 4987 (1978).
84. HASAN, N., and E.W. NESTER: Dehydroquinate Synthase in *Bacillus subtilis*, an Enzyme Associated with Chorismate Synthase and Flavin Reductase. J. Biol. Chem., **253**, 4999 (1978).
85. HASLAM, E.: The Shikimate Pathway. London: Butterworths. 1974.
86. HASLAM, E.: Shikimic Acid – Metabolism and Metabolites. Chichester: Wiley. 1993.
87. HASLAM, E., and R.J. IFE: The Shikimate Pathway, Part III: The Stereochemical Course of the L-Phenylalanine Ammonia Lyase Reaction. J. Chem. Soc. (C), 2818 (1971).
88. HAWKES, T.R., T. LEWIS, J.R. COGGINS, D.M. MOUSDALE, D.J. LOWE, and R.F. THORNELEY: Chorismate Synthase. Pre-Steady State Kinetics of Phosphate Release from 5-Enolpyruvylshikimate-3-phosphate. Biochem. J., **265**, 899 (1990).

89. HERRMANN, K., and R. SCHONER: 3-Deoxy-D-arabinoheptulosonate 7-phosphate Synthase. Purification and Molecular Characterisation of the Tyrosine Sensitive Isoenzyme from *Escherichia coli*. J. Biol. Chem., **251**, 5440 (1976).

90. HERRMANN, K., R.J. MCCANDLISS, and M.D. POLING: 3-Deoxy-D-arabinoheptulosonate 7-phosphate Synthase. Purification and Molecular Characterisation of the Phenylalanine Sensitive Isoenzyme from *Escherichia coli*. J. Biol. Chem., **253**, 4259 (1978).

91. HILVERT, D., S.H. CARPENTER, K.D. NADRED, and M.-T.M. AUDITOR: Catalysis of Concerted Reactions by Antibodies: The Claisen Rearrangement. Proc. Natl. Acad. Sci. (U.S.A.), **85**, 4953 (1988).

92. HOLTON, T.A., and Y. TANAKE: Blue Roses – A Pigment of Our Imagination? Trends in Biotechn., **12**, 40 (1994).

93. HOLTON, T.A., F. BRUGLIERA, D.R. LESTER, Y. TANAKA, C.D. HYLAND, J.G.T. MENTING, C.Y. LU, E. FARCY, T.W. STEVENSON, and E.C. CORNISH: Cloning and Expression of Cytochrome P-450 Genes Controlling Flower Colour. Nature, **366**, 276 (1993).

94. HOMMELL, U., A. LUSTIG, and K. KIRSCHNER: Purification and Characterisation of Yeast Anthranilate Phosphoribosyl Transferase. Eur. J. Biochem., **180**, 33 (1989).

95. HU, C.Y., and D.B. SPRINSON: Properties of Tyrosine-Inhibitable 3-Deoxy-D-arabinoheptulosonic Acid-7-phosphate Synthase from *Salmonella*. J. Bacteriol., **129**, 177 (1977).

96. HUANG, L., A.L. MONTOYA, and E.W. NESTER: Purification and Characterisation of Shikimate Kinase Enzyme Activity in *Bacillus subtilis*. J. Biol. Chem., **250**, 7675 (1975).

97. HUDSON, G.S., V. WONG, and B.E. DAVIDSON: Chorismate Mutase/Prephenate Dehydrogenase from *Escherichia coli*: Purification, Characterisation and Identification of a Reactive Cysteine. Biochemistry, **23**, 6240 (1984).

98. HUDSON, G.S., and B.E. DAVIDSON: Chorismate Mutase–Prephenate Dehydrogenase from *Escherichia coli*: Methods in Enzymology, **142**, 440 (1987).

99. HUTTER, R., P. NIEDERBERGER, and J.A. DEMOSS: Tryptophan Biosynthetic Genes in Eukaryotic Micro-Organisms. Ann. Rev. Microbiol., **40**, 55 (1986).

100. HYDE, C.C., S.A. AHMED, E.A. PADLAM, E.W. MILES, and D.R. DAVIES: The Three-Dimensional Structure of the Tryptophan Synthase $\alpha_2\beta_2$ Multienzyme Complex from *Salmonella typhimurium*. J. Biol. Chem., **263**, 17857 (1988).

101. JACKMAN, L.M., and J.M. EDWARDS: Chorismic Acid. A Branch Point Intermediate in Aromatic Biosynthesis. Aust. J. Chem., **18**, 1227 (1965).

102. JACKSON, D.Y., J.W. JACOBS, R. SUGASWARA, S.H. REICH, P.A. BARTLETT, and P. G. SCHULZ: An Antibody Catalysed Claisen Rearrangement. J. Amer. Chem. Soc., **110**, 4841 (1988).

103. JENSEN, R.A.: Tyrosine and Phenylalanine Biosynthesis: Relationship Between Alternative Pathways, Regulation and Subcellular Location. In: The Shikimic Acid Pathway (Recent Advances in Phytochemistry, Vol. 20), (E.E. CONN, ed.), pp. 57–81. New York: Plenum. 1986.

104. JENSEN, R.A., and R. FISCHER: The Post-Prephenate Biochemical Pathways to Phenylalanine and Tyrosine – An Overview. Methods in Enzymology, **142**, 472 (1987).

105. KAPLAN, J.B., W.K. MERKEL, and B.P. NICHOLS: Evolution of Glutamidotransferase Genes. Nucleotide Sequence of the *pabA* Genes from *Salmonella typhimurium*, *Klebsiella aerogenes* and *Serratia marcescens*. J. Mol. Biol., **183**, 327 (1985).

106. KELLER, B., E. KELLER, H. GORISCH, and F. LINGENS: Zur Biosynthese von Phenylalanin und Tyrosin in Streptomyceten. Hoppe-Seyler's Z. Physiol. Chem., **364**, 455 (1989).

107. KIRSCHNER, K., H. SZADOWSKI, T.S. JARDETZKY, and V. HAGER: Phosphoribosylanthranilate Isomerase–Indoleglycerol Phosphate Synthase from *Escherichia coli.* Methods in Enzymology, **142**, 386 (1987).

108. KLEANTHOUS, C., R. DEKA, K. DAVIS, S.M. KELLY, A. COOPER, S.E. HARDING, N.C. PRICE, A.R. HAWKINS, and J.R. COGGINS: A Comparison of the Enzymological and Biophysical Properties of Two Distinct Classes of Dehydroquinase Enzymes. Biochem. J., **282**, 687 (1992).

109. KNOWLES, J.R.: Mechanistic Ingenuity in Enzyme Catalysis. Aldrichim. Acta, **22**, 59 (1989).

110. KNOWLES, J.R., S. MEHDI, and J.W. FROST: Dehydroquinate Synthase from *Escherichia coli* and Its Substrate 3-Deoxy-D-arabinoheptulosonic Acid-7-phosphate. Methods in Enzymology, **142**, 306 (1987).

111. KOCH, G.L.E., D.C. SHAW, and F. GIBSON: The Purification and Characterisation of Chorismate Mutase-Prephenate Dehydrogenase from *Escherichia coli.* Biochim. Biophys. Acta, **229**, 795 (1971).

112. KOCH, G.L.E., D.C. SHAW, and F. GIBSON: Studies on the Relationship Between the Active Sites of Chorismate Mutase–Prephenate Dehydrogenase from *Escherichia coli* and *Aerobacter aerogenes.* Biochim. Biophys. Acta, **258**, 719 (1971).

113. KOUKOL, J., and E.E. CONN: The Metabolism of Aromatic Compounds in Higher Plants, IV: Purification and Properties of the Phenylalanine Deaminase of *Hordeum vulgare.* J. Biol. Chem., **236**, 2692 (1961).

114. VAN DER KROL, A.R., P.E. LENTING, J. VEENSTRA, I.M. VAN DER MEER, R.E. KOES, A.G.M. GERATS, J.N.M. MOL, and A.R. STUITJE: An Anti-Sense Chalcone Synthase Gene in Transgenic Plants Inhibits Flower Pigmentation. Nature, **333**, 866 (1988).

115. LAWRENCE, J., G.B. COX, and F. GIBSON: Biosynthesis of Ubiquinone in *Escherichia coli* K-12. Biochemical and Genetic Characterisation of a Mutant Unable to Convert Chorismate into 4-Hydroxybenzoate. J. Bacteriol., **118**, 41 (1974).

116. LEVIN, J.G., and D.B. SPRINSON: The Enzymatic Formation and Isolation of 3-Enolpyruvylshikimate-5-phosphate. J. Biol. Chem., **239**, 1142 (1964).

117. LEWENDON, A., and J.R. COGGINS: 3-Phosphoshikimate-1-carboxyvinyl Transferase. Methods in Enzymology, **142**, 342 (1987).

118. LIU, J., K. DUNCAN, and C.T. WALSH: Nucleotide Sequence of a Cluster of *Escherichia coli* Enterobactin Biosynthesis Genes: Identification of *entA* and Purification of Its Product 2,3-Dihydro-2,3-dihydroxybenzoate Dehydrogenase. J. Bacteriol., **171**, 791 (1989).

119. LIU, J., N. QUIN, G.A. BERCHTOLD, and C.T. WALSH: Overexpression, Purification and Characterisation of Isochorismate Synthase (*ent*C), the First Enzyme Involved in the Biosynthesis of Enterobactin from Chorismate. Biochemistry, **29**, 1417 (1990).

120. LOOMIS, L.D., and K.N. RAYMOND: Solution Equilibria of Enterobactin and Metal-Enterobactin Complexes. Inorg. Chem., **30**, 906 (1991).

121. MAITRA, U.S., and D.B. SPRINSON: 5-Dehydro-3-deoxy-D-arabinoheptulosonic Acid-7-phosphate. An Intermediate in the 3-Dehydroquinate Synthase Reaction. J. Biol. Chem., **253**, 5426 (1978).

122. MCQUADE, J.F., and T.E. CREIGHTON: Purification and Comparison of the N-(5′-Phosphoribosyl)anthranilic Acid Isomerase/Indole-3-glycerol Synthetase of Tryptophan Biosynthesis from Three Species of Enterobacteriaceae. Eur. J. Biochem., **16**, 199 (1970).

123. MEHDI, S., S.L. BENDER, and J.R. KNOWLES: Dehydroquinate Synthase: The Role of Divalent Metal Cations and of Nicotinamide Adenine Dinucleotide in Catalysis. Biochemistry, **28**, 7555 (1989).

124. MEYER, P., I. HEIDMANN, G. FORKMAN, and H. SAEDER: A New Petunia Flower Colour Generated by Transformation of a Mutant with a Maize Gene. Nature, **330**, 677 (1987).

125. MILLS, E.W., R. BAUERLE, and S.A. AHMED: Tryptophan Synthase from *Escherichia coli* and *Salmonella typhimurium*. Methods in Enzymology, **142**, 398 (1987).

126. MILLAR, G., and J.R. COGGINS: The Complete Amino Acid Sequence of 3-Dehydroquinate Synthase from *Escherichia coli* K-12. FEBS Lett., **200**, 11 (1986).

127. MOL, J.N.M., A.R. STUITJE, and A. VAN DER KROL: Genetic Manipulation of Floral Pigmentation Genes. Plant Mol. Biol. **12**, 287 (1989).

128. MORELL, H., M.J. CLARK, P.F. KNOWLES, and D.B. SPRINSON: The Enzymic Synthesis of Chorismic and Prephenic Acid from 3-Enolpyruvylshikimic Acid 5-Phosphate. J. Biol. Chem., **242**, 82 (1967).

129. MOUSDALE, D.M., and J.R. COGGINS: Detection and Sub-cellular Localisation in a Higher Plant of Chorismate Synthase. FEBS Lett., **205**, 328 (1986).

130. MOUSDALE, D.M., and J.R. COGGINS: 3-Phosphoshikimate-1-carboxyvinyl Transferase from *Pisum sativum*. Methods in Enzymology, **142**, 348 (1987).

131. MOUSDALE, D.M., M.S. CAMPBELL, and J.R. COGGINS: Purification and Characterisation of Bifunctional Dehydroquinase – Shikimate: NADP Oxidoreductase from Pea Seedlings. Phytochemistry, **26**, 2665 (1987).

132. NICHOLS, B.P., A.M. SEIBOLD, and S.K. DOKTOR: *para*-Aminobenzoate Synthesis from Chorismate Occurs in Two Steps. J. Biol. Chem., **264**, 8597 (1989).

133. OGINO, T., C. GARNER, J.L. MARKLEY, and K. HERRMANN: Biosynthesis of Aromatic Compounds: ^{13}C NMR Spectroscopy of Whole *Escherichia coli* Cells. Proc. Natl. Acad. Sci. (U.S.A), **79**, 5828 (1982).

134. OZENBERGER, B.A., T.J. BRICKMAN, and M.A. MCINTOSH: Nucleotide Sequence of *Escherichia coli* Isochorismate Synthase Gene *entC* and Evolutionary Relationships of Isochorismate Synthase and Other Chorismate Utilising Enzymes. J. Bacteriol, **171**, 775 (1989).

135. PATEL, N., S.L. STENMARK-COX, and R.A. JENSEN: Enzymological Basis of Reluctant Auxotrophy for Phenylalanine and Tyrosine in *Pseudomonas aeruginosa*. J. Biol. Chem., **253**, 2972 (1978).

136. PITTARD, A.J.: Biosynthesis of the Aromatic Amino Acids in *Escherichia coli* and *Salmonella typhimurium*: Cellular and Molecular Biology, Vol. I (F.C. Niedhart, ed.), p. 368–394. Washington: American Soc. Microbiol., 1987.

137. POULSEN, C., and R. VERPOORTE: Roles of Chorismate Mutase, Isochorismate Synthase and Anthranilate Synthase in Plants. Phytochemistry, **30**, 377 (1991).

138. PRIESTLE, J.P., M.G. GRUTTER, J.L. WHITE, M.G. VINCENT, M. KANIA, E.W. ISOU, T.S. JARDETZKY, K. KIRSCHNER, and J.N. JANSONIUS: Three-Dimensional Structure of the Bifunctional Enzyme *N*-(5′-Phosphoribosyl)-anthranilate Isomerase-Indoleglycerol-3-phosphate Synthase from *Escherichia coli*. Proc. Natl. Acad. Sci. (U.S.A), **84**, 5690 (1987).

139. RAMJEE, N., J.R. COGGINS, T.R. HAWKES, D.J. LOWE, and R.N.F. THORNELEY: Spectrophotometric Detection of a Modified Flavin Mononucleotide (FMN) Intermediate Formed During the Catalytic Cycle of Chorismate Synthase. J. Amer. Chem. Soc., **113**, 8566 (1991).

140. RAMJEE, N., S. BALASUBRAMANIAN, C. ABELL, J.R. COGGINS, G.M. DAVIES, T.R. HAWKES, D.J. LOWE, and R.N.F. THORNELEY: Reaction of (6*R*)-6-F-EPSP with Recombinant *Escherichia coli* Chorismate Synthase Generates a Stable Flavin Mononucleotide Semiquinone Radical. J. Amer. Chem. Soc., **114**, 3151 (1992).

141. RUSNAK, F., J. LIU, N. QUIN, G.A. BERCHTOLD, and C.T. WALSH: Subcloning of the Enterobactin Biosynthetic Gene *ent*B: Expression, Purification, Characterisation and Substrate Specificity of Isochorismatase. Biochemistry, **29**, 1425 (1990).
142. SAMFRATHKUMAR, P., and J.F. MORRISON: Chorismate Mutase–Prephenate Dehydrogenase from *Escherichia coli.* Purification and Properties of the Bifunctional Enzyme. Biochim. Biophys. Acta, **702**, 204 (1982).
143. SCHMIDHEIM, T., H.-U. MOSCH, J.N.S. EVANS, and G. BRAUS: Yeast Allosteric Chorismate Mutase is Locked in the Activated State by a Single Amino Acid Substitution. Biochemistry, **29**, 3660 (1990).
144. SOGO, S.G., T.S. WIDLANSKI, J.H. HOARE, C.E. GRIMSHAW, G.E. BERCHTOLD, and J.R. KNOWLES: Stereochemistry of the Rearrangement of Chorismate to Prephenate Chorismate Mutase Involves a Chair Transition State. J. Amer. Chem. Soc., **106**, 2701 (1984).
145. SMITH, D.S.S., and J.R. COGGINS: Isolation of a Bifunctional Domain from the Pentafunctional *arom* Complex of *Neurospora crassa.* Biochem. J., **213**, 405 (1983).
146. SRINIVASAN, P.R., J. ROTHSCHILD, and D.B. SPRINSON: The Enzyme Conversion of 3-Deoxy-D-arabinoheptulosonic Acid-7-phosphate to 5-Dehydroquinate. J. Biol. Chem., **238**, 3176 (1976).
147. STEINRUCKEN, H.C., A. SCHULZ, N. AMRHEIN, C.A. PORTER, and R.T. FRALEY: Overproduction of 5-Enolpyruvylshikimate-3-phosphate Synthase in Glyphosate Tolerant *Petunia hybrida* Cell Line. Arch. Biochem. Biophys., **244**, 169 (1986).
148. STENMARK, S.L., D.L. PIERSON, G.I. GLOVER, and R.A. JENSEN: Blue-Green Bacteria Synthesise L-Tyrosine by the Pretyrosine Pathway. Nature, **247**, 290 (1974).
149. STEWART, J., D.B. WILSON, and B. GANEM: A Genetically Engineered Monofunctional Chorismate Mutase. J. Amer. Chem. Soc., **112**, 4582 (1991).
150. STEWART, J., D.B. WILSON, and B. GANEM: Chorismate Mutase/Prephenate Dehydratase from *Escherichia coli*: Subcloning, Overproduction and Purification. Tetrahedron, **47**, 2573 (1991).
151. TENG, C.-T., and B. GANEM: Shikimate Derived Metabolites, 13: A Key Intermediate in the Biosynthesis of Anthranilate from Chorismate. J. Amer. Chem. Soc., **106**, 2463 (1984).
152. WALSH, C.T., M.D. ERION, A.E. WATTS, J.J. DELANY, and G.A. BERCHTOLD: Chorismate Aminations: Partial Purification of *Escherichia coli* PABA Synthase and Mechanistic Comparison with Anthranilate Synthase. Biochemistry, **26**, 4734 (1987).
153. WALSH, C.T., J. LIU, F. RUSNAK, and M. SAKAITANI: Molecular Studies on Enzymes in Chorismate Metabolism and the Enterobactin Biosynthetic Pathway. Chem. Rev., **90**, 1105 (1990).
154. WEAVER, L.M., and K. HERRMANN: Cloning of an *aro*F Allele Encoding a Tyrosine-Insensitive 3-Deoxy-D-arabinoheptulosonate-7-phosphate Synthase. J. Bacteriol., **172**, 6581 (1990).
155. WEISS, U., and J.M. EDWARDS: Biosynthesis of Aromatic Compounds. New-York: Wiley-Interscience. 1980.
156. WHITE, P.J., G. MILLAR, and J.R. COGGINS: The Overexpression, Purification and Complete Amino Acid Sequence of Chorismate Synthase from *Escherichia coli* K-12 and Its Comparison with the Enzyme from *Neurospora crassa.* Biochem. J., **251**, 313 (1988).
157. WIGHTMAN, R.H., J. STAUNTON, A.R. BATTERSBY, and K.R. HANSON: Studies of Enzyme Mediated Reactions, Part 1: Synthesis of Deuterium- or Tritium-Labelled (3*S*)- and (3*R*)-Phenylalanines: Stereochemical Course of the Elimination Catalysed by L-Phenylalanine Ammonia Lyase. J. Chem. Soc. (Perkin Trans. I), 2355 (1972);

ELLIS, B.E., M.H. ZENK, G.W. KIRBY, J. MICHAEL, and H.G. FLOSS: Steric Course of the Tyrosine Ammonia-Lyase Reaction. Phytochem. **12**, 1057 (1973).

158. YANIV, H., and C. GILVARG: Aromatic Biosynthesis, XIV: 5-Dehydroshikimic Reductase. J. Biol. Chem., **213**, 787 (1955).

159. YOUNG, I.G., and F. GIBSON: Regulation of the Enzymes Involved in the Biosynthesis of 2,3-Dihydroxybenzoic Acid in *Aerobacter aerogenes* and *Escherichia coli*. Biochim. Biophys. Acta, **177**, 401 (1969).

160. YOUNG, I.G., T.J. BATTERHAM, and F. GIBSON: The Isolation, Identification and Properties of Isochorismic Acid, an Intermediate in the Biosynthesis of 2,3-Dihydroxybenzoic Acid. Biochim. Biophys. Acta, **177**, 389 (1969).

161. YOUNG, I.G., L.M. JACKMAN, and F. GIBSON: The Isolation, Identification and Properties of 2,3-Dihydro-2,3-dihydroxybenzoic Acid, an Intermediate in the Biosynthesis of 2,3-Dihydroxybenzoic Acid. Biochim. Biophys. Acta, **177**, 401 (1969).

162. ZALKIN, H.: Anthranilate Synthase. Methods in Enzymology, **113**, 287 (1985).

163. ZIMMERMAN, A., and K. HAHLBROCK: Light Induced Changes of Enzyme Activities in Parsley Cell Cultures. Purification and Properties of Phenylalanine Ammonia Lyase (EC 4.3.1.5). Arch. Biochem. Biophys., **166**, 54 (1975).

(*Received March 13, 1995*)

Author Index

Page numbers printed in *italics* refer to References

Subject Index

Springer-Verlag and the Environment

WE AT SPRINGER-VERLAG FIRMLY BELIEVE THAT AN international science publisher has a special obligation to the environment, and our corporate policies consistently reflect this conviction.

WE ALSO EXPECT OUR BUSINESS PARTNERS – PRINTERS, paper mills, packaging manufacturers, etc. – to commit themselves to using environmentally friendly materials and production processes.

THE PAPER IN THIS BOOK IS MADE FROM NO-CHLORINE pulp and is acid free, in conformance with international standards for paper permanency.

MIX
Papier aus verantwortungsvollen Quellen
Paper from responsible sources
FSC® C105338

If you have any concerns about our products,
you can contact us on
ProductSafety@springernature.com

In case Publisher is established outside the EU,
the EU authorized representative is:
Springer Nature Customer Service Center GmbH
Europaplatz 3, 69115 Heidelberg, Germany

Printed by Libri Plureos GmbH
in Hamburg, Germany